Organic and Hybrid Materials for Photovoltaic and Photonic Applications

Organic and Hybrid Materials for Photovoltaic and Photonic Applications

Editors

Tersilla Virgili
Mariacecilia Pasini

MDPI • Basel • Beijing • Wuhan • Barcelona • Belgrade • Manchester • Tokyo • Cluj • Tianjin

Editors
Tersilla Virgili
Istituto di Fotonica e Nanotecnologie
National Research Council (CNR)
Milano
Italy

Mariacecilia Pasini
SCITEC
CNR
Milano
Italy

Editorial Office
MDPI
St. Alban-Anlage 66
4052 Basel, Switzerland

This is a reprint of articles from the Special Issue published online in the open access journal *Molecules* (ISSN 1420-3049) (available at: www.mdpi.com/journal/molecules/special_issues/Photovoltaic_Application).

For citation purposes, cite each article independently as indicated on the article page online and as indicated below:

LastName, A.A.; LastName, B.B.; LastName, C.C. Article Title. *Journal Name* **Year**, *Volume Number*, Page Range.

ISBN 978-3-0365-6879-9 (Hbk)
ISBN 978-3-0365-6878-2 (PDF)

© 2023 by the authors. Articles in this book are Open Access and distributed under the Creative Commons Attribution (CC BY) license, which allows users to download, copy and build upon published articles, as long as the author and publisher are properly credited, which ensures maximum dissemination and a wider impact of our publications.

The book as a whole is distributed by MDPI under the terms and conditions of the Creative Commons license CC BY NC-ND.

Contents

About the Editors vii

Preface to "Organic and Hybrid Materials for Photovoltaic and Photonic Applications" ix

Safa Shoaee, Anna Laura Sanna and Giuseppe Sforazzini
Elucidating Charge Generation in Green-Solvent Processed Organic Solar Cells
Reprinted from: *Molecules* 2021, 26, 7439, doi:10.3390/molecules26247439 1

Javier Álvarez-Conde, Eva M. García-Frutos and Juan Cabanillas-Gonzalez
Organic Semiconductor Micro/Nanocrystals for Laser Applications
Reprinted from: *Molecules* 2021, 26, 958, doi:10.3390/molecules26040958 15

Benedetta Maria Squeo, Lucia Ganzer, Tersilla Virgili and Mariacecilia Pasini
BODIPY-Based Molecules, a Platform for Photonic and Solar Cells
Reprinted from: *Molecules* 2020, 26, 153, doi:10.3390/molecules26010153 43

Elisa Lassi, Benedetta Maria Squeo, Roberto Sorrentino, Guido Scavia, Simona Mrakic-Sposta and Maristella Gussoni et al.
Sulfonate-Conjugated Polyelectrolytes as Anode Interfacial Layers in Inverted Organic Solar Cells
Reprinted from: *Molecules* 2021, 26, 763, doi:10.3390/molecules26030763 73

Siyang Liu, Shuwang Yi, Peiling Qing, Weijun Li, Bin Gu and Zhicai He et al.
Molecular Engineering Enhances the Charge Carriers Transport in Wide Band-Gap Polymer Donors Based Polymer Solar Cells
Reprinted from: *Molecules* 2020, 25, 4101, doi:10.3390/molecules25184101 91

Shujahadeen B. Aziz, Muaffaq M. Nofal, Mohamad A. Brza, Niyaz M. Sadiq, Elham M. A. Dannoun and Khayal K. Ahmed et al.
Innovative Green Chemistry Approach to Synthesis of Sn^{2+}-Metal Complex and Design of Polymer Composites with Small Optical Band Gaps
Reprinted from: *Molecules* 2022, 27, 1965, doi:10.3390/molecules27061965 105

Rui Feng, Jia-Hui Fan, Kai Li, Zhi-Gang Li, Yan Qin and Zi-Ying Li et al.
Temperature-Responsive Photoluminescence and Elastic Properties of 1D Lead Halide Perovskites *R*- and *S*-(Methylbenzylamine)$PbBr_3$
Reprinted from: *Molecules* 2022, 27, 728, doi:10.3390/molecules27030728 127

Omar Hassan Omar, Rosa Giannelli, Erica Colaprico, Laura Capodieci, Francesco Babudri and Alessandra Operamolla
Reductive Amination Reaction for the Functionalization of Cellulose Nanocrystals
Reprinted from: *Molecules* 2021, 26, 5032, doi:10.3390/molecules26165032 137

Mark Geoghegan, Marta M. Mróz, Chiara Botta, Laurie Parrenin, Cyril Brochon and Eric Cloutet et al.
Optical Gain in Semiconducting Polymer Nano and Mesoparticles
Reprinted from: *Molecules* 2021, 26, 1138, doi:10.3390/molecules26041138 153

Stefania Milanese, Maria Luisa De Giorgi and Marco Anni
Determination of the Best Empiric Method to Quantify the Amplified Spontaneous Emission Threshold in Polymeric Active Waveguides
Reprinted from: *Molecules* 2020, 25, 2992, doi:10.3390/molecules25132992 163

About the Editors

Tersilla Virgili

Tersilla Virgili took her Ph.D. in 2000 at the Physics Department of The University of Sheffield (UK). Since 2001, she has been a permanent researcher of the Institute of Photonics and Nanotechnogies (IFN-CNR) in Milan (Italy). Her research is based on organic materials. She is an expert of ultrafast spectroscopy, organic microcavity (weak and strong coupling regime) and of organic devices. She is co-author of more than 100 international publications, three book chapters, and she is co-inventor of two international patents.

Mariacecilia Pasini

Mariacecilia Pasini took her master degree in Chemistry in 1997. Since 2001, she has been a permanent researcher of the Istituto di Scienze e Tecnologie Chimiche "Giulio Natta"(SCITEC) of the Consiglio Nazionale delle Ricerche (National Research Council—CNR) with headquarters in Milano. Her research activity is focused on the development of advanced conjugated organic and hybrid materials for smart applications in electronics, photonics, energy and sensing, with a particular interest in green organic electronics, water soluble conjugated polyelectrolytes, nanomaterials, and sustainable materials, and processes for semiconducting organic materials. She is co-author of more than 100 international publications, three book chapters, and she is co-inventor of one international patent.

Preface to "Organic and Hybrid Materials for Photovoltaic and Photonic Applications"

This book, with a collection of seven original contributions and three literature reviews, offers select examples of photovoltaic and photonic applications of organic and hybrid materials. The first review by Prof. Sforazzini G. et al. presents a brief introduction to organic solar cells, discussing the principles for formation of free charges, and the requirement for a favorable morphology of the active layer. The second review by Prof. Cabanillas Gonzalez J. et al. discusses the main developments in the field of organic crystal lasers, describing briefly the photophysics and figures-of-merit of organic semiconductor micro/nanocrystals in terms of optical gain properties. The third review by Dr Pasini M. et al. focuses to the use of an emerging class of molecules based on BODIPY in the two main applications described in this Special Issue. The original contributions address our topic by different points of view: from chemical development of new materials, photophysical characterisations, and final applications.

Tersilla Virgili and Mariacecilia Pasini
Editors

Review

Elucidating Charge Generation in Green-Solvent Processed Organic Solar Cells

Safa Shoaee [1,*], Anna Laura Sanna [2] and Giuseppe Sforazzini [2,*]

[1] Disordered Semiconductor Optoelectronics, Institute of Physics and Astronomy, University of Potsdam, Karl-Liebknecht-Str. 24-25, 14476 Potsdam-Golm, Germany
[2] Dipartimento di Scienze Chimiche e Geologiche, Università degli Studi di Cagliari, Complesso Universitario di Monserrato, S.S. 554, Bivio per Sestu, I-09042 Monserrato, Italy; annal.sanna96@unica.it
* Correspondence: shoai@uni-potsdam.de (S.S.); giuseppe.sforazzini@unica.it (G.S.)

Abstract: Organic solar cells have the potential to become the cheapest form of electricity. Rapid increase in the power conversion efficiency of organic solar cells (OSCs) has been achieved with the development of non-fullerene small-molecule acceptors. Next generation photovoltaics based upon environmentally benign "green solvent" processing of organic semiconductors promise a step-change in the adaptability and versatility of solar technologies and promote sustainable development. However, high-performing OSCs are still processed by halogenated (non-environmentally friendly) solvents, so hindering their large-scale manufacture. In this perspective, we discuss the recent progress in developing highly efficient OSCs processed from eco-compatible solvents, and highlight research challenges that should be addressed for the future development of high power conversion efficiencies devices.

Keywords: organic solar cells; green solvents; non-halogenated solvents; exaction diffusion; photoluminescence quenching

1. Introduction

One of the greatest challenges to modern science is the search for new, clean and stable sources of energy which can provide for the growing requirements of an increasingly populated planet. Simultaneously, reducing damage to our natural environmental and halting man-made climate change is imperative. Indeed, one of the most pressing challenges of the 21st century is to reduce our net carbon emissions towards net zero to slow the effect of global warming. Thus, finding ways to generate electricity from sustainable sources and in a sustainable manner is pivotal to reaching this goal. To date, organic solar cells have shown an enormous potential as a viable candidate for green energy production. For instance bulk heterojunction (BHJ) organic photovoltaic (OPV), can be fabricated from solution at low-temperatures [1,2]. As a result, BHJ OPVs can be produced with a modest energy consumption [3,4] with respect to the high temperature approaches currently employed to process inorganic semiconductor materials (such as crystalline silicon). In addition, over the last decade, there has been substantial progress in the development of OPVs driven by advances in molecular design, material processing, and device engineering. Power conversion efficiencies (PCE) for the state of the art laboratory-scale devices exceed 18% for single junction cells [5] and are now approaching commercial viability. Thus, OPV is a promising technology with the potential to constitute a large portion of the future energy generation capability, but only if gains are made in performance, lifetime and eco-friendly fabrication.

The validated arrival of OPVs into the array of next-generation green-tech was promoted through tantalisingly attractive cost per power ($/W) values. These values would be achieved through leveraging the solubility of organic materials to diminish processing costs and fast production capacity through solution processes. To this date, approximately 30 years later, this plausibility has not been fully realized. However, given today's public awareness on the health impact of product manufacturing, and the consciousness to prevent further damage to the environment and ecosystems, eco-friendly organic solar cells grant such attention [6]. Utilization of organic materials can guarantee an economically and environmentally sustainable life-cycle of the device if use of halogenated solvents in fabrication is removed. For instance, organic photovoltaics processed from green solvents demonstrate great potential for indoor photovoltaic applications, where toxicity should be eliminated. They can provide a self-suitable power source for Internet-of-Things (IoT) systems under indoor or low light intensity conditions [7,8]. Whilst crystalline silicon cells show low PCEs under indoor light, on the contrary, the highly tunable light-absorption properties of OPV photoactive materials combined with the environmentally friendly fabrication process, make them promising candidates for indoor applications. For this application, environmentally benign OPV address several inherent weaknesses in PV systems by enhancing design flexibility, environmental consideration, aesthetic demands and increasing efficiency under low-light conditions, leading to a smaller system footprint, particularly required by product designers and architects. Reporting in nature energy, Cui et al. demonstrated that 26% PCE can be obtained under a light emitting diode illumination [9]. Among a few studies [10–12]. Cutting et al. reported that the PCE of OPV devices can be increased up to 350% under LED light intensity relative to outdoor conditions [13]. This performance is significantly higher than that of competing Si-technologies or Perovskite-type cells under the same irradiance.

Today's cutting-edge PCE organic solar cells are commonly processed from halogenated and halogenated aromatic solvents, such as chloroform (CF) and chlorobenzene (CB), and might also include additive such as 1,8-diiodooctane (DIO) to achieve favorable film morphology of the light-harvesting active layer. Despite the worldwide spread of efforts to reach PCEs comparable to those obtained from halogenated-processed OSCs, to the best of our knowledge, eco-friendly organic solar cells are not there yet and the structure-function relationship is not explored or discussed. To date, several reviews have reported different chemistry aspects of green solvents [14–16] (Figure 1), and the impact of green solvents on the morphology of the OSC's active layer [14,17–19]. Underpinning these structural considerations are the fundamentals of charge generation and collection. However, discussions on how these morphologies can be inherently linked with charge generation and recombination are currently lacking.

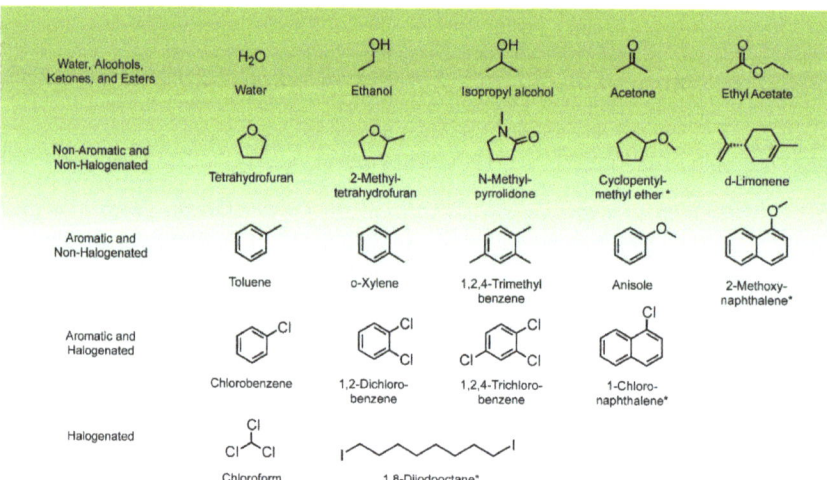

Figure 1. Representative examples of solvents used for BHJ solar cells organized in five general categories. Solvent hazard increases with the fading of green. Reviews about the toxicity and the environmental impact of the solvents have been recently published [16,20]. The mark ∗ indicates solvent additives.

In this Perspective, we begin with a brief introduction to OSCs, discussing the principles for formation of free charges, and the requirement for a favorable morphology of the active layer. Then we will discuss the recent advances on the molecular design of organic semiconductors with enhanced solubility in eco-compatible solvents, and how the implemented molecular changes affect the film morphology and the efficiency of the device. Finally, we analyze the PCE in view of the photoluminescence quenching of OSCs processed with halogenated and non-halogenated solvent. The premise behind donor and acceptor materials and their role in exciton dissociation, and the boundary conditions that exciton diffusion lengths require for effective dissociation will also be explained. Finally, we highlight the critical aspects that should be solved for the future development of OSCs processed with sustainable solvents.

2. Introduction to OSCs

Organic solar cells with a bulk heterojunction architecture consist of an active layer with phase-segregated domains of the electron donor and the acceptor components [21]. Such BHJ solar cells are commonly prepared by combining conjugated donor (D) polymers with electron-accepting (A) molecules. This leads to an interpenetrating network with a large D-A contact area, where the absorbing site is within a few nanometers of the donor-acceptor interface. When light is absorbed by the donor (or the acceptor) an 'exciton' (LE) is generated, which can be regarded as an electron-hole pair bound together by electrostatic interactions. The first step for efficient energy transduction requires dissociation of the neutral excited state, localized on a donor or acceptor component, into a charge-transfer exciton that is localized on adjacent donor and acceptor components. and subsequently into long-lived free charges with a high quantum yield and minimal loss of free energy. Thus, a potential concern is that the electron and hole must overcome their mutual Coulomb attraction, V.

$$V = \frac{e^2}{4\pi r \varepsilon_r \varepsilon_0} \tag{1}$$

where e is the charge of an electron, ε_r is the dielectric constant of the surrounding medium, ε_0 is the permittivity of vacuum, and r is the electron-hole separation distance. For organic materials, overcoming this Coulomb attraction is demanding due to both small dielectric constant ($\varepsilon_r \approx 2$–4) and the localized nature of the electronic states involved. As such, achieving efficient charge photogeneration is a key challenge for solar energy conversion technologies based upon molecular materials.

The events that occur in organic solar cells upon light illumination can be summarized as (Figure 2):

i. Local exciton generation;
ii. Exciton diffusion to the donor/acceptor interface within its lifetime;
iii. Exciton dissociation at the D/A interface and charge transfer (CT) state formation;
iv. Separation of the CT state into free carriers;
v. Charge carrier collection in the selective electrodes.

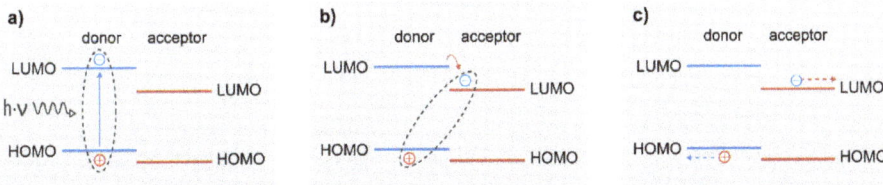

Figure 2. Description of the fundamental processes in OPV devices: (**a**) local exciton generation in the donor, (**b**) charge transfer state formation, and (**c**) free carrier formation and charge collection.

Bulk heterojunctions utilise pre-deposition solution-based mixing, with the objective of forming a randomly configured arrangement with length scales of the exciton diffusion length-order. The convoluted networks do not guarantee a percolating pathway to the collection points. Whilst largely unpredictable and susceptible to reproduction difficulties, this approach has become a field-wide standard and BHJs have consistently produced leading efficiency devices. Until recently only a small number of acceptors have proven capable of delivering high power conversion efficiencies. In particular until recently (2015) the fullerenes dominated the landscape. However since, non-fullerene acceptors (NFA) have delivered advances in cell efficiencies [22]. One of the most efficient systems, discovered in 2019, relies on the combination of the donor polymer PM6 with the small molecule NFA Y6 [23].

3. Morphology of the Photoactive Layer

The key to making efficient OSCs is to ensure that the two materials are intermixed at a length scale less than the exciton diffusion length (typically 5–10 nm) so that every LE formed can reach an interface to undergo charge transfer. At the same time, the morphology has to enable charge-carrier transport in the two different phases to minimize recombination (Figure 3). This *morphology* is determined by the *processing solvent*, concentration, miscibility and crystallinity of D and A, as well as other parameters. Depending on the degree of miscibility and crystallinity between donor and acceptor, 2D or 3D microstructures (the pure D and or A phases, and the D/A amorphous intermixed phase) describe the morphology of the active layer [24]. A certain amount of mixed amorphous phases is crucial for efficient charge generation, while the phase-separated morphology is known to be critical for charges to be both stabilized and extracted. The combination of both factors will determine the final photocurrent of the device. When the donor–acceptor miscibility is too high (hyper-miscible) it can lead to performance deteriorations due to insufficient phase separation to allow stabilization of carriers (to avoid recombination). On the other hand, a miscibility that is too low (hypo-miscible) leads to limitation in charge generation [25]. In this regard, the choice of solvent, which influences the solubility as well as miscibility, can play a significant role on the morphology. Since organic semiconductors are typically made from extended aromatic sub-units, most high performing devices are processed from halogenated solvents (e.g., chloroform, o-dichlorobenzene) which provide good solubility and thereby miscibility of D and A.

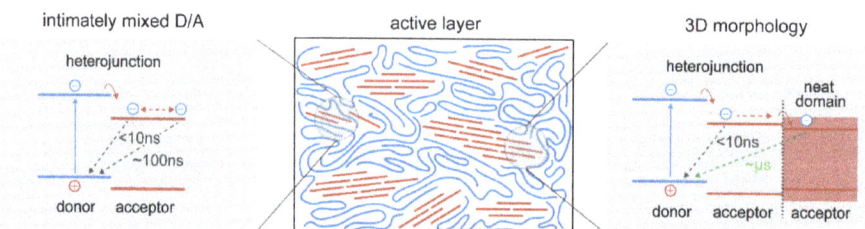

Figure 3. Illustration of the functional model for charge generation and recombination.

There have been a lot of successful attempt in device efficiencies and numerous studies of photoinduced charge separation of fullerene and non-fullerene acceptor based photoactive layers processed from halogenated solvents [26–30]. Extensive studies of these systems have led to a detailed understanding of their structure/function relationship in terms of nonadiabatic electron transfer theory [31]. It has been shown, for example, that charge photogeneration in most systems is dependent upon sufficient domain size and miscibility, whilst the use of energetic cascade (due to aggregation) is required to increase their spatial separation and avoid undesired recombination [24].

The use of halogenated solvents allows, in principle, through good miscibility, the formation of percolation pathways to achieve the electrical 'wiring' of charge photogeneration at the donor/acceptor interface to external device electrodes. Active layers processed from green solvents on the other hand, typically do not exhibit high miscibility between the donor and the acceptor that is present in aromatic halogenated solvents. The mechanism by which active layers processed from green solvents can overcome the Coulomb attraction of the photogenerated electron-hole pair, and in particular achieve this with high quantum efficiency, is central to the development of green organic solar cells.

To date, only a few publications have attempted to use green solvents (Figure 1), and until recently most non-halogenated solvent processed devices deliver inferior performance to the halogenated solvent processed ones (Figure 4) [14,32–35]. This is due to either solubility limitations and/or other morphological aspects, such as interactions of the solvents and components, as well as the rate of solvent evaporation. These have been addressed extensively in recent reviews [19]. In this perspective we focus on the implication of the morphology on charge generation.

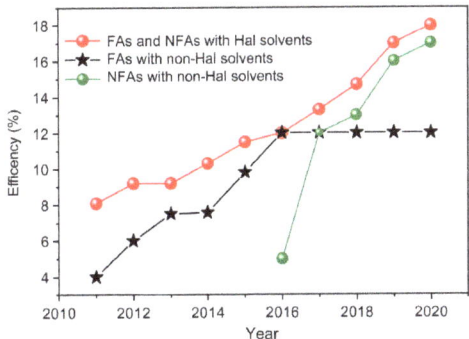

Figure 4. Trends in performance between halogenated and non-halogenated processed organic solar cells.

One limitation with respect to OSCs efficiency processed from green solvents stems from the short exciton diffusion length. The problem with morphology due to miscibility is typically expected to reflect on inefficient exciton dissociation. Efficient device performance relies upon the photogenerated exciton moving to a donor/acceptor interface so that exciton dissociation can occur [36,37]. Due to their electrical neutrality, the motion of excitons is not affected by electric fields, and thus they diffuse through the blend randomly. In this regard, an important parameter of excitons is the diffusion length: the distance an exciton can migrate before relaxing back to the ground state [31,38]. Dissociation of the exciton into charges must therefore occur within this distance. Clearly, this will limit the extent of phase segregation possible in a bulk heterojunction blend morphology for efficient device performance. In general, phase segregation on the order of the exciton diffusion length is desired. Measurements of exciton diffusion length have yielded values of 5-14 nm [37,39,40] for most conjugated polymers and [6,6]-phenyl-C61-butyric acid methyl ester (PCBM), but some recent NFAs exhibit values up to 50 nm long [41].

4. Organic Semiconductors Design towards Green Solvent Processing OSCs

To prepare high performance OSCs using green solvents it is necessary to achieve a film morphology that is favorable for charge separation and transport. This requires the coexistence of both intimately mixed D and A regions as well as phase segregated domains [22,42]. Thus, it is imperative to finely tune the design of semiconductors so as to allow for a good miscibility, and at the same time to govern the packing of the molecules in the solid state. However, finding the balance among molecule solubility, material crystallization and phase separation is not trivial. As solubilizing chains are pivotal for the processing of the materials, the intermolecular arrangement of the semiconductors is strongly influenced by them when processed into films. As a result, the electronic properties of the materials are affected by the choice of the side chains. Moreover, to achieve high

PCEs the semiconductors constituting the D/A active layer should have complementary absorption to harvest most of the solar light, whilst at the same time exhibiting favorable molecular energy levels. Fullerene-based electron acceptors commonly suffer from poor absorption in the visible and NIR portion of the solar spectrum, tunability of its energy level, and solubility in polar media. Attempts to improve these chemical and physical properties by dedicated molecular engineering usually result in materials that underperform with respect to the classic PCBM. On the other hand, NFA have raised attention due to their molecular design tailoring, which allows for improvement of both processability and optoelectronic properties. Nonfullerene acceptors are commonly designed on an acceptor–donor–acceptor (A–D–A) type structure [23]. NFA backbones consist of electron-donating fused-polycyclic systems with terminal electron-accepting units, and solubilizing chains on sides (Figure 5). Such a molecular architecture is responsible for the rapid progress on the tuning of both molecular electronic properties and processability in non-halogenated solvents. In this context, two of the most commonly adopted approaches rely on the introduction of dedicated side chains and the enhancement of the dihedral angle between the aromatic subunits. Following the increasing interest in NFAs, considerable research effort has been also devoted to design and synthesis of polymer donors with an improved solubility in green solvent. Randomizing of copolymer backbone by incorporating a third monomeric unit to design terpolymer, as well as varying the regioregularity or tuning the steric interactions of lateral substituents have been proven to be among the most successful strategies (Figure 5).

An NFA electron acceptor processable from green solvent was synthesized by Hong et al. by replacing the 2-ethylhexyl side chains of Y6 with longer 2-butyloctyl groups, so as to design BTP-4F-12. Such a change allowed for a good solubility in non-halogenated solvents such as o-xylene (o-XY), 1,2,4-trimethylbenzene (TMB), and tetrahydrofuran (THF), as well as for an improved in-plane intermolecular stacking which facilitates charge transport. The same group prepared a modified version of PBDB-TF by introducing an ester-substituted thiophene as a third polymer repeating unit. The resulting polymer, named T1, exhibits an improved solubility in non-chlorinated solvents due to the twisting of the backbone induced by steric interactions of its lateral substituents. OSCs fabricated from blends of BTP-4F-12:T1 processed in THF exhibit PCE of 16.1% which is comparable to the values obtained for CF-processed devices [34]. Recently, Jia et al. conducted a chemical modification for Y6 by inserting a second ethylene double bond π-bridge between the central fuse ring and the terminal indane derivative [43]. The resulting structure BTPV-4F exhibits an absorption that covers a larger portion of the solar spectrum than the original Y6, so as to improve its absorption also in the NIR. To render BTPV-4F suitable for non-halogenated solvents, Qin et al. replaced the 2-ethylhexyl group with 2-butyloctyl side chains, to design BTPV-4F-eC9 [44]. OSCs made from THF-processed binary blends of the latter with PTB7-Th exhibit PCEs of 12.77% that are higher than those obtained using chloroform and 1-chloronaphthalene additive. Chen et al., designed a dissymmetric version of Y6 by replacing the two fluorine atoms with two of chlorine in one terminal indane derivative, and by introducing a trifluoromethyl group in the other side of the molecule, instead of the two fluorine atoms. The corresponding compound, named BTIC-2Cl-γCF$_3$, exhibits an enhanced solubility in toluene and an ability to form a well-organized packing network in the solid state. Binary blend of BTIC-2Cl-γCF$_3$ and PBDB-TF processed from toluene afford devices with power conversion efficiency as high as 16.31% [45].

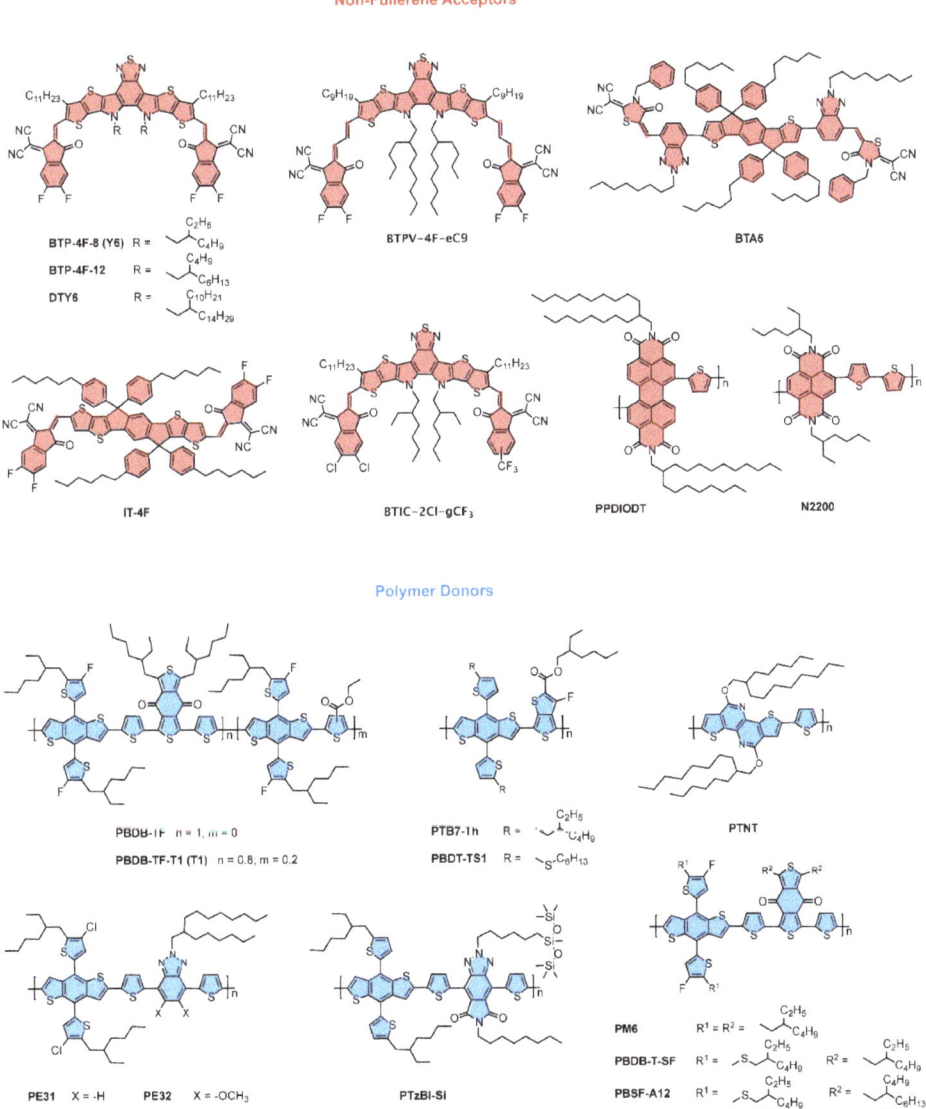

Figure 5. Selected representative examples of NFAs and polymer donors used in BHJ solar cells processed from non-halogenated solvents.

In order to increase the solubility of PBDB-T-SF, an analogue of electron donating polymer PM6, Wang et al. extended the side chains on the donor and acceptor repeating units from 2-ethylhexyl to 2-butyloctyl in order to design PBSF-D12 and PBSF-A12, respectively. As a result, the new wide band-gap polymers have a good solubility in toluene. Films PBSF-D12 exhibit red-shifted absorption spectra due to a high degree of molecular organization which is beneficial for the charge mobility. Toluene processed the OSCs based on PBSF-D12:IT-4F exhibited a good photovoltaic performance with a PCE of 13.4% [46]. Dai et al. conducted modifications for J52-C by replacing the fluorine atoms of the benzotriazole unit with hydrogen atoms, methoxy groups, chlorine atoms. The resulting compounds PE31, PE32, PE33 were tested in THF-processed OSCs made from binary blends with BTA5. Devices resulting from the unsubstituted polymer PE31 achieved the highest PCE of 10.08%. Such a result is ascribed to a better packing of the compounds leading to a good level of intermolecular organization. In contrast, due to the weak crystallinity of the methoxy-substituted polymer PE32, the resulting device shows the lowest PCE of 7.40% [47]. All polymer solar cells were prepared by Sunsun et al. using a perylenediimide (PDI)-bithiophene-based polymer acceptor PPDIODT, and PBDT-TS1 as polymer donor. The former polymeric NFA is designed to have large dihedral angles between its thiophene and PDIs subunit, and to hold a 2-octyldodecyl chain on the PDI motif. As a result, PPDIODT exhibits a good solubility in various green solvents, such as toluene and anisole. OSCs fabricated from binary blend of PPDIODT:PBDT-TS1 processed from anisole have recorded a very good PCE of 6.58% [48]. Polymer donors with enhanced solubility in non-halogenated solvents were also prepared using siloxane-functionalized side-chains. Fan et al. designed and synthesized PTzBI-Si, an electron-donating polymer containing amide-functionalized benzotriazole inclusive of the siloxane-based solubilizing group. The resulting polymer has good solubility in green solvent such as tetrahydrofuran (THF), 2-methyl-tetrafuran (2-MeTHF), and cyclopentyl methyl ether (CPME). Moreover, at the solid state, PTzBI-Si assumes preferentially face-on orientation and forms optimal morphology so as to facilitate carrier mobility in devices. All-PSCs were fabricated using binary blends of PTzBI-Si and the polymer acceptor N2200, processed from THF, 2-MeTHf and CPME, delivering the a slightly higher PCE of 11.0% with the latest solvent [33].

Besides the aforementioned examples, another approach to fabricate OSC from environmentally benign solvents is to anchor the organic semiconductors onto the surface of nanoparticles [49,50]. The later enables pre-aggregation of the donor and the acceptor domains and achieves phase separation, forming a beneficial BHJ morphology [51]. However this technology is underdeveloped and only a few cases of donor and acceptor combinations that can use this method have been reported so far [52].

The large number of variable involved in the fabrication of a solar cell, such as the different chemical structure of the donor and acceptor, the presence of various solubilizing side groups, the use of different processing techniques, as well as the different nature of media used to cast the active layer, render it difficult to quickly assess the structure-property-performance relationships of the devices. Thus, to deepen the origin of the better performances of halogenated-processed OSCs over their eco-compatible counterpart it is essential to analyze charge generation.

5. Photoluminescence Assay of Exciton Dissociation

The primary experimental technique employed to assay the efficiency of exciton dissociation in excitonic D/A blend films is photoluminescence (PL). This is a straightforward technique that monitors the yield of emission of excitons in the blend compared to the pristine material. Quenching of the radiative emission in the blend insinuates exciton dissociation.

Whilst in halogenated processed active layers domain size smaller than the order of exciton diffusion length has been achieved, for non-halogenated and green solvent processed active layers, with hypo-miscible morphology, domains of D and A can be larger than the exciton diffusion length. Thus, the exciton which is photogenerated in one domain, cannot completely diffuse to an interface for dissociation, before relaxing back to the ground state. In this case, only the fraction of excitons generated within the exciton diffusion length will dissociate to form an interfacial CT state and potentially contribute to the photocurrent.

This is exemplified in the work by Lanzi et al. who designed and synthetized water-soluble polythiophenes by introducing hexyl side chains with terminal aminium group (PT6NEt$^+$), and with pyridine unit (PT6Pir), to be blended together with PCBM. From photoluminescence studies it was shown that only about 70% of the excitons reach a D/A interface to dissociate, thus limiting charge generation [53]. This cannot, in general, be attributed to unfavorable energetics. Rather, this efficiency loss appears to derive from the tendency of the blend to form D or A aggregates (large domains) on the length scale or larger than their exciton diffusion lengths. Such a scenario is consistent with what has also been observed in D:A blends with high crystalline domains. In this regard it is anticipated that with the new generation of NFAs which have exceeded previous numbers, such statements and questions of the processes limiting device efficiency are susceptible to revision. Indeed, there are a handful of papers using green solvents, which have investigated the efficiency of exciton diffusion to the interface and thereby its dissociation. Li et al., reported an all polymer solar cell using perylene diimides PPDIODT blended with PBDT-TS1, processed from anisole. In these samples, strong PL quenching of the blend compared to the neat was observed; indicating domain size compatible with exciton diffusion length [48]. Similar observation was reported for the study of benzotriazole (BTA)-based p-type polymers (PE31, PE32, PE33, and J52-Cl) when blended with a BTA-based small molecule BTA5 using THF [47]. It was shown that PE31:BTA5 exhibited the highest PCE of 10.08%, which correlated with the highest PL quenching, whilst the other blends had reduced PL quenching; insinuating less optimum morphology of domain size. One of the most attractive systems reported yet is the PM6:DTY6 blend which shows close to 100% PL quenching and an efficiency of 16% when processed from xylene. Figure 6a plots photocurrent at short circuit against photoluminescence quenching. Correlations between two parameters suggest that the miscibility and molecular organization between the polymer and the acceptor are a key consideration for optimization of photocurrent generation and indicative of exciton diffusion limitations.

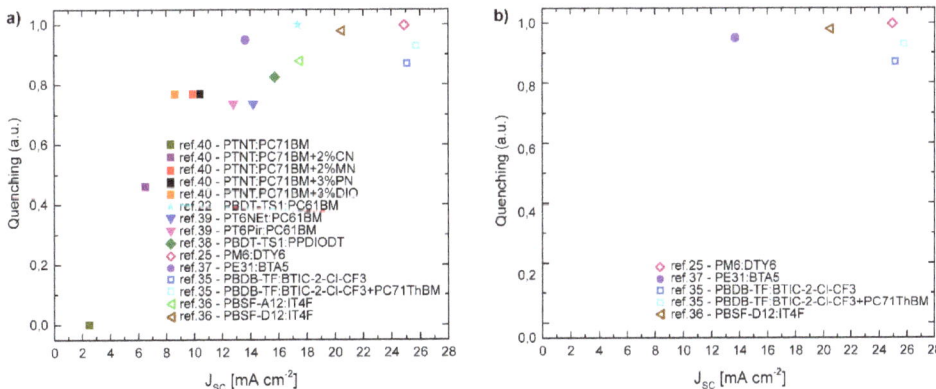

Figure 6. (a) Photoluminescence quenching as a function of photocurrent at short circuit, J_{SC}, for a number of fullerene and non-fullerene acceptor based solar cells processed from non-halogenated solvents [32,35,45,46,48,53–55]. Halogenated and non-halogenated additives: 1-Methoxynaphthalene (MN), 1-Phenylnaphthalene (PN), 1-Cloronaphthalene (CN), 1,8-Diiodooctane (DIO). (b) A selection of data from panel (a) which exhibit high quenching.

Charge separation in donor/acceptor blend systems can be most simply described as electron transfer from excitons to generate free charges. In this simple picture, the efficiency of exciton quenching at the D/A interface should correlate directly with the yield of photogenerated charges. Whilst the general trend in Figure 6a indicates that the photocurrent is limited by exciton quenching (due to too large domains), however upon a closer examination of the systems and categorization of the different class of acceptors, in Figure 6b we observe photocurrent to be independent of the exciton quenching values in the NFA based systems.

Indeed the fused-ring NFA materials benefit from higher film crystallinity (due to better π–π molecular packing) and lower energetic disorder [56]. The higher crystallinity has synergy in enhancing exciton and of free carrier diffusion length (to reach an interface and thereby dissociate) as well as suppressed recombination of free carriers due to crystalline pure phases [57,58]. The lower energetic disorder can also aid exciton dissociation. Excitons are electrically neutral and diffuse through the blend randomly. This diffusion is typically described as a Forster-type incoherent energy transfer process, which can be either intramolecular or intermolecular and usually acts to lower the energy of the exciton. This downhill energy transfer can result in trapping of the exciton in the tail of the inhomogeneously broadened density of states, where the trap sites are often associated with defects and aggregates. At this point, any further exciton migration will rely on thermal fluctuations. In the new class of NFAs, these defects are typically much smaller; 60 meV compared to 120 meV in the PCBM systems [59], thus allowing for more excitons to be able to benefit from thermal fluctuations in order to achieve exciton dissociation at an interface.

Figure 6b implies that, for this materials series, exciton quenching is not the limiting factor for charge photogeneration: the efficiency of the CT-state dissociation into free charges is instead. Thus in addition to the exciton dissociation, the degree of miscibility between the donor and the acceptor still plays a role in CT dissociation and recombination of the free carriers. It is well established that a 3D morphology consisting of a mixed phase as well as pure phases is required to initially generate carriers but to also spatially stabilise the free carriers from one another [24,60]. Thus, in the cases where green solvent prevents coexistence of intimately mixed as well as aggregated phases, the CT state dissociation efficiency is either inefficient or requires a new mechanism for generation of free carriers. Whilst the field is being driven forward in terms of numbers, there is however very limited fundamental studies, elucidating the loss and working mechanisms in green solvent processed organic solar cells. Whilst the initial reports of non-halogenated solvent processed solar cells is encouraging; however, further and general enhancement warrants a deeper understanding of the limiting steps.

6. Concluding Remarks

The proceeding sections showcased the constant development of novel materials rapidly pushing forward the PCEs of OSCs. However, the efficiencies of devices processed via eco-compatible solvents have not yet coherently reached the performance of the OPVs processed by halogenated solvents. To reach the commercialization of OSCs, eco-compatible solvents have to replace toxic organic solvents. Herein, we reviewed current state of the art efficiencies of organic solar cells when processed from non-halogenated solvents. In addition, we discussed the chemical structure and some of the performance-limiting steps of OSCs processed from eco-compatible solvents. A recurrent theme in the current literature is the role of morphology for device performance. Many studies and other reviews have addressed the influence of blend nanomorphology on the device performance. However, studies relating morphology to charge photogeneration remain rather indirect and very few. Nevertheless, from the few studies, experimental evidence is slowly accumulating that exciton dissociation is not the limiting photocurrent as exemplified by PM6:DY6 [35]. However, in order to get an accurate and reliable predictive model to relate materials' structures to photovoltaic device performance understanding the limiting steps in charge generation and recombination is deemed necessary.

Author Contributions: S.S., A.L.S. and G.S. collected and analysed the data. S.S. and G.S. wrote the manuscript with input from all authors. All authors have read and agreed to the published version of the manuscript.

Funding: Financial support from The Alexander von Humboldt Foundation (Sofja Kovalevskaja prize) and PON Ricerca e Innovazione 2014-2020 grant (DOT1304455-2) are gratefully acknowledged.

Institutional Review Board Statement: Not applicable.

Informed Consent Statement: Not applicable.

Data Availability Statement: Not applicable.

Conflicts of Interest: There is no conflict of interest.

References

1. Etxebarria, I.; Ajuria, J.; Pacios, R. Solution-Processable Polymeric Solar Cells: A Review on Materials, Strategies and Cell Architectures to Overcome 10%. *Org. Electron.* **2015**, *19*, 34–60. [CrossRef]
2. Burgués-Ceballos, I.; Stella, M.; Lacharmoise, P.; Martínez-Ferrero, E. Towards Industrialization of Polymer Solar Cells: Material Processing for Upscaling. *J. Mater. Chem. A* **2014**, *2*, 17711–17722. [CrossRef]
3. Brabec, C.J.; Hauch, J.A.; Schilinsky, P.; Waldauf, C. Production Aspects of Organic Photovoltaics and Their Impact on the Commercialization of Devices. *MRS Bull.* **2005**, *30*, 50–52. [CrossRef]
4. Shaheen, S.E.; Ginley, D.S.; Jabbour, G.E. Organic-Based Photovoltaics: Toward Low-Cost Power Generation. *MRS Bull.* **2005**, *30*, 10–19. [CrossRef]
5. Zhang, M.; Zhu, L.; Zhou, G.; Hao, T.; Qiu, C.; Zhao, Z.; Hu, Q.; Larson, B.W.; Zhu, H.; Ma, Z.; et al. Single-Layered Organic Photovoltaics with Double Cascading Charge Transport Pathways: 18% Efficiencies. *Nat. Commun.* **2021**, *12*, 309. [CrossRef] [PubMed]
6. March, W. Organic Photovoltaics—Truly Green Energy: "Ultra-Low Carbon Footprint"; 2020. Available online: https://www.heliatek.com/fileadmin/user_upload/pdf_documents/2020-10_Heliatek_Truly_Green_Energy_Ultra_Low_Carbon_Footprint_REV2.pdf (accessed on 15 November 2021).
7. Ahsan Saeed, M.; Hyeon Kim, S.; Baek, K.; Hyun, J.K.; Youn Lee, S.; Won Shim, J. PEDOT:PSS: CuNW-Based Transparent Composite Electrodes for High-Performance and Flexible Organic Photovoltaics under Indoor Lighting. *Appl. Surf. Sci.* **2021**, *567*, 150852. [CrossRef]
8. You, Y.-J.; Saeed, M.A.; Shafian, S.; Kim, J.; Hyeon Kim, S.; Kim, S.H.; Kim, K.; Shim, J.W. Energy Recycling under Ambient Illumination for Internet-of-Things Using Metal/Oxide/Metal-Based Colorful Organic Photovoltaics. *Nanotechnology* **2021**, *32*, 465401. [CrossRef] [PubMed]
9. Cui, Y.; Wang, Y.; Bergqvist, J.; Yao, H.; Xu, Y.; Gao, B.; Yang, C.; Zhang, S.; Inganäs, O.; Gao, F.; et al. Wide-Gap Non-Fullerene Acceptor Enabling High-Performance Organic Photovoltaic Cells for Indoor Applications. *Nat. Energy* **2019**, *4*, 768–775. [CrossRef]
10. Lee, S.; Lee, Y.; Park, J.; Choi, D. Stitchable Organic Photovoltaic Cells with Textile Electrodes. *Nano Energy* **2014**, *9*, 88–93. [CrossRef]
11. De Rossi, F.; Pontecorvo, T.; Brown, T.M. Characterization of Photovoltaic Devices for Indoor Light Harvesting and Customization of Flexible Dye Solar Cells to Deliver Superior Efficiency under Artificial Lighting. *Appl. Energy* **2015**, *156*, 413–422. [CrossRef]
12. Lechêne, B.P.; Cowell, M.; Pierre, A.; Evans, J.W.; Wright, P.K.; Arias, A.C. Organic Solar Cells and Fully Printed Super-Capacitors Optimized for Indoor Light Energy Harvesting. *Nano Energy* **2016**, *26*, 631–640. [CrossRef]
13. Cutting, C.L.; Bag, M.; Venkataraman, D. Indoor Light Recycling: A New Home for Organic Photovoltaics. *J. Mater. Chem. C* **2016**, *4*, 10367–10370. [CrossRef]
14. Hong, L.; Yao, H.; Cui, Y.; Ge, Z.; Hou, J. Recent Advances in High-Efficiency Organic Solar Cells Fabricated by Eco-Compatible Solvents at Relatively Large-Area Scale. *APL Mater.* **2020**, *8*, 120901. [CrossRef]
15. Yang, Y. The Original Design Principles of the Y-Series Nonfullerene Acceptors, from Y1 to Y6. *ACS Nano* **2021**. [CrossRef]
16. Capello, C.; Fischer, U.; Hungerbühler, K. What Is a Green Solvent? A Comprehensive Framework for the Environmental Assessment of Solvents. *Green Chem.* **2007**, *9*, 927–934. [CrossRef]
17. Burgués-Ceballos, I.; Machui, F.; Min, J.; Ameri, T.; Voigt, M.M.; Luponosov, Y.N.; Ponomarenko, S.A.; Lacharmoise, P.D.; Campoy-Quiles, M.; Brabec, C.J. Solubility Based Identification of Green Solvents for Small Molecule Organic Solar Cells. *Adv. Funct. Mater.* **2014**, *24*, 1449–1457. [CrossRef]
18. McDowell, C.; Bazan, G.C. Organic Solar Cells Processed from Green Solvents Formation of the Bulk Heterojunction from Solution. *Curr. Opin. Green Sustain. Chem.* **2017**, *5*, 49–54. [CrossRef]
19. Lee, S.; Jeong, D.; Kim, C.; Lee, C.; Kang, H.; Woo, H.Y.; Kim, B.J. Eco-Friendly Polymer Solar Cells: Advances in Green-Solvent Processing and Material Design. *ACS Nano* **2020**, *14*, 14493–14527. [CrossRef]
20. Jessop, P.G. Searching for Green Solvents. *Green Chem.* **2011**, *13*, 1391–1398. [CrossRef]
21. Brabec, C.J.; Durrant, J.R. Solution-Processed Organic Solar Cells. *MRS Bull.* **2008**, *33*, 670. [CrossRef]
22. Armin, A.; Li, W.; Sandberg, O.J.; Xiao, Z.; Ding, L.; Nelson, J.; Neher, D.; Vandewal, K.; Shoaee, S.; Wang, T.; et al. A History and Perspective of Non-Fullerene Electron Acceptors for Organic Solar Cells. *Adv. Energy Mater.* **2021**, *11*, 2003570. [CrossRef]
23. Yuan, J.; Zhang, Y.; Yuan, J.; Zhang, Y.; Zhou, L.; Zhang, G.; Yip, H.; Lau, T.; Lu, X.; Li, Y.; et al. Single-Junction Organic Solar Cell with over 15 % Efficiency Using Fused-Ring Acceptor with Electron-Deficient Core. *Joule* **2019**, *3*, 1140–1151. [CrossRef]
24. Shoaee, S.; Subramaniyan, S.; Xin, H.; Keiderling, C.; Tuladhar, P.S.; Jamieson, F.; Jenekhe, S.A.; Durrant, J.R. Charge Photogeneration for a Series of Thiazolo-Thiazole Donor Polymers Blended with the Fullerene Electron Acceptors PCBM and ICBA. *Adv. Funct. Mater.* **2013**, *23*, 3286–3298. [CrossRef]
25. Ghasemi, M.; Balar, N.; Peng, Z.; Hu, H.; Qin, Y.; Kim, T.; Rech, J.J.; Bidwell, M.; Mask, W.; McCulloch, I.; et al. A Molecular Interaction–Diffusion Framework for Predicting Organic Solar Cell Stability. *Nat. Mater.* **2021**, *20*, 525–532. [CrossRef] [PubMed]
26. Shoaee, S.; Clarke, T.; Huang, C.; Barlow, S.; Marder, S.; Heeney, M.; McCulloch, I.; Durrant, J.R. Acceptor Energy Level Control of Charge Photogeneration in Organic Donor/Acceptor Blends. *J. Am. Chem. Soc.* **2010**, *132*, 12919–12926. [CrossRef] [PubMed]
27. Shoaee, S.; An, Z.; Zhang, X.; Barlow, S.; Marder, S.R.; Duffy, W.; Heeney, M.; McCulloch, I.; Durrant, J.R. Charge Photogeneration in Polythiophene–Perylene Diimide Blend Films. *Chem. Commun.* **2009**, *36*, 5445–5447. [CrossRef] [PubMed]

28. Wang, X.; Sun, Q.; Gao, J.; Wang, J.; Xu, C.; Ma, X.; Zhang, F. Recent Progress of Organic Photovoltaics with Efficiency over 17%. *Energies* **2021**, *14*, 4200. [CrossRef]
29. Cheng, P.; Yang, Y. Narrowing the Band Gap: The Key to High-Performance Organic Photovoltaics. *Acc. Chem. Res.* **2020**, *53*, 1218–1228. [CrossRef]
30. Cheng, P.; Li, G.; Zhan, X.; Yang, Y. Next-Generation Organic Photovoltaics Based on Non-Fullerene Acceptors. *Nat. Photonics* **2018**, *12*, 131–142. [CrossRef]
31. Clarke, M.T.; Durrant, R.J. Charge Photogeneration in Organic Solar Cells. *Chem. Rev.* **2010**, *110*, 6736–6767. [CrossRef]
32. Zhang, H.; Yao, H.; Zhao, W.; Ye, L.; Hou, J. High-Efficiency Polymer Solar Cells Enabled by Environment-Friendly Single-Solvent Processing. *Adv. Energy Mater.* **2016**, *6*, 1502177. [CrossRef]
33. Li, Z.; Ying, L.; Zhu, P.; Zhong, W.; Li, N.; Liu, F.; Huang, F.; Cao, Y. A Generic Green Solvent Concept Boosting the Power Conversion Efficiency of All-Polymer Solar Cells to 11%. *Energy Environ. Sci.* **2019**, *12*, 157–163. [CrossRef]
34. Hong, L.; Yao, H.; Wu, Z.; Cui, Y.; Zhang, T.; Xu, Y.; Yu, R.; Liao, Q.; Gao, B.; Xian, K.; et al. Eco-Compatible Solvent-Processed Organic Photovoltaic Cells with Over 16% Efficiency. *Adv. Mater.* **2019**, *31*, 1903441. [CrossRef] [PubMed]
35. Dong, S.; Jia, T.; Zhang, K.; Jing, J.; Huang, F. Single-Component Non-Halogen Solvent-Processed High-Performance Organic Solar Cell Module with Efficiency over 14%. *Joule* **2020**, *4*, 2004–2016. [CrossRef]
36. Bittner, E.R.; Santos, R.J.G.; Karabunarliev, S. Exciton Dissociation Dynamics in Model Donor-Acceptor Polymer Heterojunctions. I. Energetics and Spectra. *J. Chem. Phys.* **2005**, *122*, 214719. [CrossRef]
37. Tamai, Y.; Ohkita, H.; Benten, H.; Ito, S. Exciton Diffusion in Conjugated Polymers: From Fundamental Understanding to Improvement in Photovoltaic Conversion Efficiency. *J. Phys. Chem. Lett.* **2015**, *6*, 3417–3428. [CrossRef]
38. Burlakov, V.M.; Kawata, K.; Assender, H.E.; Briggs, G.A.D.; Ruseckas, A.; Samuel, I.D.W. Discrete Hopping Model of Exciton Transport in Disordered Media. *Phys. Rev. B* **2005**, *72*, 75206. [CrossRef]
39. Sajjad, M.T.; Ward, A.J.; Kästner, C.; Ruseckas, A.; Hoppe, H.; Samuel, I.D.W. Controlling Exciton Diffusion and Fullerene Distribution in Photovoltaic Blends by Side Chain Modification. *J. Phys. Chem. Lett.* **2015**, *6*, 3054–3060. [CrossRef] [PubMed]
40. Mikhnenko, O.V.; Blom, P.W.M.; Nguyen, T.-Q. Exciton Diffusion in Organic Semiconductors. *Energy Environ. Sci.* **2015**, *8*, 1867–1888. [CrossRef]
41. Firdaus, Y.; Le Corre, V.M.; Karuthedath, S.; Liu, W.; Markina, A.; Huang, W.; Chattopadhyay, S.; Nahid, M.M.; Nugraha, M.I.; Lin, Y.; et al. Long-Range Exciton Diffusion in Molecular Non-Fullerene Acceptors. *Nat. Commun.* **2020**, *11*. [CrossRef]
42. Collins, B.A.; Tumbleston, J.R.; Ade, H. Miscibility, Crystallinity, and Phase Development in P3HT/PCBM Solar Cells: Toward an Enlightened Understanding of Device Morphology and Stability. *J. Phys. Chem. Lett.* **2011**, *2*, 3135–3145. [CrossRef]
43. Jia, Z.; Qin, S.; Meng, L.; Ma, Q.; Angunawela, I.; Zhang, J.; Li, X.; He, Y.; Lai, W.; Li, N.; et al. High Performance Tandem Organic Solar Cells via a Strongly Infrared-Absorbing Narrow Bandgap Acceptor. *Nat. Commun.* **2021**, *12*, 178. [CrossRef] [PubMed]
44. Qin, S.; Jia, Z.; Meng, L.; Zhu, C.; Lai, W.; Zhang, J.; Huang, W.; Sun, C.; Qiu, B.; Li, Y. Non-Halogenated-Solvent Processed and Additive-Free Tandem Organic Solar Cell with Efficiency Reaching 16.67%. *Adv. Funct. Mater.* **2021**, *31*, 2102361. [CrossRef]
45. Chen, H.; Lai, H.; Chen, Z.; Zhu, Y.; Wang, H.; Han, L.; Zhang, Y.; He, F. 17.1%-Efficient Eco-Compatible Organic Solar Cells from a Dissymmetric 3D Network Acceptor. *Angew. Chem. Int. Ed. Engl.* **2021**, *60*, 3238–3246. [CrossRef]
46. Wang, K.; Li, W.; Guo, X.; Zhu, Q.; Fan, Q.; Guo, Q.; Ma, W.; Zhang, M. Optimizing the Alkyl Side-Chain Design of a Wide Band-Gap Polymer Donor for Attaining Nonfullerene Organic Solar Cells with High Efficiency Using a Nonhalogenated Solvent. *Chem. Mater.* **2021**, *33*, 5981–5990. [CrossRef]
47. Dai, T.; Lei, P.; Zhang, B.; Tang, A.; Geng, Y.; Zeng, Q.; Zhou, E. Fabrication of High VOC Organic Solar Cells with a Non-Halogenated Solvent and the Effect of Substituted Groups for "Same-A-Strategy" Material Combinations. *ACS Appl. Mater. Interfaces* **2021**, *13*, 21556–21564. [CrossRef] [PubMed]
48. Li, S.; Zhang, H.; Zhao, W.; Ye, L.; Yao, H.; Yang, B.; Zhang, S.; Hou, J. Green-Solvent-Processed All-Polymer Solar Cells Containing a Perylene Diimide-Based Acceptor with an Efficiency over 6.5%. *Adv. Energy Mater.* **2016**, *6*, 1501991. [CrossRef]
49. Yamamoto, N.A.D.; Payne, M.E.; Koehler, M.; Facchetti, A.; Roman, L.S.; Arias, A.C. Charge Transport Model for Photovoltaic Devices Based on Printed Polymer: Fullerene Nanoparticles. *Sol. Energy Mater. Sol. Cells* **2015**, *141*, 171–177. [CrossRef]
50. Shimizu, H.; Yamada, M.; Wada, R.; Okabe, M. Preparation and Characterization of Water Self-Dispersible Poly(3-Hexylthiophene) Particles. *Polym. J.* **2008**, *40*, 33–36. [CrossRef]
51. Gärtner, S.; Christmann, M.; Sankaran, S.; Röhm, H.; Prinz, E.-M.; Penth, F.; Pütz, A.; Türeli, A.E.; Penth, B.; Baumstümmler, B.; et al. Eco-Friendly Fabrication of 4% Efficient Organic Solar Cells from Surfactant-Free P3HT:ICBA Nanoparticle Dispersions. *Adv. Mater.* **2014**, *26*, 6653–6657. [CrossRef]
52. Xie, C.; Heumüller, T.; Gruber, W.; Tang, X.; Classen, A.; Schuldes, I.; Bidwell, M.; Späth, A.; Fink, R.H.; Unruh, T.; et al. Overcoming Efficiency and Stability Limits in Water-Processing Nanoparticular Organic Photovoltaics by Minimizing Microstructure Defects. *Nat. Commun.* **2018**, *9*, 5335. [CrossRef] [PubMed]
53. Lanzi, M.; Salatelli, E.; Giorgini, L.; Marinelli, M.; Pierini, F. Effect of the Incorporation of an Ag Nanoparticle Interlayer on the Photovoltaic Performance of Green Bulk Heterojunction Water-Soluble Polythiophene Solar Cells. *Polymer (Guildf)*. **2018**, *149*, 273–285. [CrossRef]
54. Kocak, G.; Gedefaw, D.; Andersson, M.R. Optimizing Polymer Solar Cells Using Non-Halogenated Solvent Blends. *Polymers* **2019**, *11*, 1–11. [CrossRef] [PubMed]

55. Liao, C.; Zhang, M.; Xu, X.; Liu, F.; Li, Y.; Peng, Q. Green Solvent-Processed Efficient Non-Fullerene Organic Solar Cells Enabled by Low-Bandgap Copolymer Donors with EDOT Side Chains. *J. Mater. Chem. A* **2019**, *7*, 716–726. [CrossRef]
56. Zhang, G.; Chen, X.-K.; Xiao, J.; Chow, P.C.Y.; Ren, M.; Kupgan, G.; Jiao, X.; Chan, C.C.S.; Du, X.; Xia, R.; et al. Delocalization of Exciton and Electron Wavefunction in Non-Fullerene Acceptor Molecules Enables Efficient Organic Solar Cells. *Nat. Commun.* **2020**, *11*, 3943. [CrossRef] [PubMed]
57. Hosseini, S.M.; Tokmoldin, N.; Lee, Y.W.; Zou, Y.; Young, H.; Neher, D.; Shoaee, S. Putting Order into PM6:Y6 Solar Cells to Reduce the Langevin Recombination in 400 Nm Thick Junction. *Sol. RRL* **2020**, *4*, 2000498. [CrossRef]
58. Tokmoldin, N.; Hosseini, S.M.; Raoufi, M.; Phuong, L.Q.; Sandberg, O.; Guan, H.; Zou, Y.; Neher, D.; Shoaee, S. Extraordinarily Long Diffusion Length In PM6:Y6 Organic Solar Cells. *J. Mater. Chem. A* **2020**, *8*, 7854–7860. [CrossRef]
59. Kaiser, C.; Sandberg, O.J.; Zarrabi, N.; Li, W.; Meredith, P.; Armin, A. A Universal Urbach Rule for Disordered Organic Semiconductors. *Nat. Commun.* **2021**, *12*, 3988. [CrossRef]
60. He, X.; Collins, B.A.; Watts, B.; Ade, H.; McNeill, C.R. Studying Polymer/Fullerene Intermixing and Miscibility in Laterally Patterned Films with X-Ray Spectromicroscopy. *Small* **2012**, *8*, 1920–1927. [CrossRef]

Review

Organic Semiconductor Micro/Nanocrystals for Laser Applications

Javier Álvarez-Conde [1,2], Eva M. García-Frutos [1,*] and Juan Cabanillas-Gonzalez [2,*]

[1] Instituto de Ciencia de Materiales de Madrid (ICMM), CSIC, Cantoblanco, E-28049 Madrid, Spain; javier.alvarez@imdea.org

[2] IMDEA Nanociencia, Calle Faraday 9, Ciudad Universitaria de Cantoblanco, E-28049 Madrid, Spain

* Correspondence: emgfrutos@icmm.csic.es (E.M.G.-F.); juan.cabanillas@imdea.org (J.C-G); Tel.: +34-91-334-9038 (E.M.G.-F.); +34 912998784 (J.C.-G.)

Abstract: Organic semiconductor micro/nanocrystals (OSMCs) have attracted great attention due to their numerous advantages such us free grain boundaries, minimal defects and traps, molecular diversity, low cost, flexibility and solution processability. Due to all these characteristics, they are strong candidates for the next generation of electronic and optoelectronic devices. In this review, we present a comprehensive overview of these OSMCs, discussing molecular packing, the methods to control crystallization and their applications to the area of organic solid-state lasers. Special emphasis is given to OSMC lasers which self-assemble into geometrically defined optical resonators owing to their attractive prospects for tuning/control of light emission properties through geometrical resonator design. The most recent developments together with novel strategies for light emission tuning and effective light extraction are presented.

Keywords: organic molecules; single crystals; molecular packing; lasers; optical resonators

1. Introduction

Since the turn of century, organic semiconductor micro/nanocrystals (OSMCs) [1–7] have attracted continuous attention as a promising research topic with promising electronic and optoelectronic applications, including organic field-effect transistors (OFETs) [8–11], organic photovoltaics (OPVs), organic light-emitting diodes (OLEDs) [12,13], photodetectors (PDs) [14,15], and lasers [16–20]. A huge amount of OSMCs have been developed, with several assets over their inorganic counterparts, such us a plethora of molecular structures with diverse properties, low-cost device fabrication, compatibility with stretchy, flexible, and lightweight moldable substrates, etc. [21]. Consequently, the low thickness, light weight, foldability, and stretchability of OSMCs make them suitable, for instance, for flexible or miniaturized organic electronics and optoelectronic applications. On the other hand, OSMCs offer the possibility to be shaped into diverse molecular assemblies, which are finely tuned by crystal engineering [22]. Single crystals of high quality offer long exciton diffusion length and long lifetimes which are attractive for light-to-energy conversion [23,24]. Finally, they present large compatibility with nonexpensive solution-processed methods which can be easily scaled-up for device fabrication [25]. Different techniques such as spin-coating, drop-casting or ink-jet printing can be used with organic single crystals for the development of electronic devices. The high purity and low density of defects in organic single crystals is crucial for the development of high-performance devices and circuits. To optimize device performance, it is necessary to avoid grain boundaries, defects, impurities, and dislocations.

OSMCs can be prepared into different micro/nanostructures with controlled fabrication procedures, enabling for different properties, depending on the nucleation, molecular packing, and assembly [22]. The formation of OSMCs is determined by regular and stretched intermolecular packing among neighboring molecules with moderately weak noncovalent bonds, such as Van der Waals forces, hydrogen bonds, and the π-π interactions. This affords different packing modes that are decisive in the formation of organic micro/nanocrystals which highly influence their optoelectronic properties. These intermolecular interactions are also easily influenced by the external conditions like the light, solvents and the temperature.

The study of both crystal growth and engineering for the preparation of high-quality OSMCs has been widely addressed. An in-depth analysis of the crystal engineering allows getting a variety of different micro/nanocrystals with various physical properties. Herein, one of the most important aspects is the chemical versatility and modular nature of organic materials, allowing for modulation and change in the intermolecular interactions through subtle changes in the molecular structure. Moreover, physical properties such us melting point, sublimation temperature, or solubility are also important aspects to consider for OSMC growth. Thus, many organic semiconductors with diverse molecular structures have been synthesized and described.

Therefore, taking into account the dimensionality and the shape of the microstructure/nanostructure, a control of the different properties can be achieved. OSMCs can be prepared into 1D wires, tubes, 2D-sheets, belts or discs, affording different crystals morphologies [5,7].

Meanwhile, OSMCs development has not only restricted to achieving randomly dispersed organic crystals, but also there have been huge efforts in developing methods for alignment and patterning of OSMCs into ordered arrays, with the goal of obtaining much better device performance [4,26].

There is an important relation between the molecular structures, packing modes, crystal morphologies, and optoelectronic properties (Scheme 1), achieving the final and complex property of the material.

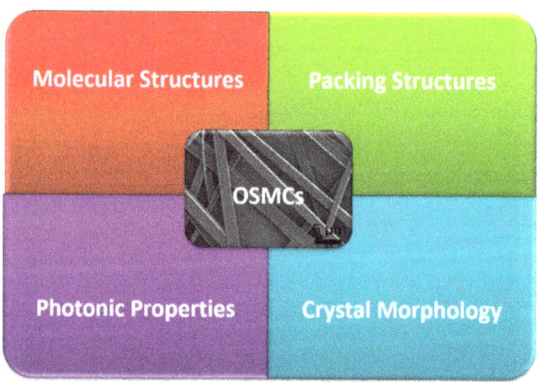

Scheme 1. Relationships in OSMCs among crystal morphology, molecular structures, packing structures and photonic properties.

In this review, we firstly introduce the importance of molecular packing in the crystals. The different noncovalent intermolecular interactions such as hydrogen bonding, π-π stacking, van der Walls forces amongst others, govern the packing arrangement of the molecules, aside from electronic and optoelectronic properties. Subsequently, we discuss the different large number of organic crystal growth techniques for controlling the crystallization of organic semiconductors to get OSMCs. Finally, we discuss the main developments in the field of organic crystal lasers, describing briefly the photophysics and figures-of-merit of OSMCs in terms of optical gain properties, the types of crystalline optical resonators more commonly reported and recent relevant examples of laser cavities based on the different types of crystal resonators.

2. Molecular Packing in Crystals

In organic semiconductors absorption and emission of light implicates an electronic transfer between the highest occupied molecular orbital (HOMO) and the lowest unoccupied molecular orbital (LUMO) which forms and removes an exciton. Optical properties of these semiconductors differ in solution and in the solid state (Figure 1a). In solution organic molecules are surrounded by solvent molecules which implies no disruption in exciton formation, whereas in solid state molecules of the semiconductor are close to each other involving a direct overlap of the molecular orbitals (MOs) of neighboring molecules, creating excitonic couplings. One example is the interaction of transition dipole densities, affecting the optical and electrical properties of the material [27–31].

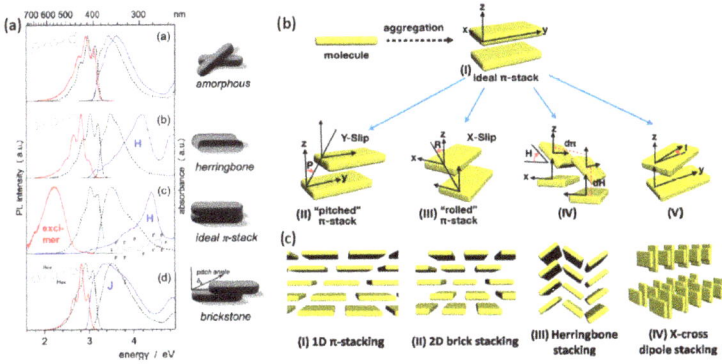

Figure 1. (a) Absorption and emission of distyrylbenzene (DSB)-based materials. Reproduced with permission from ref. [27]. Copyright 2013 RSC. (b–c) Schematic of molecular aggregation for dimers. Reproduced with permission from ref. [32]. Copyright 2018, Wiley-VCH.

As mentioned before semiconductor molecules aggregate by noncovalent, weak intermolecular interactions as hydrogen bonds, π-π stacking, Van der Walls forces amongst others. These forces are the ones that govern the molecular packing the molecule undergoes. There are four typical packings [32] (Figure 1b,c), the first one consisting of two adjacent molecules that are arranged completely face to face, called an ideal π stacking (Figure 1b I), and stands to an H-aggregate in relation to their optical properties. This arrangement gives the largest intermolecular overlap which leads to a decline in the optical properties due to the strong π-π overlap; on the other hand, it is the packing that allows for efficient charge transfer and high mobility alongside the stacking direction. To achieve this packing is hindered by the high electrostatic repulsion of the neighboring molecules [33]. Generally, a face-to-face arrangement involves a slight translation along the Y plane between the two molecules. This situation is referred as a pitched π-stack which is defined by the pitch (P) angle [34,35] (Figure 1b II, c I). By increasing the pitch angle by 50% or more slipping, the H-aggregation changes to a J-aggregation and the overlapping of the orbitals and splitting energies are reduced, providing better emission properties [36–38]. Another way of achieving this is by moving the stack on the X plane, achieving a rolled π-stack (Figure 1b III,c II), defined by a rolled (R) angle. When the π-π overlap is decreased, the exciton created localizes quickly, making the intermolecular vibrations barely participative in the emission spectra. Contrarily, when the π-π overlap increases the charge transfer (CT) increases, promoting an intermolecular separation upon electronic de-excitation which implies a loss of vibronic structure, red-shift and excimer emission features [17]. One example of the pitched and rolled arrangements is 1,4-bis(R-cyano-4-diphenylaminostyryl)-2,5-diphenylbenzene(CNDPASDB) [39–41] (Figure 2a) in which the molecules are stacked but shifted along the X or Y plane. Another arrangement we can encounter is the herringbone (Figure 1b IV,c III) in which one of the molecules is rotated along its long axis with an angle (H), getting an edge to face alignment. This packing decreases the π-π stacking as in the J-aggregates allowing better emission properties. Pentacene and 1,4-bis(4-methylstyryl)benzene (p-MSB) crystals are examples of a herringbone motif [42–44] (Figure 2c). The last arrangement we can encounter is the X-aggregation (Figure 1b V, c IV), when one molecule rotates around the stacking axis but retaining the molecular planes parallel with each other. It is the arrangement that in theory should give rise to the strongest fluorescence properties due to the least π-overlapping and large molecular distance and high carrier mobility, depending on the rotating angle [45–47]. An example of the X-aggregate takes place in perylene-3,4,9,10-tetracarboxylic tetrabutylester (PTE) depicted in Figure 2e wherein one molecule is rotated with respect of the other 70.2° [48].

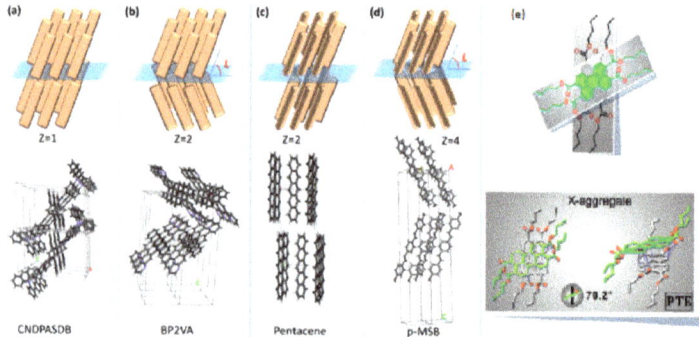

Figure 2. (**a**–**d**). Schematic representation of molecular packings. Reproduced with permission from ref. 50. Copyright 2014, Wiley-VCH. (**e**) Schematic representation of X-aggregate of PTE molecule. Reproduced with permission from ref. [48]. Copyright 2018, Wiley-VCH.

Another aspect to bear in mind is the presence of intralayer molecular interactions which are much weaker than the face-to-face interactions; these interactions produce a tilting in the molecular layer which can be measured by an angle (L) between the normal of the bottom crystal plane and the molecular long axis (Figure 2). Depending on the interactions between the layers there can be a "zig-zag" disposition (Figure 2b,d). Figure 2b for instance displays 9,10-bis((*E*)-2-(pyrid-2-yl)vinyl)anthracene (BP2VA) molecules having a pitch or roll packing but with a "zig zag" arrangement between layers [49,50]. In addition, p-MSB crystal has a herringbone motif but with a "zig zag" arrangement between layers [51].

The different H-, J- and X-aggregate arrangements give raise to changes in the electrical and optical properties depending on the exciton and splitting energy. (Figure 3). Focusing on the optical properties, H-aggregates lead to an absorbance blue shift (hypsochromic) respect to solution concomitant with a low radiative constant (K_r), whereas J-aggregates absorbance exhibit a red shift (bathochromic) and a high K_r [36,52]. In the case of X-aggregates generally the absorbance in solution and in the aggregate itself is similar. As aforementioned H-aggregates often causes quenching in the solid state due to the strong π-overlap, but in the case of J-aggregate or herringbone packing the π-overlap decreases, so the optical properties improve. A strategy to achieve good mobility and emission is to join simultaneously these two, J-aggregate and a herringbone packing, or the use of the X-aggregates which reduces the π-overlap but maintains the planarity [53–56].

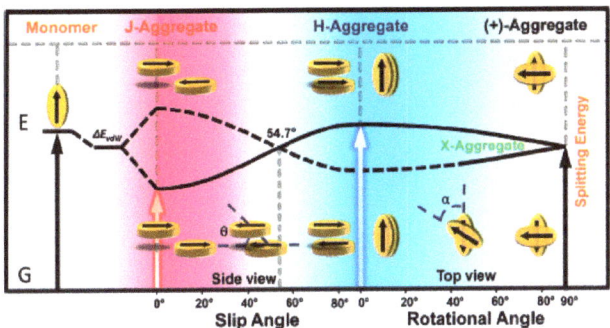

Figure 3. Schematic representation of splitting energy of different aggregates. Reproduced with permission from ref. [48]. Copyright 2018, Wiley-VCH.

3. The Growth Techniques for the Preparation of Organic Semiconductor Micro-/Nanocrystals

The methods used nowadays for inorganic crystal formation are not adequate for organic single molecules due to the harsh conditions used such as high pressure and temperature, hard reaction conditions and numerous solvents used [57].

The preparation methods of OSMCs can be classified into three categories: solution, melting and vapor processing. Solution processing is often used for nonthermally stable materials whereas melting and vapor processing can be used for materials with high thermal stability and low solubility[2].

A wide variety of different techniques for controlling the crystallization of OSMCs have been developed during the last decades [2,3]. A precise control of the crystallization process is key to achieve high quality crystals, therefore the growth method and conditions are essential to the morphologies and the molecular arrangements. The morphology and the stacking of these OSMCs depend on the different conditions in the growth methods. A control in these growth methods should afford a better crystal quality providing better device performance.

3.1. Solution-Processing Techniques

These are the most simple and effective approaches to grow organic crystals because most of organic molecules are soluble in a multitude of organic solvents in a wide range of temperatures and pressures. The concentration in solution increases upon solvent evaporation, reaching a point of saturation where molecules self-assemble creating complex micro/nanocrystalline structures [58–60].

3.1.1. Drop-Casting

The drop-casting method is the most efficient approach to grow OSMCs by self-assembly of organic molecules. The self-assembly process depends on the intermolecular interactions between solvent molecules, organic molecules and organic-solvent molecules. The growth of the crystals with this method requires control of different conditions such as solvent, concentration, atmosphere and temperature. The growth condition is key to optimize the crystal quality. The method consists of dropping a volume of organic semiconductor solution onto a substrate and let the solvent evaporate for several hours or days. Precise control of concentration, temperature, atmosphere and substrate surface enables for the formation of high-quality crystals [61–63]. One example is 9,10-bis(phenylethynyl)anthracene (BPEA), which led to different crystal phases depending on the solvent used (chlorobenzene or dichloromethane) because of the evaporation velocity, which is determined by the interaction between the molecule and the solvent, and whether it is an open system or a quasi-closed system [64]. Another example is diphenylfluorenone (DPFO): adding a solution of this molecule in THF to a substrate leads to microfibers, whereas adding more solution on top caused microfiber redissolution and subsequent formation of microplates upon solvent evaporation (Figure 4a–c)[65].

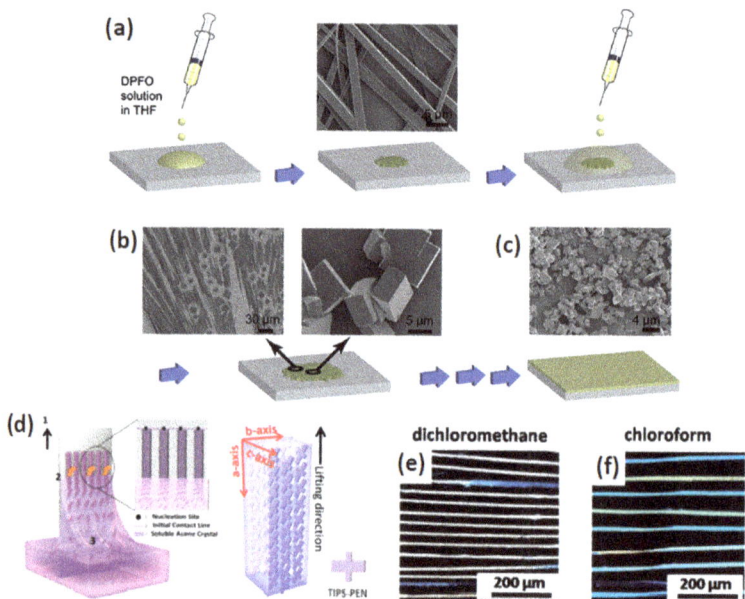

Figure 4. (**a**–**c**) Representation of the steps involved in the drop-casting method of microsized DPFO self-assembled crystals. Reproduced with permission from ref. [65]. Copyright 2015, Wiley-VCH. (**d**) Graphic illustration of dip coating process. (**e**–**f**) Cross polarized optical microscopy images of dip-coated TIPS-PEN crystals from (**e**) dichloromethane and (**f**) chloroform. Reproduced with permission from ref. [4]. Copyright 2016, Wiley-VCH.

3.1.2. Dip-Coating

This technique consists of pulling out a substrate that is immersed in a solution of the organic molecule. This technique allows obtaining organized pattern crystals. Firstly, molecules crystallize in the substrate due to solvent evaporation. Owing to the concentration gradient and capillary forces, more molecules from the solution will move to the contact line depositing more material, crystallizing opposite to the pulling direction. The main parameter to control is the dip-coating speed which strongly influences the morphologies of the crystals formed. Nanoribbon arrays were obtained for instance from a solution of BPEA and triisopropylsilyethynyl pentacene (TIPS-PEN) with this method. Applying pulling speeds higher than 80 μm s^{-1} led to individual nanoribbons whereas lowering the speed below 30 μm s^{-1} led to a conglomerate of nanoribbons (Figure 4d–f) [4,26].

3.1.3. Solvent Exchange

A commonly used method consists of making the solution saturated or hypersaturated by adding an antisolvent. Through diffusion of the solvent and antisolvent the molecule precipitates and self-assembles [66]. In order to implement this method, a few conditions have to be fulfilled: (i) the organic molecule must be soluble in one solvent and insoluble in the other; (ii) both solvents ought to be miscible with one another; and (iii) both solvents should have different densities in order form an interface and avoid rapid mixing which would lead to fast nucleation [67–70]. An example of this process is depicted in Figure 5a. This method led to C_{60} crystals with different plate or rod morphologies by changing the solvents and solvent ratios of CCl_4, m-xylene and isopropanol [71].

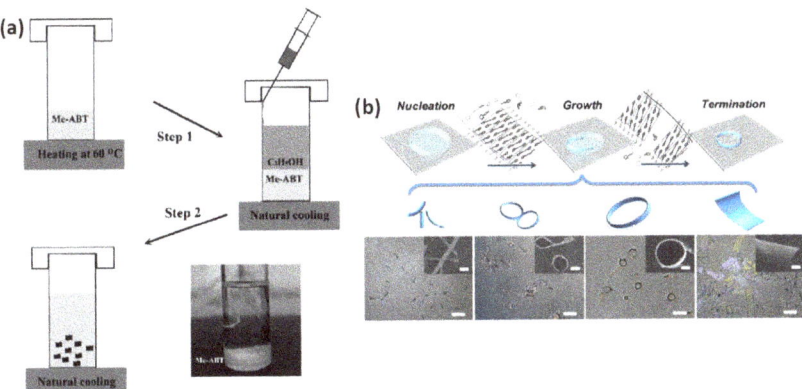

Figure 5. (**a**) Illustration of self-assembly process of 6-methyl-anthra[2,3-*b*]benzo[*d*]thiophene (Me-ABT) through solvent exchange method. Reproduced with permission from ref. [69]. Copyright 2010, Wiley-VCH. (**b**) Diagram of the mechanism form which micro-rings are self-assembled by interfacial tension of liquid drops. Reproduced with permission from ref. [72]. Copyright 2013, Wiley-VCH.

Another interesting example stands for the use of this method to obtain micro-rings; this technique uses the solvent exchange method along with the interfacial tension (Figure 5b). Micro-rings are achieved by adding a droplet of a solution of 1,5-diphenyl-1,4-penta-dien-3-one (DPPDO) (a flexible compound) in ethanol/water into a substrate, the ethanol evaporation triggers precipitation of the compound on the water droplet. Nucleation starts preferentially at the drop edge, whereas the water tension and weak intermolecular interactions enable the wires to bend into micro-rings. In addition, if the concentration is increased to 10 mmol L^{-1}, microtiles are obtained [72].

3.1.4. Solvent Vapor Diffusion (SVD)

When avoidance of solvent mixing is an obstacle or solvents have same densities, solvent-vapor diffusion (SVD) can be used. It is a variant of the solvent exchange method with the difference that the antisolvent is placed outside rather than inside the solution. Slow evaporation of the antisolvent leads to its gradual diffusion inside solutions to mix with the solvent, provoking molecular precipitation and self-assembly. This method reduces the mixing speed of the solvents being able to achieve higher and better quality crystals [73–75]. High quality 2,2′,7,7′-tetrakis(*N*,*N*-di-*p*-methoxyphenyl-amine) 9,9′-spirobifluorene (spiro-OMeTAD) crystals were developed with this method (Figure 6) [75]. As shown in Figure 6a, the inner vial contains the spiro-OMeTAD solution in DMSO at a concentration of 1 mg/mL, whereby the outer contains the nondissolving methanol. When the diffusion of methanol vapor proceeds slowly to the inner vial at room temperature, it provokes a sustained reduction of solubility of spiro-OMeTAD in the methanol-enriching solution, finishing in a supersaturation state that triggers its crystallization (Figure 6b,c).

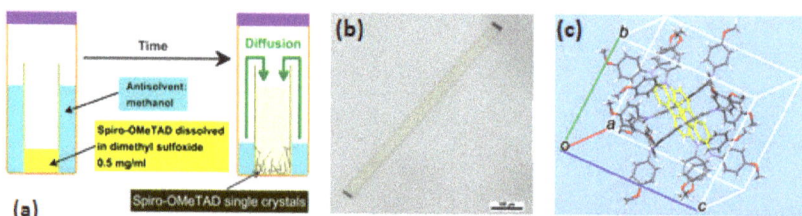

Figure 6. Crystal growth, shape, and crystallography. (**a**) Schematic diagram of the crystallization process. (**b**) Confocal optical microscopy image of a spiro-OMeTAD single crystal. (**c**) Unit cell of the single-crystal structure of spiro-OMeTAD (the fluorene fragments are highlighted in yellow). Reproduced with permission from ref. [75]. Copyright 2016, AAAS.

Single crystals of diperylene bisimide were also grown through this strategy. In this case the inner vial contains the solution of the bisimide in toluene and the outer vial contains methanol as the antisolvent [76].

3.2. Melting Processed Crystals

The melt crystal growth methods, such as Bridgman and Stockbarger, Czochralski, or floating zone methods, are often used for growing large crystals of inorganic semiconductors, however these methods have also been used for organic single crystal growth (Figure 7) [77]. Normally, these methods are known as zone refining, zone melting or zone freezing technique. They have been less employed for growing organic crystals due to the high vapor pressure and chemical stability of organic molecules around melting temperatures.

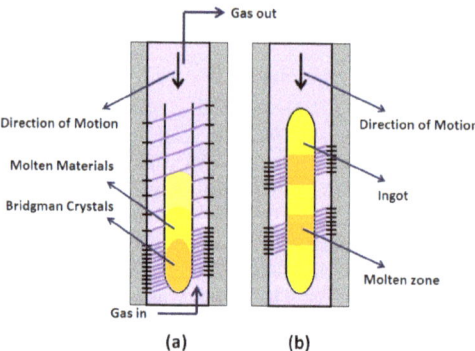

Figure 7. (**a**) Bridgman and (**b**) zone melting method utilized for organic single crystals growth.

This method, however, has some constraints: (i) molecules must have a well-defined melting point; (ii) large thermal stability at fusion temperature is required; (iii) molecules must have low chemical activity; and (iv) they must be extremely pure because this technique is highly sensitive to impurities [57].

Among these requirements, the main challenge concerns with the thermal stability of the organic compound. In addition, these methods require a large amount of material, and relatively expensive apparatus. A way to circumvent this problem consists of growing the crystals between two glass or quartz slides [78]. Thiophene-phenyl-pyrrole (TPP) crystals were grown by this method upon placing the material between two quartz substrates in a hot stage, obtaining flat crystals that show waveguiding properties (Figure 8a) [79].

3.3. Vapor Processed Crystals

As mentioned above this method is used primarily for molecules with low solubility and high thermal stability. It involves phase transitions between solid, liquid and vapor phases.

3.3.1. Physical Vapor Transport (PVT)

The most common technique is physical vapor transport (PVT). First proposed by Kloc and Laudise et al. [80,81] became one of the most popular methods for growing organic crystals. It consists of heating the material, most of the cases under vacuum so that the boiling point of the material lowers. The sublimated material is then transported to a lower temperature zone by an inert gas where it crystallizes. By this method impurities also crystallize in front or behind the crystallization zone. The control parameters in this technique are carrier gas flow, temperature gradient, and vacuum level (Figure 8b) [82–84]. Doped crystals of BSB-Me with tetracene and pentacene were grown with PVT using just the doped powder, to obtain crystals thicknesses of less than 400 nm and length of several millimetres (Figure 8c) [85].

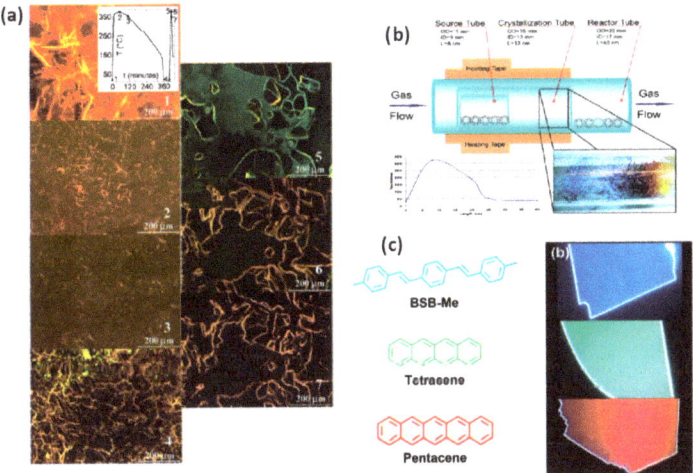

Figure 8. (a) Crystalization process from melted TPP. Reproduced with permission from ref. [79]. Copyright 2010, American Chemical Society. (b) Illustration from PVT and crystal growth of pentacene. Reproduced with permission from ref. [84]. Copyright 2005, American Chemical Society. (c) Doped crystals of BSB-Me with tetracene and pentacene grown by PVT. Reproduced with permission from ref. [85]. Copyright 2017, Wiley-VCH.

3.3.2. Microspacing Sublimation

Crystals grown from PVT are of high quality, but the growth requires high vacuum environment, inert flowing gas, high control of the temperature and expensive apparatus. In this sense Xutang Tao et al. reported a new method to obtain organic crystals through sublimation based on microspacing distance, which is low cost and requires less parameter control (Figure 9) [86]. It consists of heating the organic molecule in a hot stage deposited in a substrate until sublimation to the upper substrate, separated around 400 μm. The evaporated material condenses in droplets on the upper substrate, from which crystals grow. They obtained high quality crystals of anthracene, perylene, pentacene, pyrene among others with sizes from 10 to 50 μm [87,88].

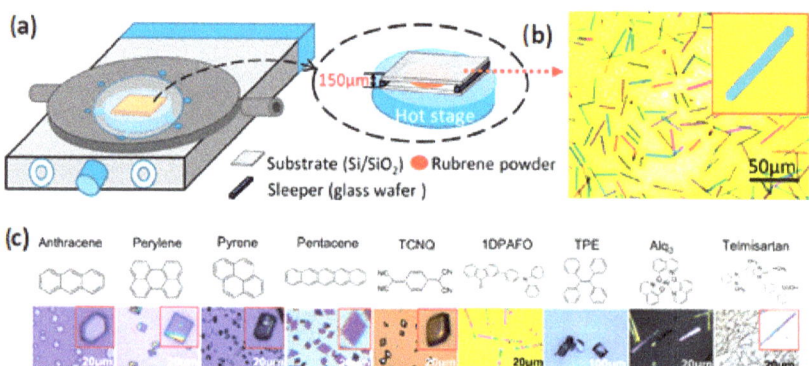

Figure 9. (a) Schematic representation of the microspacing sublimation technique. (b) Rubrene crystals grown from microspacing technique. (c) Crystals grown by microspacing sublimation. Reproduced with permission from ref. [86]. Copyright 2018, American Chemical Society.

4. Organic Solid Lasers

A laser consists of an optical gain medium located in an optical cavity providing optical feedback in one, two, or three directions. Light is generated inside the medium by electrical (electrically-pumped lasers) or optical (optically-pumped lasers) stimuli and amplified by stimulated emission. Organic π-conjugated materials exhibit very interesting features as active media in laser devices. They show high room-temperature photoluminescence quantum efficiencies (PLQE) [89,90] which translates into large stimulated emission (SE) cross-section values and their photoluminescence (PL) spectra can be tuned through chemical functionalization [91]. Moreover, organic lasers can be processed by cost-effective solution-based methods [92], they have capabilities to confine and guide light due to their elevated refractive indexes [93] and they possess unique mechanical properties (flexibility and light weight) leading to new potential market niches for organic laser devices. From a more intrinsic point of view, organic semiconductors are potential four-level laser systems, (see scheme in Figure 10a). In this configuration excited states generated by an electrical or optical perturbation relax efficiently to the lowest excited state. Depending on the nature of the organic molecule such state can have a local excited [94,95], or intramolecular charge transfer character [96]. In all cases, fast internal conversion and vibrational cooling leads to fast pumping of the lowest excited state from which, radiative emission to the upper vibrational levels of the ground state occurs. Given that these upper vibrational levels are empty at room temperature, stimulated emission takes place without competition with resonant ground state absorption. This lack of competition is what endorses conjugated molecules with unique features for optical gain, thus guaranteeing laser action at low optical pumping thresholds. Another interesting feature relies on the fact that emission is displaced from the absorption spectral region by the so-called Stokes shift, minimizing re-absorption. Furthermore, spectral displacement can be achieved through the realization of co-crystals based on donor-acceptor moieties which also enable to span the luminescence across the visible [97,98]. A myriad of organic crystals are found to combine high PLQE and outstanding charge transport [85,99–101], paving the way for the realization of electrically-pumped lasers, the first demonstration being recently reported in 2019 [102].

A basic example of an organic laser cavity is composed by and organic semiconductor film located between two metallic or dielectric mirrors or gratings, either in external [103] or in an integrated microcavity geometry [104]. Other types of organic laser geometries are mirror-free surface-emitting distributed feedback (DFBs) cavities [105,106] where a grating is imprinted on (or formed by refractive index variation of) the gain medium, producing feedback in one direction. In both cases waveguiding by total internal reflection due to the difference in refractive indices at the organic layer boundaries is required to confine the light in the direction perpendicular to the layer.

Organic crystals provide the possibility to merge optical gain and optical feedback due to the presence of well-defined crystal faces enabling optical confinement through total internal reflection inside the crystal cavity. Despite the vast number of reports on optical gain from organic crystal lasers, the majority of them refer to processes which are not supported by cavity geometrical resonances, (e.g., random lasing or amplified spontaneous emission (ASE)). ASE does not require optical feedback because light amplification takes place by a single pass along the optical gain medium. ASE is typical of organic waveguides (slabs, 1D planar waveguides or optical fibers) [107–112] but can also be supported by organic crystals [68,113,114]. The ASE output is constituted by a spectrally broader emission linewidth (~10 nm) corresponding to the amplified waveguided mode along the crystal. Random lasing is instead associated to multiple scattering effects given raise to closed loops and optical feedback. This process is often observed in crystals which present refractive index inhomogeneities including crystal dislocations, different morphology areas or impurities. These inhomogeneities lead to multiple scattering and formation of resonances via random walk. They often display multiple lasing modes and their emission cannot be controlled though crystal design owing to their disorder nature [115–117].

Hereafter, we will address organic crystal lasers constituted by optical microresonators formed by boundaries of the organic gain medium itself and whose resonances are determined by the microresonator geometry and the refractive index of the medium. Different organic crystal microresonators with defined geometries have been reported, ranging from wires [118,119], fibers [120], rings [121,122], polygonal cavities [123,124], slab crystals [125] or disks [126].

4.1. 1D Fabry-Perot Resonators

Fabry-Perot (FP) cavities are typical of one-dimensional microstructures such as wires, fibres and hollow fibres, where photons undergo total internal reflections at the cavity walls and bounce back and forth at the cavity facets, leading to a periodical interference pattern (Figure 10b). Light confinement becomes observable when the diameter of the cavity approaches the wavelength of the light. Dimensions below the wavelength give raise of strong diffraction effects lowering considerably the confinement efficiency, which is given by the fractional mode power within the core.

$$\eta = 1 - \left(2.405 \exp\left[-\frac{1}{V}\right]\right)^2 V^{-3} \quad (1)$$

where $V = \pi d/\lambda \, (n^2 - n^2_0)^{0.5}$, d is the wire diameter, and n and n_0 stand for the refractive index of the wire and surrounding medium (air). For n ~ 1.7, characteristic of an organic medium, and λ = 460 nm, a confinement efficiency of > 85% is expected when r amounts to 300 nm [127]. Spontaneous emission gives rise to a given spectral distribution $I(\lambda)$, the light outcoupled from the Fabry-Perot (FP) resonator being given by:

$$I_t = I(\lambda) \frac{(1-R)^2}{(1-R)^2 + 4R\sin^2\left(\frac{2\pi}{\lambda}L\right)} \quad (2)$$

where R and L stand here for the reflectance at the cavity facets and the length of the cavity respectively. The light outcoming the FP displays characteristic periodical resonances arising from longitudinal modes, spectrally separated by $\Delta\nu \sim c/2\,nL$. The number of longitudinal modes that a certain cavity can support is given by the ratio $\Delta\nu_{sp}/\Delta\nu$, where $\Delta\nu_{sp}$ stands for the spectral bandwidth of spontaneous emission.

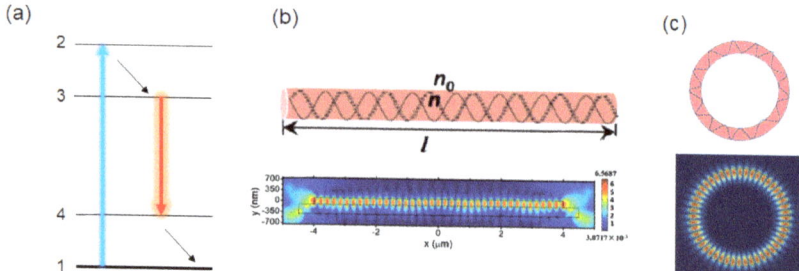

Figure 10. (**a**) Simplified 4-level laser system diagram involving photoexcitation (1→2), fast relaxation to the lowest excited state (2→3), spontaneous and stimulated emission (3→4), and ground state recovery (4→1). (**b**) Optical waveguiding in a FP wire resonator showing optical confinement due to reflection at the wires tips (up) and distribution of the internal electric field (down). Reproduced with permission from ref. [119]. Copyright 2017 Wiley-VCH. (**c**) Optical confinement on a WGM ring resonator (up) and internal distribution of the electric field (down). Reproduced with permission from ref. [128]. Copyright 2020 Wiley-VCH.

4.2. Whispering Gallery Mode Resonators

Whispering gallery mode (WGM) resonances are a tangible phenomenon when referred to sound waves. Many of us has surely experienced before the capability of curved walls from arches or domes to propagate sounds as weak as whispers. Like sound, electromagnetic waves experience an analogous effect. In curved surfaces based on highly dense medium, light experiences multiple total internal reflections at the medium-air interface leading to closed loops where light interferes constructively. The result is a standing wave pattern distributed along the curved surface (Figure 10c) [122]. WGMs are present in spheric and hemispheric cavities and cavities with circular surfaces like cylinders, or rings. Cavities of polygonal shape can also support WGMs. The number of loops that photons undergo before leaving the cavity is given by the quality factor (Q). This magnitude is defined as:

$$Q = \frac{\lambda}{\Delta \lambda} \quad (3)$$

and expresses the capability of the cavity to trap light. Q factors in WGM resonators can be as high as 10^{11} although in organic WGM resonators values typically range between 10^3 and 10^4 [128]. A direct consequence of a high Q-factor is the concentration of high optical power in the resonator giving raise to strong-light matter interactions. WGM resonators can be applied to the development of lasers with low pumping thresholds and very narrow linewidth. In these cavities the spectrum of the confined light experiences a rippling ascribed to the multiple confined modes. The spacing between these resonances is inversely proportional to the cavity diameter within the geometrical approximation:

$$\Delta \lambda = \frac{\lambda^2}{\left(n - \lambda \frac{dn}{d\lambda}\right) L} \quad (4)$$

A direct implication of EQ.3 and EQ.4 is that the Q factor increases with increasing the cavity diameter. An exact treatment requires the solution of the spherical or cylindrical vectorial electromagnetic boundary problem, i.e., Mie theory or the equivalent for cylindrical structures. Transverse electric (TE) and transverse magnetic (TM) solutions result from these calculations. Whereas microresonators generally support various lasing modes within the spectral range of the gain material, it is possible to achieve single mode lasing by coupling two such resonators of different sizes [129]. An interesting feature of WGM cavities is that they have extremely high Q-factors, meaning that photons in these structures undergo many round trips, which makes them interesting for low threshold lasing [130] and highly sensitive transduction of physical or chemical perturbations [131]. Consequently, WGM cavities have received an ever-increasing interest for biosensing applications [132], imaging in biological media [133], and physical sensing [131].

4.3. Lasers Based on 1D Fabry-Perot Resonators

Organic molecules arranged in one-dimensional structures supported by intermolecular interactions, such as hydrogen bonding, π-π or halogenated bonds display unique photonic properties in terms of light transport and optical amplification. Structures with highly defined flat end faces can behave as efficient FP cavities producing the required feedback to achieve laser action (Figure 11a,b). Crystalline one-dimensional wires are one example [134–136]. The length and width of the wire as well as the molecular orientation determine whether the FP cavity is constituted along the wire, or between the lateral faces of the wire. A clear indication of the resonance direction comes from analysis of wires with different lengths (L) and the assessment of the inversely proportional relation with mode spacing ($\Delta\lambda$) given by EQ.4 (Figure 11c,d). Crystalline wires with longitudinal FP modes (i.e., those yielding from oscillations between the wire end tips) were for instance reported by Wang et al. based on a methoxyphenyl-hydroxynaphthalen (DMHP) derivative [119]. Single crystals of this compound obtained with the solvent-exchange method were composed of molecules arranged in orthorhombic structure with a unit cell composed of four symmetrically equivalent molecules, growing in one direction. The resulting wires had lengths ranging from 5 to 100 μm and widths from 50 to 250 nm with very smooth lateral faces of few nanometers roughness which displayed red emission and PLQEs of 32%. The PL spectrum was decorated by the characteristic periodically displaced modes attributed to FP cavities and the separation between the modes followed a linear relation with $1/L$. Multimode lasing centered at 720 nm was observed upon pumping with fluences above the 1.4 μJ cm^{-2} lasing threshold.

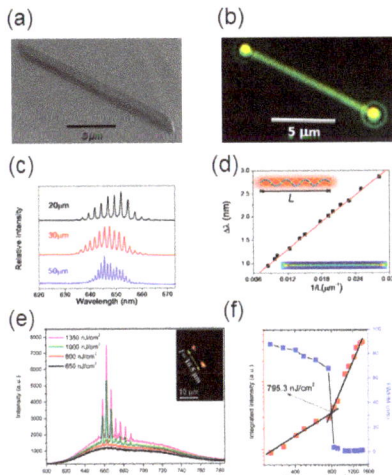

Figure 11. (a) Scanning electron microscope image of a BP2T-CN crystal wire. (b) Fluorescence microscope image of same type of wire. The intense luminescence emitted at the wire tips is the result of effective fluorescence waveguiding. Reproduced with permission from ref. [135]. Copyright 2017 Wiley-VCH. (c) Typical FP modes emitted by S-BF$_2$ nanowires of different lengths. The effect of length is manifested in an increase in the spacing between the FP modes. (d) Dependence of optical mode spacing with the wire length. Reproduced with permission ref. [136]. Copyright 2017, American Chemical Society]. (e) Multimode laser action in DMHP crystal wires. (f) Integrated emission output and full width half maximum of the emission spectrum as a function of pump fluence showing the lasing threshold. Reproduced with permission from ref. [119] Copyright 2017 Wiley-VCH.

Wires with large widths can in some cases support FP cavities defined between the lateral faces. An example is the work by Fu and co-workers on 1,4-dimethoxy-2,5-di[4′-(methylthio)styryl]benzene (TDSB) [118]. Single crystals of TDSB arrange into H-aggregates forming a monoclinic structure. The resulting wires have a characteristic rectangular shape of 0.5–2 m width and variable lengths up to 100 μm. Herein, the tight molecular co-facial packing and strong electron-phonon coupling in the TDSB crystals enhance the oscillator strength of the H-aggregate allowed transition leading to a PLQE of 81%. Differing from the previous example, the cavity mode spacing arising from these wires was independent of the wire length but followed a linear inverse dependence with the cavity width, confirming that the FP cavity is built-in between the wire lateral faces. Restricting the dimensions of the same wires in the transversal direction can be an effective tool to design the resonant cavity geometry. This can be achieved for instance with capillary bridge lithography [123,137]. This method exploits de-wetting of a liquid film deposited on a substrate by placing the solution in contact with a template of periodically arranged ribs with their surface modified with heptadecafluorodecyltrimethoxysilane (Figure 12a–c). Capillary forces drive the solution underneath the pillars where slow solvent evaporation triggers molecular nucleation leading to crystal patterns which follow the rib shape. Capillary bridge lithography was employed to obtain TDSB wires of same crystal structure and rectangular section as those obtained through self-assembling in solution but of lesser width (0.5 × 0.5 μm) [123]. Detailed analysis of the PL spectra and mode spacing of wires with different length confirmed that the resonant modes arose from longitudinal optical oscillations between the wire tips.

FP wires typically exhibit multimode lasing although single mode lasing can be achieved in short length cavities where the mode spectral separation $\Delta\lambda$ is larger than the optical gain spectral bandwidth of the organic crystal. Liao et al. reported multimode lasing in microbelt-shaped crystals of 1,4-dimethoxy-2,5-di[4′-(cyano)styryl]benzene, an organic compound which displays J-aggregation [138]. Microbelts supported FP resonant modes between the lateral faces and displayed different mode spacing dependent on the microbelt width. Some of these crystals with the smallest widths (625 nm) gave rise to single mode lasing at 531 nm. In principle, single mode lasing is particularly interesting because it lacks from competition between multiple modes, which a priori would lead to lowering of the lasing thresholds. Nevertheless, the threshold achieved in these single mode cavities were somewhat larger than similar multimode cavities, a result which could be explained by the individual characteristics of the measured microbelt.

Regarding the laser emission characteristics of organic FP resonators, the different works report laser lines which span from 500–750 nm depending on the compound, with lasing thresholds which are typically in the 0.1–10 μJ cm^{-2} range achieved by pumping with 100–300 fs pulses at 1 kHz repetition rate, (see Table 1 and Figure 13 for a detailed description). The spectral position of the lasing lines is determined by the gain spectral bandwidth, which typically corresponds to the bandwidth of the 0–1 vibronic PL peak, and the cavity modes supported within this bandwidth. Interestingly, in some cases optical gain from 0–1 and 0–2 can coexist leading to simultaneous lasing at two spectral regions. Wang et al. demonstrated that short wires of DMHP exhibited multimode lasing centered at 660 nm whereas long wires exhibited a shift of the peaks towards 720 nm, coinciding with the 0–1 and 0–2 vibrational PL peaks of DMHP [119]. This effect was explained as due to the long DMHP absorption tail extending down to the 0–1 PL spectral region. Under these circumstances, losses by self-absorption at 660 nm become important and lasing can only be achieved through stimulating the emission from the 0–2 transition, although at expenses of a significant increase in lasing threshold. Wires of 30.7 μm length exhibited dual laser emission centered in these two spectral regions.

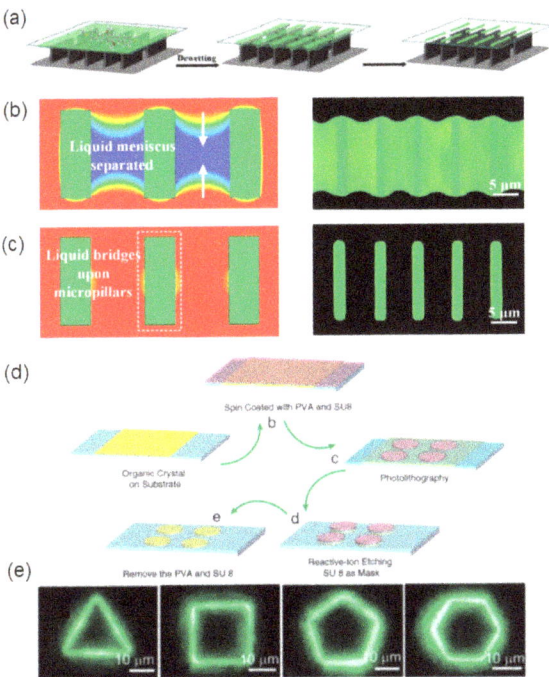

Figure 12. (**a**) Description of the capillary bridged lithography method. A template composed of ribs is placed in contact with a liquid film deposited on a substrate. (**b**) Simulation (left) and fluorescence image (right) of the liquid undergoing de-wetting. (**c**) Asymmetric wettability between the substrate and the ribs drives the liquid underneath the ribs (left). Subsequent solidification leads to crystalline geometrical patterns which reproduce the template (right). Reproduced with permission from ref. [123]. Copyright 2017 Wiley-VCH. (**d**) Mask-assisted photolithography following five steps: (i) physical vapor deposition of the organic film, (ii) subsequent coating with PVA and SU8, (iii) space-selective light irradiation, (iv) reactive ion etching of SU8 nonprotected areas and v) PVA lift-off. (**e**) Fluorescence image of different geometries which can be deployed with this method. Reproduced with permission from ref. [121]. Copyright 2012 Wiley-VCH.

Table 1. Crystallographic and photophysical emission data of organic π-conjugated materials.

Molecule	Crystal Structure	Type of Cavity	Dimensions (W × L)	Lasing Threshold (µJ cm^{-2})	Q-factor	Lasing Wavelength (nm)	PLQE (%)	Ref
TDSB	H-aggregates monoclinic, space group of P 2 1/c	wires, slab FP	0.5–2µm (rectangular section) × 10–30 µm	0.1	1000	500	81	[118]
DMHP	orthorhombic, space group of Pca2$_1$	wires, FP	0.05–0.25 µm × 5–100 µm	1.4	-	660–720	30	[119]
TDSB	H-aggregates, monoclinic, space group of P 2 1/c	patterned wires, FP	0.5 µm (square section) × 10–30 µm	0.4	-	500	-	[123]
S-BF$_2$	monoclinic	wires, FP	2 µm × 20–100 µm	12.8	850	645	10	[136]
COPV	J-aggregates, monoclinic	microbelts, slab FP	0.3–0.6 µm × 1–10 µm	1.09	868	525	58	[138]
DMHC	Anhortic, space group P-1	patterned wires, FP	0.3 µm × 1 µm	9.9	2340	775	2	[137]
BP2T (BP2T-CN)	monoclinic P21/c (P21)	microbelts, FP	1 µm × 5-100 µm	11	354 (377)	535	-	[138]
HDMAC	monoclinic	wires FP; square microdisks WGM*	500 × 10–30 µm; 2–10 µm, 600 nm height	1.05; 0.43	~7500 in microdisks	650	30	[139]
DASB	Monoclinic P2$_1$/c	octahedron crystals WGM	> 1 µm side of lateral faces	6.9	>1500	531	30	[124]
DSB	H-aggregates herringbone, orthorhombic	hexagonal microdisks WGM	1–5 µm edge	0.79	210	440	65	[126]
BP1T, (BP2T)	-	patterned multishape microdisks WGM	Variable circular radii/ polygonal > 1 µm edge	88 nJ	2030	500, (525)	36, (59)	[121]
HDFMAC		micro-rings WGM	10–30 µm diameter, 250 nm width, 2.5 µm height	14.2	2000–4000	647	23	[122]
COPV2, (COPV3)	monoclinic P2$_1$/n	rhombic microdisks quasi-WGM	4 × 35 µm Edge	18	3400	460, 490, 510	65–76	[140]

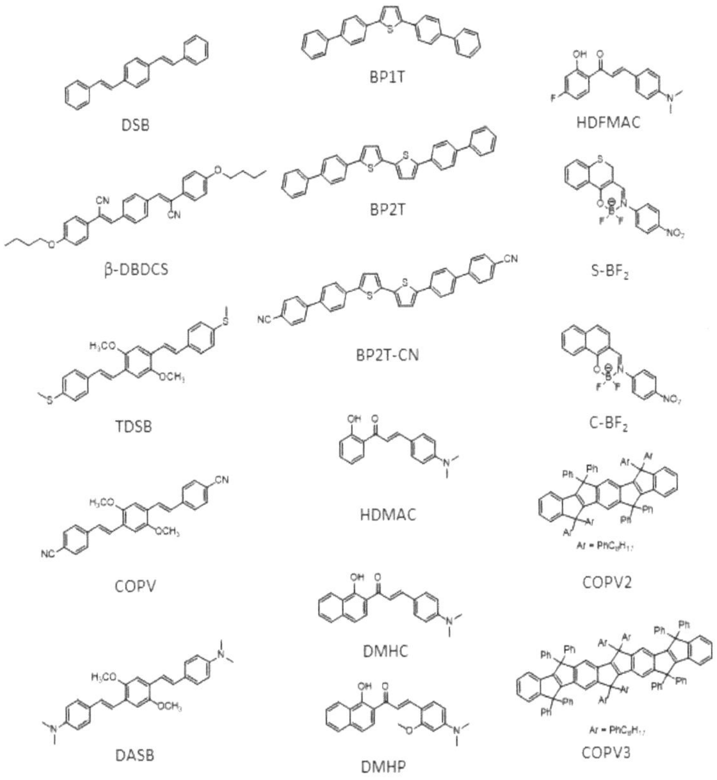

Figure 13. Chemical structures of compounds featuring in Table 1.

1.4. Lasers Based on Whispering Gallery Mode Resonators

Organic molecules with tendency to self-assemble along two crystal growth directions rather than one preferential direction offer the possibility to achieve 2D and 3D crystalline geometries with capability to behave as WGM optical resonators [128]. This is the case for instance of p-distyrylbenzene (DSB) or 1,4-dimethoxy-2,5-di[4′-dimethylaminostyryl]benzene (DASB). DSB self-assemble into lamellar herringbone aggregates leading to an orthorhombic lattice with perfect square symmetry, depicting sharp spots in the selective area diffraction pattern characteristic of a single crystal [124]. The resulting crystals are hexagonal microdisks with edges of 1–5 µm size and thicknesses of about 350 nm (Figure 14a). These crystals are highly fluorescent with typical PLQEs of 65% arising from the large oscillator strength of the transition between the upper level in the H-band and the ground state. The PL spectrum of a single hexagonal disk appears decorated with resonant modes, their density becoming larger as the size of the hexagon increases. Information into the exact geometry of these resonances inside the hexagons is inferred from the single crystal fluorescence images and from the refractive index group estimated from mode spacing analysis, assuming three possible resonance geometries. The fluorescence image of a given microdisk (Figure 14b) depicts an alternate bright-dark edge, thus confirming that a closed loop involves only three total internal reflections (3-WGM or 3D-WGM) at alternate faces, (Figure 14c). Among these two resonance geometries, D3-WGM is more plausible based on the feasible group refractive index value estimated from EQ.4. The resulting hexagonal microdisks behave as laser cavities with lasing thresholds as low as 0.79 µJcm^{-2} and the possibility to transit from multimode to single mode lasing by shrinking the hexagon edges from 5 to 1 µm (Figure 14d). More complex octahedron microcrystals were obtained from self-assembling of DASB which showed PLQE values of 30%, and visible WGMs in the PL spectrum which were rationalized as closed loop formed by six total internal reflections at six different crystal edges [126]. In line with DSB hexagonal disks, DASB octahedrons also led to multimode lasing with the possibility of achieving single mode lasing by restricting the cavity size. The Q-factors of these crystals ranged from 1500 to 7900. The Q-factors of WGM resonators are typically larger than those from FP resonators. This is basically a consequence of the WGM resonator geometry which facilitates total internal reflections because light travels more parallel compared with close to normal incidence at the FP ends. Following this same argument, larger Q-factors are expected from WGM resonators of smaller curvatures (e.g., polygonal cavities of high order) or of larger sizes [121,139].

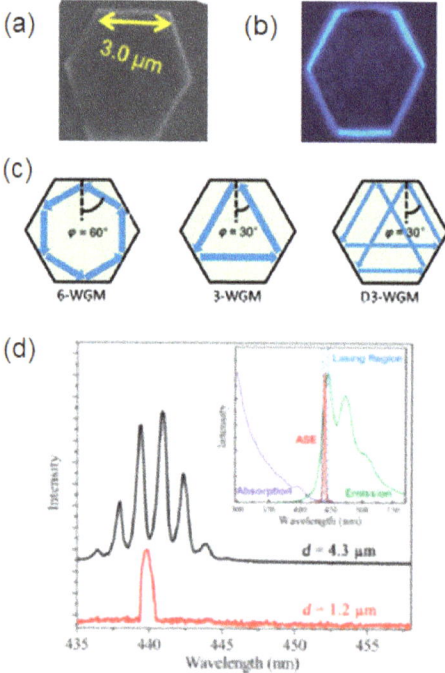

Figure 14. (a) Scanning electron microscope image of a DSB hexagonal disk and (b) corresponding fluorescence image. The alternate bright edges support the existence of only three reflecting edges. (c) Possible geometric resonances that the disk can support. (d) Multimode and single mode lasing in disks of 4.3 μm and 1.2 μm perimeter. Reproduced with permission from ref. [126]. Copyright 2014, Wiley-VCH.

Control of light emission in these cavities has also been achieved by patterning [121,123] and solvent-assisted methods [139]. Growth of a OH-substituted 3-[4-(dimethylamino)phenyl]-1-(2-hy-droxyphenyl)prop-2-en-1-on (HDMAC) into 1D-wires or 2D-microdisks was achieved through self-assembling from protic or aprotic solvents respectively [139]. The molecules pack along the [010] and [002] crystal direction with the OH- groups piling along this plane. Hydrogen bonding of protic solvent molecules at the OH position interferes with [002] growth leading to 1D wires along the [010] direction. The use of aprotic solvents enables instead the formation of rectangular microdisks with WGM resonance loops defined by four reflections at each of the disk faces. TE and TM polarized modes characteristic of WGM resonators were clearly visible in the PL spectrum. Large resonators of about 55 μm round trip exhibited the largest Q-factors (~7000). The lowest laser thresholds achieved were 0.4 μJcm^{-2}.

Control of sizes and shapes in WGM resonators has also been achieved by lithography methods. Capillary bridged lithography was applied by Honbing Fu and co-workers to obtain indistinctly WGM and FP resonators [123]. They obtained micro-rings with different number of WGMs upon tuning the diameter of the pillar template which determines the ring diameter. Thus, 8, 12 and 16 μm diameter rings were achieved which led to 3, 5 and 7 optical modes in the PL spectra. Multimode lasing was achieved with the lowest threshold at 0.3 μJcm^{-2}. The validity of this method to achieve small diameter WGM resonators capable to support single mode lasing remains nevertheless under question due to the low mechanical resilience of organic crystals which could develop cracks when patterned into high curvature surfaces. A different patterning method to achieve WGM resonators of controlled size was reported by Fang et al. based on reactive-ion etching (Figure 12d) [121]. A typical realization consisted of depositing either a 5,5'-bis(4-biphenylyl)-2,2'-thiophene (BP1T) or a 5,5''-bis(biphenyl-4-yl)-2,2':5',2''-bithiophene (BP2T) layer of nanometer roughness by PVD onto a substrate and spin-coating subsequently two layers of PVA and SU8 photoresist. The desired motifs were transferred into SU8 by UV lithography and temperature annealing leading to coated patterns of PVA and irradiated SU8 on top of the crystalline film. This method required applying baking temperatures below 70 °C to avoid organic film cracking. The interstitial film between the motifs was then removed by reactive ion etching. Finally, the PVA/SU8 capping was lift-off using an appropriate orthogonal solvent. This method enables the development of a wide range of polygonal disks with triangular, square, circular, pentagonal, hexagonal or star-shape sections and of different sizes, (Figure 12e). Although it offers an attractive way to shape the resonator, perhaps the most interesting result in practical terms is the possibility to control the resonator size, because the standard self-assembling methods give raise to a wide distribution of sizes. The best performing WGM resonators were those patterned into rings showing Q-factors of 2030 and narrower lasing modes respect to polygonal cavities.

An important aspect of optical resonators is the development of strategies to enhance light outcoupling. In WGM resonators this has been envisaged by means of using external photonic structures (generally wires). The main challenge here is to avoid interference of the outcoupling structure on the waveguiding properties of the WGM resonator. Lv et al. demonstrated that 3-[4-(dimethylamino)phenyl]-1-(2-hydroxy-4-fluorophenyl)-2-propen-1-one (HDFMAC) self-assembles into thin micro rings with a width smaller than the propagating wavelength and very high aspect ratio (250 nm width, 2.5 m height) [122]. This geometry leads to considerably leakage of the transversal electric (TE) mode, polarized parallel to the substrate, whilst the transverse magnetic (TM) mode locates far away from the substrate, avoiding substrate-induced propagation losses and enabling light outcoupling to an external photonic structure.

By changing the proportion of ethanol:dichloromethane in the master solution, HDF-MAC was observed to assemble into micro-rings tailed with microbelts (Figure 15a,b). Microbelts outcoupled efficiently the from the resonator. The PL collected from the microbelt tip displayed characteristic WGMs and the lasing threshold was weakly altered by the presence of the microbelt tail (14.2 and 15.9 μJcm^{-2} without and with microbelt) (Figure 15c,d). A different strategy exploited plasmon-photon coupling to create heterostructures composed of dye-doped polystyrene microspheres with 4–20 mm diameter with tangentially connected Ag nanowires of 170 nm diameter (Figure 15e,f) [141]. The surface plasmon polaritons (SPP) of the nanowires were efficiently launched by the optical modes confined in the microdisk. High photon-plasmon coupling efficiency between 1 to 3.5% were achieved as the result of WGM-SPP modes momentum match at the interface due to the tangential connection between the resonator and the nanowire. Interestingly, the mode distribution at the nanowire output was strongly dependent of the nanowire length, owing to the different dampings of modes located at low and high frequencies. This strategy can serve as a way to exploit the nanowires as frequency filters.

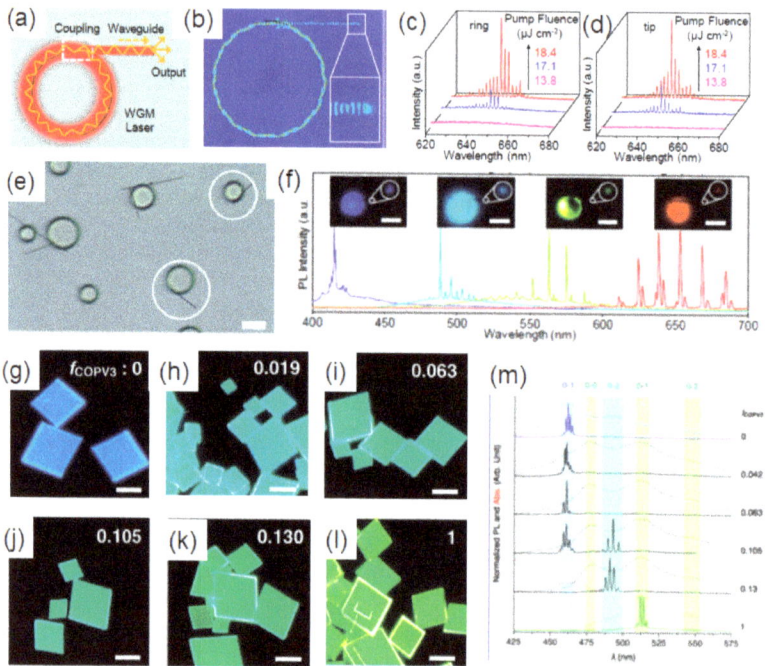

Figure 15. (**a**) Micro-ring coupled to a microbelt. (**b**) Optical simulation of the electric field inside the structure showing its distribution along the ring and the microbelt. (**c**) Laser emission detected from the ring and (**d**) from the microbelt tip. Reproduced with permission from ref. [122]. Copyright 2019, American Chemical Society. (**e**) Scanning electron microscopy images of microdisks coupled to plasmonic Ag wires (scale bar 5 mm) and (**f**) laser emission output detected from the nanowire tip. Reproduced with permission from ref. [141]. Copyright 2017, American Chemical Society. (**g–l**) Fluorescence image of rhombic COPV2:COPV3 co-crystals with different COPV3 content (scale bars 10 mm) and (**m**) tuning of laser emission by FRET in the co-crystals. Reproduced with permission from ref. [140]. Copyright 2018, American Chemical Society.

Reducing the surface contact of the resonator with the substrate is an interesting way to favor light outcoupling and reduce optical losses induced by substrate leakage. Okada et al. demonstrated how the use of silver-coated substrate can favor the formation of edge on instead of face on crystals [140]. Enhanced optical confinement in this case enabled to reduce the lasing threshold by a four-fold. In this work, tuning of laser emission output was also achieved by developing WGM resonators based on rhombic co-crystals of blue and green emitting COPVs coupled by Förster resonance energy transfer (FRET) (Figure 15g–l). By tuning the concentration of the blue (donor) and green (acceptor) compound in the cocrystal, the FRET rate was tuned above or below the lasing rate of the donor, enabling for green or blue lasing. At a critical donor:acceptor concentration simultaneous blue and green lasing was observed (Figure 15m).

5. Conclusions

Over the recent few decades, OMSCs have involved increasing attention as a promising field of knowledge and technology that involves physics, chemistry and materials science. The chemical versatility and flexible nature of the organic materials afford different nucleation, molecular packing and assembly between the organic compounds, allowing a control in the crystal growth. The intrinsic properties, molecular arrangement and the structure–property relationships of the OSMCs allow various crystal morphologies that can possess different electronic and optoelectronic properties. The current progress of OSMCs includes a wide range of techniques for achieving high quality OSMCs, without grain boundaries, defects, impurities and dislocations. The optimization of different methods for the crystal growth such ss solution-processing techniques (drop-casting, dip-coating, solvent exchange, solvent vapor diffusion) or vapor process crystals (PVT, microspacing sublimation) allows a control in the crystal morphologies. Solution-based methods are frequently utilized for organic molecules that exhibit large solubility in a wide range of organic solvents whereas PVT is the technique of choice for organic compounds that sublime without decomposition. Consequently, a huge effort in the optimization and exploration of single-crystals growth techniques have been developed to obtain high purity and diverse morphologies in the single crystals. These efforts have boosted the investigations into light emission and light amplification properties in the recent years. The important developments done in the field of OSMC lasers will pave the way to integrate them as basic elements in complex photonic circuits and optical logic gates. Such objectives will be enabled by exploiting lithography methods developed to control resonator size and shape. In these complex circuits, light coupling between the optical resonators and other optical elements will play a crucial role, so new strategies to boost light management in multicomponent structures will be of interest. OSMCs will also play an important role on the development of electrically-pumped lasers, owing to their outstanding charge transport properties. In this respect, a major goal will be the integration of the crystal on an LED-type sandwich structure with charge injecting electrodes without interfering with the lasing properties of the OSMC. Exploiting OSMC resonators for low-end optical pumped lasing (i.e., pumping with cost-efficient LED laser diode sources) is another promising milestone yet to be accomplished. This pumping scheme already demonstrated in conjugated-polymer-based DFBs should be on reach in OMSCs given their superior Q-factor. This result could be of interest for applications in solid state lighting as well as in the sensing field, where WGM and FP optical resonators are already employed as high sensitivity platforms to reveal changes in physical parameters or presence of chemical analytes through the shift of the resonator modes. In summary, OSMCs lasers will continue to attract attention in the next year due to their fascinating photonic and optoelectronic properties.

Author Contributions: Conceptualization, methodology, writing, supervising, funding (J.C.-G.). Conceptualization, methodology, writing-original draft preparation, writing-review and editing, supervision, funding acquisition (E.M.G.-F.). Writing-original draft preparation, writing-review and editing, visualization (J.Á.-C.). All authors have read and agreed to the published version of the manuscript.

Funding: We are grateful for support from the Agencia Estatal de Investigación (AEI, Spain) (project PID2019-105479RB-I00). IMDEA Nanociencia acknowledges support from the "Severo Ochoa" Programme for Centres of Excellence in R&D (MINECO, grant SEV-2016-0686). J. C.-G. acknowledges financial support from the Spanish Ministry of Science, Innovation and Universities (RTI2018-097508-B-I00), and from the Regional Government of Madrid (NMAT2D S2018/NMT4511).

Data Availability Statement: No new data were created or analyzed in this study. Data sharing is not applicable to this article.

Acknowledgments: J.Á.-C. acknowledges financial support from the European Social fund through "Programa de empleo juvenil y la iniciativa de empleo juvenil" (PEJD-2017-PRE/IND-4536) of the Government of Madrid.

Conflicts of Interest: The authors declare no conflict of interest

References

1. Deng, W.; Zhang, X.; Zhang, X.; Guo, J.; Jie, J. Ordered and Patterned Assembly of Organic Micro/Nanocrystals for Flexible Electronic and Optoelectronic Devices. *Adv. Mater. Technol.* **2017**, *2*, 1600280. [CrossRef]
2. Wang, Y.; Sun, L.; Wang, C.; Yang, F.; Ren, X.; Zhang, X.; Dong, H.; Hu, W. Organic Crystalline Materials in Flexible Electronics. *Chem. Soc. Rev.* **2019**, *48*, 1492–1530. [CrossRef] [PubMed]
3. Wang, C.; Dong, H.; Jiang, L.; Hu, W. Organic Semiconductor Crystals. *Chem. Soc. Rev.* **2018**, *47*, 422–500. [CrossRef] [PubMed]

4. Zhang, X.J.; Jie, J.S.; Deng, W.; Shang, Q.X.; Wang, J.C.; Wang, H.; Chen, X.F.; Zhang, X.H. Alignment and Patterning of Ordered Small-Molecule Organic Semiconductor Micro-/Nanocrystals for Device Applications. *Adv. Mater.* **2016**, *28*, 2475–2503. [CrossRef]
5. Li, R.; Hu, W.; Liu, Y.; Zhu, D. Micro- and Nanocrystals of Organic Semiconductors. *Acc. Chem. Res.* **2010**, *43*, 529–540. [CrossRef]
6. Garcia-Frutos, E.M. Small Organic Single-Crystalline One-Dimensional Micro- and Nanostructures for Miniaturized Devices. *J. Mater. Chem. C* **2013**, *1*, 3633–3645. [CrossRef]
7. Zhao, Y.S.; Fu, H.; Peng, A.; Ma, Y.; Liao, Q.; Yao, J. Construction and Optoelectronic Properties of Organic One-Dimensional Nanostructures. *Acc. Chem. Res.* **2010**, *43*, 409–418. [CrossRef] [PubMed]
8. Park, K.S.; Cho, B.; Baek, J.; Hwang, J.K.; Lee, H.; Sung, M.M. Single-Crystal Organic Nanowire Electronics by Direct Printing from Molecular Solutions. *Adv. Funct. Mater.* **2013**, *23*, 4776–4784. [CrossRef]
9. Deng, W.; Zhang, X.; Gong, C.; Zhang, Q.; Xing, Y.; Wu, Y.; Zhang, X.; Jie, J. Aligned Nanowire Arrays on Thin Flexible Substrates for Organic Transistors with High Bending Stability. *J. Mater. Chem. C* **2014**, *2*, 1314–1320. [CrossRef]
10. Yang, Y.S.; Yasuda, T.; Kakizoe, H.; Mieno, H.; Kino, H.; Tateyama, Y.; Adachi, C. High Performance Organic Field-Effect Transistors Based on Single-Crystal Microribbons and Microsheets of Solution-Processed Dithieno[3{,}2-b:2′{,}3′-d]Thiophene Derivatives. *Chem. Commun.* **2013**, *49*, 6483–6485. [CrossRef]
11. Briseno, A.L.; Mannsfeld, S.C.B.; Jenekhe, S.A.; Bao, Z.; Xia, Y. Introducing Organic Nanowire Transistors. *Mater. Today* **2008**, *11*, 38–47. [CrossRef]
12. Martín, C.; Kennes, K.; der Auweraer, M.; Hofkens, J.; de Miguel, G.; García-Frutos, E.M. Self-Assembling Azaindole Organogel for Organic Light-Emitting Devices (OLEDs). *Adv. Funct. Mater.* **2017**, *27*, 1702176. [CrossRef]
13. Mei, J.; Hong, Y.; Lam, J.W.Y.; Qin, A.; Tang, Y.; Tang, B.Z. Aggregation-Induced Emission: The Whole Is More Brilliant than the Parts. *Adv. Mater.* **2014**, *26*, 5429–5479. [CrossRef] [PubMed]
14. Wei, L.; Yao, J.; Fu, H. Solvent-Assisted Self-Assembly of Fullerene into Single-Crystal Ultrathin Microribbons as Highly Sensitive UV–Visible Photodetectors. *ACS Nano* **2013**, *7*, 7573–7582. [CrossRef]
15. Zhang, X.; Jie, J.; Zhang, W.; Zhang, C.; Luo, L.; He, Z.; Zhang, X.; Zhang, W.; Lee, C.; Lee, S. Photoconductivity of a Single Small-Molecule Organic Nanowire. *Adv. Mater.* **2008**, *20*, 2427–2432. [CrossRef]
16. Zhao, F.; Liu, C.; Sun, Y.; Li, Q.; Zhao, J.; Li, Z.; Zhang, B.; Lu, C.; Li, Q.; Qiao, S.; et al. Controlled Self-Assembly of Triazatruxene Overlength Microwires for Optical Waveguide. *Org. Electron.* **2019**, *74*, 276–281. [CrossRef]
17. Gierschner, J.; Varghese, S.; Park, S.Y. Organic Single Crystal Lasers: A Materials View. *Adv. Opt. Mater.* **2016**, *4*, 348–364. [CrossRef]
18. Hayashi, S.; Koizumi, T.; Kamiya, N. 2,5-Dimethoxybenzene-1,4-Dicarboxaldehyde: An Emissive Organic Crystal and Highly Efficient Fluorescent Waveguide. *Chempluschem* **2019**, *84*, 247–251. [CrossRef]
19. Zhang, W.; Yao, J.; Zhao, Y.S. Organic Micro/Nanoscale Lasers. *Acc. Chem. Res.* **2016**, *49*, 1691–1700. [CrossRef]
20. Jiang, Y.; Liu, Y.-Y.; Liu, X.; Lin, H.; Gao, K.; Lai, W.-Y.; Huang, W. Organic Solid-State Lasers: A Materials View and Future Development. *Chem. Soc. Rev.* **2020**, *49*, 5885–5944. [CrossRef] [PubMed]
21. Briseno, A.L.; Tseng, R.J.; Ling, M.-M.; Falcao, E.H.L.; Yang, Y.; Wudl, F.; Bao, Z. High-Performance Organic Single-Crystal Transistors on Flexible Substrates. *Adv. Mater.* **2006**, *18*, 2320–2324. [CrossRef]
22. Yu, P.; Zhen, Y.; Dong, H.; Hu, W. Crystal Engineering of Organic Optoelectronic Materials. *Chem* **2019**, *5*, 2814–2853. [CrossRef]
23. Mikhnenko, O.V.; Blom, P.W.M.; Nguyen, T.-Q. Exciton Diffusion in Organic Semiconductors. *Energy Environ. Sci.* **2015**, *8*, 1867–1888. [CrossRef]
24. Lunt, R.R.; Benziger, J.B.; Forrest, S.R. Relationship between Crystalline Order and Exciton Diffusion Length in Molecular Organic Semiconductors. *Adv. Mater.* **2010**, *22*, 1233–1236. [CrossRef] [PubMed]
25. Hasegawa, T. 18-Advances in Device Fabrication Scale-up Methods. In *Handbook of Organic Materials for Electronic and Photonic Devices (Second Edition)*; Ostroverkhova, O., Ed.; Woodhead Publishing Series in Electronic and Optical Materials; Woodhead Publishing: Sawston, UK; Cambridge, UK, 2019; pp. 579–597. [CrossRef]
26. Wang, W.; Wang, L.; Dai, G.; Deng, W.; Zhang, X.; Jie, J.; Zhang, X. Controlled Growth of Large-Area Aligned Single-Crystalline Organic Nanoribbon Arrays for Transistors and Light-Emitting Diodes Driving. *Nano-Micro Lett.* **2017**, *9*, 1–11. [CrossRef] [PubMed]
27. Gierschner, J.; Park, S.Y. Luminescent Distyrylbenzenes: Tailoring Molecular Structure and Crystalline Morphology. *J. Mater. Chem. C* **2013**, *1*, 5818–5832. [CrossRef]
28. Gierschner, J.; Huang, Y.S.; Van Averbeke, B.; Cornil, J.; Friend, R.H.; Beljonne, D. Excitonic versus Electronic Couplings in Molecular Assemblies: The Importance of Non-Nearest Neighbor Interactions. *J. Chem. Phys.* **2009**, *130*. [CrossRef] [PubMed]
29. Gierschner, J.; Lüer, L.; Milián-Medina, B.; Oelkrug, D.; Egelhaaf, H.J. Highly Emissive H-Aggregates or Aggregation-Induced Emission Quenching? The Photophysics of All-Trans Para-Distyrylbenzene. *J. Phys. Chem. Lett.* **2013**, *4*, 2686–2697. [CrossRef]
30. Coropceanu, V.; Cornil, J.; da Silva Filho, D.A.; Olivier, Y.; Silbey, R.; Brédas, J.L. Charge Transport in Organic Semiconductors. *Chem. Rev.* **2007**, *107*, 926–952. [CrossRef] [PubMed]
31. Guo, Q.; Wang, L.; Bai, F.; Jiang, Y.; Guo, J.; Xu, B.; Tian, W. Polymorphism Dependent Charge Transport Property of 9,10-Bis((E)-2-(Pyrid-2-Yl)Vinyl)Anthracene: A Theoretical Study. *RSC Adv.* **2015**, *5*, 18875–18880. [CrossRef]
32. Zhang, X.; Dong, H.; Hu, W. Organic Semiconductor Single Crystals for Electronics and Photonics. *Adv. Mater.* **2018**, *30*, 1–34. [CrossRef]

33. Kaufmann, C.; Bialas, D.; Stolte, M.; Würthner, F. Discrete π-Stacks of Perylene Bisimide Dyes within Folda-Dimers: Insight into Long- and Short-Range Exciton Coupling. *J. Am. Chem. Soc.* **2018**, *140*, 9986–9995. [CrossRef]
34. Curtis, M.D.; Cao, J.; Kampf, J.W. Solid-State Packing of Conjugated Oligomers: From π-Stacks to the Herringbone Structure. *J. Am. Chem. Soc.* **2004**, *126*, 4318–4328. [CrossRef] [PubMed]
35. Cornil, J.; Beljonne, D.; Calbert, J.P.; Brédas, J.L. Interchain Interactions in Organic π-Conjugated Materials: Impact on Electronic Structure, Optical Response, and Charge Transport. *Adv. Mater.* **2001**, *13*, 1053–1067. [CrossRef]
36. Spano, F.C. The Spectral Signatures of Frenkel Polarons in H- And J-Aggregates. *Acc. Chem. Res.* **2010**, *43*, 429–439. [CrossRef]
37. Zhou, J.; Zhang, W.; Jiang, X.F.; Wang, C.; Zhou, X.; Xu, B.; Liu, L.; Xie, Z.; Ma, Y. Magic-Angle Stacking and Strong Intermolecular π-π Interaction in a Perylene Bisimide Crystal: An Approach for Efficient Near-Infrared (NIR) Emission and High Electron Mobility. *J. Phys. Chem. Lett.* **2018**, *9*, 596–600. [CrossRef]
38. Hestand, N.J.; Spano, F.C. Interference between Coulombic and CT-Mediated Couplings in Molecular Aggregates: H- to J-Aggregate Transformation in Perylene-Based π-Stacks. *J. Chem. Phys.* **2015**, *143*. [CrossRef] [PubMed]
39. Li, Y.; Shen, F.; Wang, H.; He, F.; Xie, Z.; Zhang, H.; Wang, Z.; Liu, L.; Li, F.; Hanif, M.; et al. Supramolecular Network Conducting the Formation of Uniaxially Oriented Molecular Crystal of Cyano Substituted Oligo(p-Phenylene Vinylene) and Its Amplified Spontaneous Emission (ASE) Behavior. *Chem. Mater.* **2008**, *20*, 7312–7318. [CrossRef]
40. Li, C.; Hanif, M.; Li, X.; Zhang, S.; Xie, Z.; Liu, L.; Yang, B.; Su, S.; Ma, Y. Effect of Cyano-Substitution in Distyrylbenzene Derivatives on Their Fluorescence and Electroluminescence Properties. *J. Mater. Chem. C* **2016**, *4*, 7478–7484. [CrossRef]
41. Fang, H.-H.; Chen, Q.-D.; Yang, J.; Xia, H.; Ma, Y.-G.; Wang, H.-Y.; Sun, H.-B. Two-Photon Excited Highly Polarized and Directional Upconversion Emission from Slab Organic Crystals. *Opt. Lett.* **2010**, *35*. [CrossRef]
42. Mattheus, C.C.; Dros, A.B.; Baas, J.; Meetsma, A.; De Boer, J.L.; Palstra, T.T.M. Polymorphism in Pentacene. *Acta Crystallogr. Sect. C Cryst. Struct. Commun.* **2001**, *57*, 939–941. [CrossRef]
43. Cao, M.; Zhang, C.; Cai, Z.; Xiao, C.; Chen, X.; Yi, K.; Yang, Y.; Lu, Y.; Wei, D. Enhanced Photoelectrical Response of Thermodynamically Epitaxial Organic Crystals at the Two-Dimensional Limit. *Nat. Commun.* **2019**, *10*, 1–11. [CrossRef]
44. Guerrini, M.; Calzolari, A.; Corni, S. Solid-State Effects on the Optical Excitation of Push-Pull Molecular J-Aggregates by First-Principles Simulations. *ACS Omega* **2018**, *3*, 10481–10486. [CrossRef]
45. Liu, J.; Meng, L.; Zhu, W.; Zhang, C.; Zhang, H.; Yao, Y.; Wang, Z.; He, P.; Zhang, X.; Wang, Y.; et al. A Cross-Dipole Stacking Molecule of an Anthracene Derivative: Integrating Optical and Electrical Properties. *J. Mater. Chem. C* **2015**, *3*, 3068–3071. [CrossRef]
46. Ramakrishnan, R.; Niyas, M.A.; Lijina, M.P.; Hariharan, M. Distinct Crystalline Aromatic Structural Motifs: Identification, Classification, and Implications. *Acc. Chem. Res.* **2019**, *52*, 3075–3086. [CrossRef] [PubMed]
47. Ge, Y.; Wen, Y.; Liu, H.; Lu, T.; Yu, Y.; Zhang, X.; Li, B.; Zhang, S.T.; Li, W.; Yang, B. A Key Stacking Factor for the Effective Formation of Pyrene Excimer in Crystals: Degree of π-π Overlap. *J. Mater. Chem. C* **2020**, *8*, 11830–11838. [CrossRef]
48. Sebastian, E.; Philip, A.M.; Benny, A.; Hariharan, M. Null Exciton Splitting in Chromophoric Greek Cross (+) Aggregate. *Angew. Chemie Int. Ed.* **2018**, *57*, 15696–15701. [CrossRef]
49. Dong, Y.; Xu, B.; Zhang, J.; Tan, X.; Wang, L.; Chen, J.; Lv, H.; Wen, S.; Li, B.; Ye, L.; et al. Piezochromic Luminescence Based on the Molecular Aggregation of 9,10-Bis((E)-2-(Pyrid-2-Yl)Vinyl)Anthracene. *Angew. Chemie-Int. Ed.* **2012**, *51*, 10782–10785. [CrossRef] [PubMed]
50. Fang, H.H.; Yang, J.; Feng, J.; Yamao, T.; Hotta, S.; Sun, H.B. Functional Organic Single Crystals for Solid-State Laser Applications. *Laser Photonics Rev.* **2014**, *8*, 687–715. [CrossRef]
51. Kabe, R.; Nakanotani, H.; Sakanoue, T.; Yahiro, M.; Adachi, C. Effect of Molecular Morphology on Amplified Spontaneous Emission of Bis-Styrylbenzene Derivatives. *Adv. Mater.* **2009**, *21*, 4034–4038. [CrossRef]
52. Rhodes, S.; Liang, W.; Wang, X.; Reddy, N.R.; Fang, J. Transition from H-Aggregate Nanotubes to J-Aggregate Nanoribbons. *J. Phys. Chem. C* **2020**, *124*, 11722–11729. [CrossRef]
53. Li, J.; Zhou, K.; Liu, J.; Zhen, Y.; Liu, L.; Zhang, J.; Dong, H.; Zhang, X.; Jiang, L.; Hu, W. Aromatic Extension at 2,6-Positions of Anthracene toward an Elegant Strategy for Organic Semiconductors with Efficient Charge Transport and Strong Solid State Emission. *J. Am. Chem. Soc.* **2017**, *139*, 17261–17264. [CrossRef] [PubMed]
54. Liu, J.; Zhang, H.; Dong, H.; Meng, L.; Jiang, L.; Jiang, L.; Wang, Y.; Yu, J.; Sun, Y.; Hu, W.; et al. High Mobility Emissive Organic Semiconductor. *Nat. Commun.* **2015**, *6*, 7. [CrossRef] [PubMed]
55. Li, Z.Z.; Liao, L.S.; Wang, X.D. Controllable Synthesis of Organic Microcrystals with Tunable Emission Color and Morphology Based on Molecular Packing Mode. *Small* **2018**, *14*, 1–8. [CrossRef] [PubMed]
56. Ma, S.; Zhou, K.; Hu, M.; Li, Q.; Liu, Y.; Zhang, H.; Jing, J.; Dong, H.; Xu, B.; Hu, W.; et al. Integrating Efficient Optical Gain in High-Mobility Organic Semiconductors for Multifunctional Optoelectronic Applications. *Adv. Funct. Mater.* **2018**, *28*, 1–8. [CrossRef]
57. Kloc, C.; Siegrist, T.; Pflaum, J. Growth of Single-Crystal Organic Semiconductors. In *Springer Handbook of Crystal Growth*; Dhanaraj, G., Byrappa, K., Prasad, V., Dudley, M., Eds.; Springer Berlin Heidelberg: Berlin, Heidelberg, 2010; pp. 845–867.
58. Dunitz, J.D.; Gavezzotti, A. How Molecules Stick Together in Organic Crystals: Weak Intermolecular Interactions. *Chem. Soc. Rev.* **2009**, *38*, 2622–2633. [CrossRef] [PubMed]
59. Jiang, L.; Dong, H.; Meng, Q.; Li, H.; He, M.; Wei, Z.; He, Y.; Hu, W. Millimeter-Sized Molecular Monolayer Two-Dimensional Crystals. *Adv. Mater.* **2011**, *23*, 2059–2063. [CrossRef] [PubMed]

60. Ito, F.; Fujimori, J.I. Fluorescence Visualization of the Molecular Assembly Processes during Solvent Evaporation via Aggregation-Induced Emission in a Cyanostilbene Derivative. *CrystEngComm* **2014**, *16*, 9779–9782. [CrossRef]
61. Wang, C.; Liang, Z.; Liu, Y.; Wang, X.; Zhao, N.; Miao, Q.; Hu, W.; Xu, J. Single Crystal N-Channel Field Effect Transistors from Solution-Processed Silylethynylated Tetraazapentacene. *J. Mater. Chem.* **2011**, *21*, 15201–15204. [CrossRef]
62. Liu, D.; Li, C.; Niu, S.; Li, Y.; Hu, M.; Li, Q.; Zhu, W.; Zhang, X.; Dong, H.; Hu, W. A Case Study of Tuning the Crystal Polymorphs of Organic Semiconductors towards Simultaneously Improved Light Emission and Field-Effect Properties. *J. Mater. Chem. C* **2019**, *7*, 5925–5930. [CrossRef]
63. Inada, Y.; Yamao, T.; Inada, M.; Itami, T.; Hotta, S. Giant Organic Single-Crystals of a Thiophene/Phenylene Co-Oligomer toward Device Applications. *Synth. Met.* **2011**, *161*, 1869–1877. [CrossRef]
64. Wang, C.; Liu, Y.; Wei, Z.; Li, H.; Xu, W.; Hu, W. Biphase Micro/Nanometer Sized Single Crystals of Organic Semiconductors: Control Synthesis and Their Strong Phase Dependent Optoelectronic Properties. *Appl. Phys. Lett.* **2010**, *96*, 94–97. [CrossRef]
65. Xu, J.; Semin, S.; Cremers, J.; Wang, L.; Savoini, M.; Fron, E.; Coutino, E.; Chervy, T.; Wang, C.; Li, Y.; et al. Controlling Microsized Polymorphic Architectures with Distinct Linear and Nonlinear Optical Properties. *Adv. Opt. Mater.* **2015**, *3*, 948–956. [CrossRef]
66. Wakahara, T.; D'Angelo, P.; Miyazawa, K.; Nemoto, Y.; Ito, O.; Tanigaki, N.; Bradley, D.D.C.; Anthopoulos, T.D. Fullerene/Cobalt Porphyrin Hybrid Nanosheets with Ambipolar Charge Transporting Characteristics. *J. Am. Chem. Soc.* **2012**, *134*, 7204–7206. [CrossRef] [PubMed]
67. Choi, S.; Chae, S.H.; Hoang, M.H.; Kim, K.H.; Huh, J.A.; Kim, Y.; Kim, S.J.; Choi, D.H.; Lee, S.J. An Unsymmetrically π-Extended Porphyrin-Based Single-Crystal Field-Effect Transistor and Its Anisotropic Carrier-Transport Behavior. *Chem.-A Eur. J.* **2013**, *19*, 2247–2251. [CrossRef] [PubMed]
68. Zhang, Z.; Song, X.; Wang, S.; Li, F.; Zhang, H.; Ye, K.; Wang, Y. Two-Dimensional Organic Single Crystals with Scale Regulated, Phase-Switchable, Polymorphism-Dependent, and Amplified Spontaneous Emission Properties. *J. Phys. Chem. Lett.* **2016**, *7*, 1697–1702. [CrossRef] [PubMed]
69. Guo, Y.; Du, C.; Yu, C.; Di, C.A.; Jiang, S.; Xi, H.; Zheng, J.; Yan, S.; Yu, C.; Hu, W.; et al. High-Performance Phototransistors Based on Organic Microribbons Prepared by a Solution Self-Assembly Process. *Adv. Funct. Mater.* **2010**, *20*, 1019–1024. [CrossRef]
70. Zhou, C.; Worku, M.; Neu, J.; Lin, H.; Tian, Y.; Lee, S.; Zhou, Y.; Han, D.; Chen, S.; Hao, A.; et al. Facile Preparation of Light Emitting Organic Metal Halide Crystals with Near-Unity Quantum Efficiency. *Chem. Mater.* **2018**, *30*, 2374–2378. [CrossRef]
71. Wang, S.; Lai, Z.; Tran, T.H.; Han, F.; Su, D.; Wang, R.; Zhang, H.; Wang, H.; Chen, H. Solvent Exchange as a Synthetic Handle for Controlling Molecular Crystals. *Carbon, N.Y.* **2020**, *160*, 188–195. [CrossRef]
72. Zhang, C.; Zou, C.L.; Yan, Y.; Wei, C.; Cui, J.M.; Sun, F.W.; Yao, J.; Zhao, Y.S. Self-Assembled Organic Crystalline Microrings as Active Whispering-Gallery-Mode Optical Resonators. *Adv. Opt. Mater.* **2013**, *1*, 357–361. [CrossRef]
73. Wang, S.; Dössel, L.; Mavrinskiy, A.; Gao, P.; Feng, X.; Pisula, W.; Müllen, K. Self-Assembly and Microstructural Control of a Hexa-Peri-Hexabenzocoronene- Perylene Diimide Dyad by Solvent Vapor Diffusion. *Small* **2011**, *7*, 2841–2846. [CrossRef]
74. Wang, C.; Fang, Y.; Wen, L.; Zhou, M.; Xu, Y.; Zhao, H.; De Cola, L.; Hu, W.; Lei, Y. Vectorial Diffusion for Facile Solution-Processed Self-Assembly of Insoluble Semiconductors: A Case Study on Metal Phthalocyanines. *Chem.-A Eur. J.* **2014**, *20*, 10990–10995. [CrossRef] [PubMed]
75. Shi, D.; Qin, X.; Li, Y.; He, Y.; Zhong, C.; Pan, J.; Dong, H.; Xu, W.; Li, T.; Hu, W.; et al. Spiro-OMeTAD Single Crystals: Remarkably Enhanced Charge-Carrier Transport via Mesoscale Ordering. *Sci. Adv.* **2016**, *2*. [CrossRef] [PubMed]
76. Lv, A.; Puniredd, S.R.; Zhang, J.; Li, Z.; Zhu, H.; Jiang, W.; Dong, H.; He, Y.; Jiang, L.; Li, Y.; et al. High Mobility, Air Stable, Organic Single Crystal Transistors of an n-Type Diperylene Bisimide. *Adv. Mater.* **2012**, *24*, 2626–2630. [CrossRef] [PubMed]
77. Jiang, H.; Kloc, C. Single-Crystal Growth of Organic Semiconductors. *MRS Bull.* **2013**, *38*, 28–33. [CrossRef]
78. Schweicher, G.; Paquay, N.; Amato, C.; Resel, R.; Koini, M.; Talvy, S.; Lemaur, V.; Cornil, J.; Geerts, Y.; Gbabode, G. Toward Single Crystal Thin Films of Terthiophene by Directional Crystallization Using a Thermal Gradient. *Cryst. Growth Des.* **2011**, *11*, 3663–3672. [CrossRef]
79. Tavazzi, S.; Miozzo, L.; Silvestri, L.; Mora, P.; Spearman, P.; Moret, M.; Rizzato, S.; Braga, D.; Diagne Diaw, A.K.; Gningue-Sall, D.; et al. Crystal Structure and Optical Properties of N-Pyrrole End-Capped Thiophene/Phenyl Co-Oligomer: Strong h-Type Excitonic Coupling and Emission Self-Waveguiding. *Cryst. Growth Des.* **2010**, *10*, 2342–2349. [CrossRef]
80. Kloc, C.; Simpkins, P.G.; Siegrist, T.; Laudise, R.A. Physical Vapor Growth of Centimeter-Sized Crystals of α-Hexathiophene. *J. Cryst. Growth* **1997**, *182*, 416–427. [CrossRef]
81. Laudise, R.A.; Kloc, C.; Simpkins, P.G.; Siegrist, T. Physical Vapor Growth of Organic Semiconductors. *J. Cryst. Growth* **1998**, *187*, 449–454. [CrossRef]
82. Yong, B.; Zhao, S.; Fu, H.; Hu, F.; Peng, A.; Yang, W.; Yao, J. Tunable Emission from Binary Organic One-Dimensional Nanomaterials: An Alternative Approach to White-Light Emission. *Adv. Mater.* **2008**, *20*, 79–83. [CrossRef]
83. Huang, Y.; Yuan, R.; Zhou, S. Gas Phase-Based Growth of Highly Sensitive Single-Crystal Rectangular Micro- and Nanotubes. *J. Mater. Chem.* **2012**, *22*, 883–888. [CrossRef]
84. Roberson, L.B.; Kowalik, J.; Tolbert, L.M.; Kloc, C.; Zeis, R.; Chi, X.; Fleming, R.; Wilkins, C. Pentacene Disproportionation during Sublimation for Field-Effect Transistors. *J. Am. Chem. Soc.* **2005**, *127*, 3069–3075. [CrossRef]
85. Ding, R.; Feng, J.; Dong, F.X.; Zhou, W.; Liu, Y.; Zhang, X.L.; Wang, X.P.; Fang, H.H.; Xu, B.; Li, X.B.; et al. Highly Efficient Three Primary Color Organic Single-Crystal Light-Emitting Devices with Balanced Carrier Injection and Transport. *Adv. Funct. Mater.* **2017**, *27*. [CrossRef]

86. Ye, X.; Liu, Y.; Han, Q.; Ge, C.; Cui, S.; Zhang, L.; Zheng, X.; Liu, G.; Liu, J.; Liu, D.; et al. Microspacing In-Air Sublimation Growth of Organic Crystals. *Chem. Mater.* **2018**, *30*, 412–420. [CrossRef]
87. Ye, X.; Liu, Y.; Guo, Q.; Han, Q.; Ge, C.; Cui, S.; Zhang, L.; Tao, X. 1D versus 2D Cocrystals Growth via Microspacing In-Air Sublimation. *Nat. Commun.* **2019**, *10*. [CrossRef] [PubMed]
88. Guo, Q.; Ye, X.; Lin, Q.; Han, Q.; Ge, C.; Zheng, X.; Zhang, L.; Cui, S.; Wu, Y.; Li, C.; et al. Micro-Spacing In-Air Sublimation Growth of Ultrathin Organic Single Crystals. *Chem. Mater.* **2020**, *32*, 7618–7629. [CrossRef]
89. Qian, Y.; Wei, Q.; Del Pozo, G.; Mróz, M.M.; Lüer, L.; Casado, S.; Cabanillas-Gonzalez, J.; Zhang, Q.; Xie, L.; Xia, R.; et al. H-Shaped Oligofluorenes for Highly Air-Stable and Low-Threshold Non-Doped Deep Blue Lasing. *Adv. Mater.* **2014**, *26*, 2937–2942. [CrossRef]
90. Zhang, Q.; Wu, Y.; Lian, S.; Gao, J.; Zhang, S.; Hai, G.; Sun, C.; Li, X.; Xia, R.; Cabanillas-Gonzalez, J.; et al. Simultaneously Enhancing Photoluminescence Quantum Efficiency and Optical Gain of Polyfluorene via Backbone Intercalation of 2,5-Dimethyl-1,4-Phenylene. *Adv. Opt. Mater.* **2020**, *8*. [CrossRef]
91. Bai, K.; Wang, S.; Zhao, L.; Ding, J.; Wang, L. Efficient Blue, Green, and Red Electroluminescence from Carbazole-Functionalized Poly(Spirobifluorene)S. *Macromolecules* **2017**, *50*, 6945–6953. [CrossRef]
92. Wang, J.; Song, C.; Zhong, Z.; Hu, Z.; Han, S.; Xu, W.; Peng, J.; Ying, L.; Wang, J.; Cao, Y. In Situ Patterning of Microgrooves via Inkjet Etching for a Solution-Processed OLED Display. *J. Mater. Chem. C* **2017**, *5*, 5005–5009. [CrossRef]
93. Smirnov, J.R.C.; Zhang, Q.; Wannemacher, R.; Wu, L.; Casado, S.; Xia, R.; Rodriguez, I.; Cabanillas-González, J. Flexible All-Polymer Waveguide for Low Threshold Amplified Spontaneous Emission. *Sci. Rep.* **2016**, *6*, 34565. [CrossRef]
94. Zhang, Q.; Liu, J.; Wei, Q.; Guo, X.; Xu, Y.; Xia, R.; Xie, L.; Qian, Y.; Sun, C.; Lüer, L.; et al. Host Exciton Confinement for Enhanced Förster-Transfer-Blend Gain Media Yielding Highly Efficient Yellow-Green Lasers. *Adv. Funct. Mater.* **2018**, *28*, 1705824. [CrossRef]
95. Cabanillas-Gonzalez, J.; Sciascia, C.; Lanzani, G.; Toffanin, S.; Capelli, R.; Ramon, M.C.; Muccini, M.; Gierschner, J.; Hwu, T.-Y.; Wong, K.-T. Molecular Packing Effects on the Optical Spectra and Triplet Dynamics in Oligofluorene Films. *J. Phys. Chem. B* **2008**, *112*, 11605–11609. [CrossRef] [PubMed]
96. Dong, H.; Wei, Y.; Zhang, W.; Wei, C.; Zhang, C.; Yao, J.; Zhao, Y.S. Broadband Tunable Microlasers Based on Controlled Intramolecular Charge-Transfer Process in Organic Supramolecular Microcrystals. *J. Am. Chem. Soc.* **2016**, *138*, 1118–1121. [CrossRef] [PubMed]
97. Khan, A.; Wang, M.; Usman, R.; Sun, H.; Du, M.; Xu, C. Molecular Marriage via Charge Transfer Interaction in Organic Charge Transfer Co-Crystals toward Solid-State Fluorescence Modulation. *Cryst. Growth Des.* **2017**, *17*, 1251–1257. [CrossRef]
98. Yan, D.; Delori, A.; Lloyd, G.O.; Friščić, T.; Day, G.M.; Jones, W.; Lu, J.; Wei, M.; Evans, D.G.; Duan, X. A Cocrystal Strategy to Tune the Luminescent Properties of Stilbene-Type Organic Solid-State Materials. *Angew. Chemie Int. Ed.* **2011**, *50*, 12483–12486. [CrossRef]
99. Park, S.K.; Kim, J.H.; Ohto, T.; Yamada, R.; Jones, A.O.F.; Whang, D.R.; Cho, I.; Oh, S.; Hong, S.H.; Kwon, J.E.; et al. Highly Luminescent 2D-Type Slab Crystals Based on a Molecular Charge-Transfer Complex as Promising Organic Light-Emitting Transistor Materials. *Adv. Mater.* **2017**, *29*, 1701346. [CrossRef]
100. He, D.; Qiao, J.; Zhang, L.; Wang, J.; Lan, T.; Qian, J.; Li, Y.; Shi, Y.; Chai, Y.; Lan, W.; et al. Ultrahigh Mobility and Efficient Charge Injection in Monolayer Organic Thin-Film Transistors on Boron Nitride. *Sci. Adv.* **2017**, *3*, e1701186. [CrossRef]
101. Ullah, M.; Wawrzinek, R.; Nagiri, R.C.R.; Lo, S.-C.; Namdas, E.B. UV–Deep Blue–Visible Light-Emitting Organic Field Effect Transistors with High Charge Carrier Mobilities. *Adv. Opt. Mater.* **2017**, *5*, 1600973. [CrossRef]
102. Sandanayaka, A.S.D.; Matsushima, T.; Bencheikh, F.; Terakawa, S.; Potscavage, W.J.; Qin, C.; Fujihara, T.; Goushi, K.; Ribierre, J.-C.; Adachi, C. Indication of Current-Injection Lasing from an Organic Semiconductor. *Appl. Phys. Express* **2019**, *12*, 61010. [CrossRef]
103. Stagira, S.; Zavelani-Rossi, M.; Nisoli, M.; DeSilvestri, S.; Lanzani, G.; Zenz, C.; Mataloni, P.; Leising, G. Single-Mode Picosecond Blue Laser Emission from a Solid Conjugated Polymer. *Appl. Phys. Lett.* **1998**, *73*, 2860–2862. [CrossRef]
104. Canazza, G.; Scotognella, F.; Lanzani, G.; De Silvestri, S.; Zavelani-Rossi, M.; Comoretto, D. Lasing from All-Polymer Microcavities. *Laser Phys. Lett.* **2014**, *11*. [CrossRef]
105. Navarro-Fuster, V.; Calzado, E.M.; Boj, P.G.; Quintana, J.A.; Villalvilla, J.M.; Díaz-García, M.A.; Trabadelo, V.; Juarros, A.; Retolaza, A.; Merino, S. Highly Photostable Organic Distributed Feedback Laser Emitting at 573 Nm. *Appl. Phys. Lett.* **2010**, *97*, 171104. [CrossRef]
106. Xu, Y.; Hai, G.; Xu, H.; Zhang, H.; Zuo, Z.; Zhang, Q.; Xia, R.; Sun, C.; Castro-Smirnov, J.; Sousaraei, A.; et al. Efficient Optical Gain from Near-Infrared Polymer Lasers Based on Poly[N-9′-Heptadecanyl-2,7-Carbazole-Alt-5,5-(4′,7′-Di-2-Thienyl-2′,1′,3′-Benzothiadiazole)]. *Adv. Opt. Mater.* **2018**, *6*, 1800263. [CrossRef]
107. Milanese, S.; De Giorgi, M.L.; Anni, M. Determination of the Best Empiric Method to Quantify the Amplified Spontaneous Emission Threshold in Polymeric Active Waveguides. *Molecules* **2020**, *25*, 2992. [CrossRef]
108. Muñoz-Mármol, R.; Bonal, V.; Paternò, G.M.; Ross, A.M.; Boj, P.G.; Villalvilla, J.M.; Quintana, J.A.; Scotognella, F.; D'Andrea, C.; Sardar, S.; et al. Dual Amplified Spontaneous Emission and Lasing from Nanographene Films. *Nanomaterials* **2020**, *10*, 1525. [CrossRef] [PubMed]
109. Sun, C.; Mróz, M.M.; Castro Smirnov, J.R.; Lüer, L.; Hermida-Merino, D.; Zhao, C.; Takeuchi, M.; Sugiyasu, K.; Cabanillas-González, J. Amplified Spontaneous Emission in Insulated Polythiophenes. *J. Mater. Chem. C* **2018**, *6*, 6591–6596. [CrossRef]

110. Calzado, E.M.; Villalvilla, J.M.; Boj, P.G.; Quintana, J.A.; Díaz-García, M.A. Concentration Dependence of Amplified Spontaneous Emission in Organic-Based Waveguides. *Org. Electron.* **2006**, *7*, 319–329. [CrossRef]
111. Illarramendi, M.A.; Arrue, J.; Ayesta, I.; Jiménez, F.; Zubia, J.; Bikandi, I.; Tagaya, A.; Koike, Y. Amplified Spontaneous Emission in Graded-Index Polymer Optical Fibers: Theory and Experiment. *Opt. Express* **2013**, *21*, 24254–24266. [CrossRef]
112. Zhang, Q.; Wei, Q.; Guo, X.; Hai, G.; Sun, H.; Li, J.; Xia, R.; Qian, Y.; Casado, S.; Castro-Smirnov, J.R.; et al. Concurrent Optical Gain Optimization and Electrical Tuning in Novel Oligomer:Polymer Blends with Yellow-Green Laser Emission. *Adv. Sci.* **2019**, *6*, 1801455. [CrossRef]
113. Huang, R.; Wang, C.; Wang, Y.; Zhang, H. Elastic Self-Doping Organic Single Crystals Exhibiting Flexible Optical Waveguide and Amplified Spontaneous Emission. *Adv. Mater.* **2018**, *30*, 1800814. [CrossRef] [PubMed]
114. Fang, H.-H.; Ding, R.; Lu, S.-Y.; Yang, J.; Zhang, X.-L.; Yang, R.; Feng, J.; Chen, Q.-D.; Song, J.-F.; Sun, H.-B. Distributed Feedback Lasers Based on Thiophene/Phenylene Co-Oligomer Single Crystals. *Adv. Funct. Mater.* **2012**, *22*, 33–38. [CrossRef]
115. Zhu, W.-S.; Han, Y.-M.; An, X.; Weng, J.-N.; Yu, M.-N.; Bai, L.-B.; Wei, C.-X.; Lin, J.-Y.; Liu, W.; Ou, C.-J.; et al. Highly Emissive Hierarchical Uniform Dialkylfluorene-Based Dimer Microcrystals for Ultraviolet Organic Laser. *J. Phys. Chem. C* **2019**, *123*, 28881–28886. [CrossRef]
116. Ou, C.-J.; Ding, X.-H.; Li, Y.-X.; Zhu, C.; Yu, M.-N.; Xie, L.-H.; Lin, J.-Y.; Xu, C.-X.; Huang, W. Conformational Effect of Polymorphic Terfluorene on Photophysics, Crystal Morphologies, and Lasing Behaviors. *J. Phys. Chem. C* **2017**, *121*, 14803–14810. [CrossRef]
117. Varghese, S.; Park, S.K.; Casado, S.; Fischer, R.C.; Resel, R.; Milián-Medina, B.; Wannemacher, R.; Park, S.Y.; Gierschner, J. Stimulated Emission Properties of Sterically Modified Distyrylbenzene-Based H-Aggregate Single Crystals. *J. Phys. Chem. Lett.* **2013**, *4*, 1597–1602. [CrossRef]
118. Xu, Z.; Liao, Q.; Shi, Q.; Zhang, H.; Yao, J.; Fu, H. Low-Threshold Nanolasers Based on Slab-Nanocrystals of H-Aggregated Organic Semiconductors. *Adv. Mater.* **2012**, *24*, OP216–OP220. [CrossRef] [PubMed]
119. Wang, X.; Li, Z.-Z.; Zhuo, M.-P.; Wu, Y.; Chen, S.; Yao, J.; Fu, H. Tunable Near-Infrared Organic Nanowire Nanolasers. *Adv. Funct. Mater.* **2017**, *27*, 1703470. [CrossRef]
120. Duong Ta, V.; Chen, R.; Ma, L.; Jun Ying, Y.; Dong Sun, H. Whispering Gallery Mode Microlasers and Refractive Index Sensing Based on Single Polymer Fiber. *Laser Photon. Rev.* **2013**, *7*, 133–139. [CrossRef]
121. Fang, H.H.; Ding, R.; Lu, S.Y.; De Yang, Y.; Chen, Q.D.; Feng, J.; Huang, Y.Z.; Sun, H.B. Whispering-Gallery Mode Lasing from Patterned Molecular Single-Crystalline Microcavity Array. *Laser Photonics Rev.* **2013**, *7*, 281–288. [CrossRef]
122. Lv, Y.; Xiong, X.; Liu, Y.; Yao, J.; Li, Y.J.; Zhao, Y.S. Controlled Outcoupling of Whispering-Gallery-Mode Lasers Based on Self-Assembled Organic Single-Crystalline Microrings. *Nano Lett.* **2019**, *19*, 1098–1103. [CrossRef]
123. Feng, J.; Jiang, X.; Yan, X.; Wu, Y.; Su, B.; Fu, H.; Yao, J.; Jiang, L. "Capillary-Bridge Lithography" for Patterning Organic Crystals toward Mode-Tunable Microlaser Arrays. *Adv. Mater.* **2017**, *29*, 1603652. [CrossRef]
124. Xu, Z.; Liao, Q.; Wang, X.; Fu, H. Whispering Gallery Mode Laser Based on a Self-Assembled Organic Octahedron Microcrystal Microresonator. *Adv. Opt. Mater.* **2014**, *2*, 1160–1166. [CrossRef]
125. Varghese, S.; Yoon, S.-J.; Casado, S.; Fischer, R.C.; Wannemacher, R.; Park, S.Y.; Gierschner, J. Orthogonal Resonator Modes and Low Lasing Threshold in Highly Emissive Distyrylbenzene-Based Molecular Crystals. *Adv. Opt. Mater.* **2014**, *2*, 542–548. [CrossRef]
126. Wang, X.; Liao, Q.; Kong, Q.; Zhang, Y.; Xu, Z.; Lu, X.; Fu, H. Whispering-Gallery-Mode Microlaser Based on Self-Assembled Organic Single-Crystalline Hexagonal Microdisks. *Angew. Chemie Int. Ed.* **2014**, *53*, 5863–5867. [CrossRef] [PubMed]
127. O'Carroll, D.; Lieberwirth, I.; Redmond, G. Microcavity Effects and Optically Pumped Lasing in Single Conjugated Polymer Nanowires. *Nat. Nanotechnol.* **2007**, *2*, 180–184. [CrossRef] [PubMed]
128. Wei, G.-Q.; Wang, X.-D.; Liao, L.-S. Recent Advances in Organic Whispering-Gallery Mode Lasers. *Laser Photon. Rev.* **2020**, *14*, 2000257. [CrossRef]
129. Zhang, C.; Zou, C.-L.; Dong, H.; Yan, Y.; Yao, J.; Zhao, Y.S. Dual-Color Single-Mode Lasing in Axially Coupled Organic Nanowire Resonators. *Sci. Adv.* **2017**, *3*, e1700225. [CrossRef]
130. Kushida, S.; Okada, D.; Sasaki, F.; Lin, Z.-H.; Huang, J.-S.; Yamamoto, Y. Lasers: Low-Threshold Whispering Gallery Mode Lasing from Self-Assembled Microspheres of Single-Sort Conjugated Polymers. *Adv. Opt. Mater.* **2017**, *5*, 1700123. [CrossRef]
131. Foreman, M.R.; Swaim, J.D.; Vollmer, F. Whispering Gallery Mode Sensors. *Adv. Opt. Photonics* **2015**, *7*, 168–240. [CrossRef]
132. Wun, A.W.; Snee, P.T.; Chan, Y.; Bawendi, M.G.; Nocera, D.G. Non-Linear Transduction Strategies for Chemo/Biosensing on Small Length Scales. *J. Mater. Chem.* **2005**, *15*, 2697–2706. [CrossRef]
133. Humar, M.; Hyun Yun, S. Intracellular Microlasers. *Nat. Photonics* **2015**, *9*, 572–576. [CrossRef]
134. Shi, Y.-L.; Wang, X.-D. 1D Organic Micro/Nanostructures for Photonics. *Adv. Funct. Mater.* **2020**, 2008149. [CrossRef]
135. Torii, K.; Higuchi, T.; Mizuno, K.; Bando, K.; Yamashita, K.; Sasaki, F.; Yanagi, H. Organic Nanowire Lasers with Epitaxially Grown Crystals of Semiconducting Oligomers. *ChemNanoMat* **2017**, *3*, 625–631. [CrossRef]
136. Yu, Z.; Wu, Y.; Xiao, L.; Chen, J.; Liao, Q.; Yao, J.; Fu, H. Organic Phosphorescence Nanowire Lasers. *J. Am. Chem. Soc.* **2017**, *139*, 6376–6381. [CrossRef] [PubMed]
137. Wu, J.-J.; Gao, H.; Lai, R.; Zhuo, M.-P.; Feng, J.; Wang, X.-D.; Wu, Y.; Liao, L.-S.; Jiang, L. Near-Infrared Organic Single-Crystal Nanolaser Arrays Activated by Excited-State Intramolecular Proton Transfer. *Matter* **2020**, *2*, 1233–1243. [CrossRef]
138. Liao, Q.; Jin, X.; Zhang, H.; Xu, Z.; Yao, J.; Fu, H. An Organic Microlaser Array Based on a Lateral Microcavity of a Single J-Aggregation Microbelt. *Angew. Chemie Int. Ed.* **2015**, *54*, 7037–7041. [CrossRef]

139. Wang, X.; Liao, Q.; Lu, X.; Li, H.; Xu, Z.; Fu, H. Shape-Engineering of Self-Assembled Organic Single Microcrystal as Optical Microresonator for Laser Applications. *Sci. Rep.* **2014**, *4*, 7011. [CrossRef]
140. Okada, D.; Azzini, S.; Nishioka, H.; Ichimura, A.; Tsuji, H.; Nakamura, E.; Sasaki, F.; Genet, C.; Ebbesen, T.W.; Yamamoto, Y. π-Electronic Co-Crystal Microcavities with Selective Vibronic-Mode Light Amplification: Toward Förster Resonance Energy Transfer Lasing. *Nano Lett.* **2018**, *18*, 4396–4402. [CrossRef]
141. Lv, Y.; Li, Y.J.; Li, J.; Yan, Y.; Yao, J.; Zhao, Y.S. All-Color Subwavelength Output of Organic Flexible Microlasers. *J. Am. Chem. Soc.* **2017**, *139*, 11329–11332. [CrossRef]

Review

BODIPY-Based Molecules, a Platform for Photonic and Solar Cells

Benedetta Maria Squeo [1], Lucia Ganzer [2], Tersilla Virgili [2,*] and Mariacecilia Pasini [1,*]

1. Istituto di Scienze e Tecnologie Chimiche (SCITEC), Consiglio Nazionale delle Ricerche (CNR), Via A. Corti 12, 20133 Milano, Italy; benedetta.squeo@scitec.cnr.it
2. Istituto di Fotonica e Nanotecnologie (IFN), Consiglio Nazionale delle Ricerche (CNR), Dipartimento di Fisica, Politecnico di Milano, P.zza Leonardo da Vinci 32, 20132 Milano, Italy; lucia.ganzer@polimi.it
* Correspondence: tersilla.virgili@polimi.it (T.V.); mariacecilia.pasini@scitec.cnr.it (M.P.)

Abstract: The 4,4-difluoro-4-bora-3a,4a-diaza-s-indacene (BODIPY)-based molecules have emerged as interesting material for optoelectronic applications. The facile structural modification of BODIPY core provides an opportunity to fine-tune its photophysical and optoelectronic properties thanks to the presence of eight reactive sites which allows for the developing of a large number of functionalized derivatives for various applications. This review will focus on BODIPY application as solid-state active material in solar cells and in photonic devices. It has been divided into two sections dedicated to the two different applications. This review provides a concise and precise description of the experimental results, their interpretation as well as the conclusions that can be drawn. The main current research outcomes are summarized to guide the readers towards the full exploitation of the use of this material in optoelectronic applications.

Keywords: BODIPY; optoelectronics; photonics; solar cells; organic dyes

1. Introduction

The rapid appearance of organic optoelectronics as a promising alternative to conventional optoelectronics has been largely achieved through the design and development of new conjugated systems [1]. The foremost advantage of organic materials is that they are cheap, lightweight, and flexible and they can offer the opportunity to make the electronics sector greener through the use of abundant materials and manufacturing methods rely on fewer, safer, and more abundant raw materials and with cheaper processing based on solution techniques so the embodied energy for organic devices is expected to be lower than for their inorganic counterparts usually processed by vacuum and high-temperature techniques [2–7].

Since its discovery [8], the 4,4-difluoro-4-bora-3a,4a-diaza-s-indacene (BODIPY) molecule has been the subject of considerable interest thanks to its chemical stability, easy functionalization, and interesting optical and electronic properties combined with low toxicity and high biocompatibility and it has therefore emerged as a promising platform for multiple optoelectronic applications [9–12]. This type of structure is commonly described as an example of a "rigidified" monomethine cyanine dye or, as it has three p-delocalized rings, one is pyrrole and the others are azafulvene and diazaborinin-type ring systems [13], by analogy with the all-carbon tricyclic ring can be considered as a boradiazaindacene, and the numbering of any substituents follows rule setup for the carbon polycycles (see Figure 1). This dye has also been [14] called "porphyrin's little sister" and this happy definition has been so successful that in analogy with porphyrinic systems, the 8-position is often referred to as the meso site.

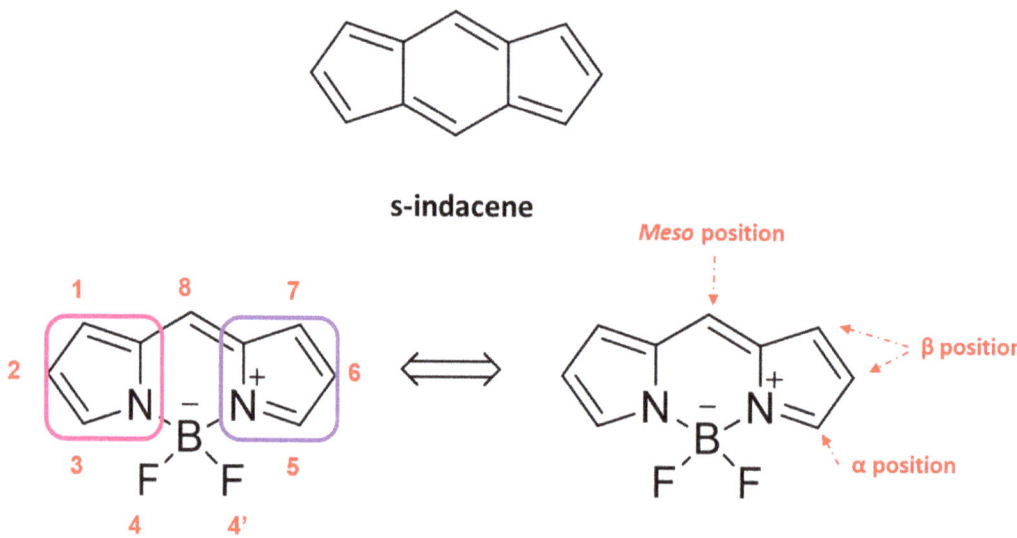

Figure 1. Naming and numbering of the 4,4-difluoro-4-bora-3a,4a-diaza-s-indacene (BODIPY)-based on s-indacene. Pyrrole ring pink boxes and azafulvene violet boxes Reproduced from Ref. [12] with permission of Taylor and Francis.

Over the years, several synthetic strategies have been implemented which have allowed the modulation of the optoelectronic properties of the BODIPY that is assumed to have two equivalent resonance forms resulting in the forming of two rings named azafulvene and pyrrole. Pyrrole is aromatic whereas azafulvene is a quinoid-type ring and it is not possible to distinguish the two forms [13]. Depending on the position and type of the substituents, it is possible to modify different parameters such as the delocalization of the π-electrons in the molecule and the supramolecular organization and the morphology of the solid thin film [15–17]. In this regard, the BODIPY core has eight reactive positions [18] that can be used to modulate its optical properties [19,20] (Figure 1): the two Boron positions, the two α-positions, the four β-positions, and the meso-position. The positions α and β are those which have the greatest influence on the electronic delocalization, while the other positions, meso and Boron, have a lower impact on electron delocalization but have a strong impact on the steric hindrance of the molecule. The new class of compounds referred to as AZA-BODIPY with a nitrogen substituted to a carbon in meso-position [21,22] will not be discussed in this review.

BODIPY-based materials are playing an increasingly important role in the field of organic semiconductors as demonstrated by the high number of citations about 11,000 achieved from 2019 (Google scholar source) of which 5300 in the year 2020.

This review intends to provide an overview of the recent progress on the design and development of BODIPY-based semiconductor molecules focusing only on molecular materials since polymeric systems require different treatment (discussion) for their application as active materials in solar cells and photonic devices [23,24].

This review is divided into two main parts:

(1) BODIPY-based active materials in OPV. Since numerous comprehensive review articles have focused on the synthetic strategy [25] and general working principles of the organic photovoltaic (OPV) devices, in this review we focus more on the analysis of the best results obtained from the first published data in 2009 [26,27] until the last one in 2020 [28,29].

(2) BODIPY-based active materials in photonics devices. In particular, we focus on the work carried out to optimize the lasing performance and photostability in solid-state, as compared to solutions.

We then conclude with the future perspectives of the use of this promising molecule for photovoltaic and photonic applications.

2. BODIPY-Based Active Materials in OPV

Renewable energy generated by solar cells is one of the potential solutions to the problem of maintaining our energy supply. Organic solar cells have the potential to be part of the next generation of low-cost solar cells. An encouraging state-of-the-art device has exhibited power conversion efficiencies (PCEs) over 17% [30], which represents a crucial step toward the commercialization of OPVs. Among the rapid development of optical-active donor/acceptor materials, interface materials, and device engineering, the properties of donors and acceptors, such as absorption, energy levels, and bandgaps, play a vital role in realizing high PCEs. Therefore, the evolution of new photovoltaic materials with a smaller optical bandgap (E_g) reasonable highest occupied molecular orbital (HOMO) and lowest unoccupied molecular orbital (LUMO) levels is crucial to further improve the PCEs of OPVs [31]. For these reasons, in the last 10 years BODIPY-based molecules have been of great interest as active materials for photovoltaic devices, thanks to their good absorbance at low energies (low optical gaps), high molar extinction coefficients, good chemical-stability, and opportune redox properties [23]. Table 1 summarizes the results in OPV where a BODIPY-based molecular material has been used as an active layer and the main parameters of a photovoltaic device are mentioned such as the current density J, the fill factor (FF), and the power conversion efficiency (PCE), where the FF is a measure of the quality of the solar cell and it is calculated by comparing the maximum power with the theoretical one [32].

The BODIPY derivatives have good solubility in common organic solvents, suitable HOMO-LUMO levels (see Table 1), and as previously mentioned, possible functionalization at several positions (Figure 1). Despite these interesting features, only recently satisfactory results with PCE above 5%, have been obtained and they are discussed in detail and reported in Table 2.

Thanks to the planar core, easy lateral functionalization, and electron-deficient characteristics BODIPY derivative is an ideal acceptor unit [33] for the design of donor–acceptor D–A type semiconducting architectures [34,35] which is one of the most used approaches to modulate the optoelectronic properties of organic semiconductors [36–38]. The BODIPY derivatives have been used both as donors and acceptors in OPV devices, in direct and inverted architecture [23]. Some of the most interesting results have recently been obtained using conjugated polyelectrolytes at the interface with the metal cathode [29,39]. This class of materials allows the engineering of the interface [40,41] between the metal and the active layer improving the extraction of the charge without complicating the preparation of the device. In fact, for the preparation of films, it is possible to use solution techniques thanks to the use of solvents orthogonal to the active layer. It is important to underline that the parameters that influence the OPV efficiency as mentioned before, are many. It is not always possible to evaluate the individual contribution because they are often intimately connected and closely related to the chemical structure and the choice of position, type, and the number of lateral substituents. In this part of the review on the OPV devices, we report most of these parameters and we try to analyze their individual contribution. All the device architectures have been reported in Figure 2.

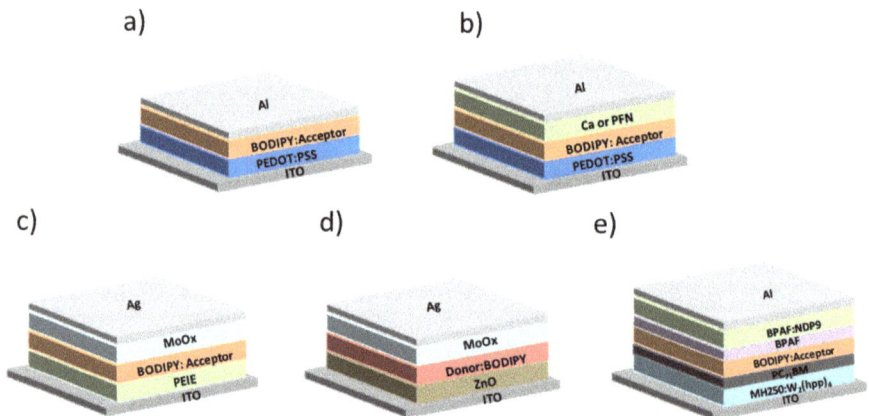

Figure 2. Different Organic Solar Cells (OSCs) architectures are presented in the work. (**a**,**b**) Direct architecture; (**c**,**d**) inverted architecture; (**e**) architecture presented in ref. [42] with an intrinsic layer between n- and p-doped active materials.

2.1. BODIPY Molecules as Electron Acceptor Materials in Organic Solar Cells

The fullerene has been, so far, the most common material used as an acceptor in an organic solar cell, however in recent years the search for non-fullerene acceptors has gained significant attention. At the same time, there has also been an increasing effort in the development of small molecules to be used as active materials for OPVs, due to the potential control of frontier orbital energy level that they offer [43]. The combination of these two interests has triggered research for small molecules, non-fullerene acceptors, for bulk heterojunction (BHJ) OPV.

To the best of our knowledge, only two research groups have reported on the synthesis of non-fullerene acceptor based on BODIPY.

In 2014, Thayumanavana et al. [44] presented a series of acceptor–donor–acceptor molecules containing terminal BODIPY unit conjugated through meso-position using a 3-hexylthiophene linker to thiophebenzodithiophene, cyclopentadithiophene, or dithienopyrrole (**1a**, **1b**, **1c**, Scheme 1). The three molecules have deep LUMO energy levels of −3.79 eV, −3.82 eV, and −3.74 eV, and good visible absorption, with absorption maxima in thin-film at 532 nm, 530 nm, and 532 nm, respectively. The molecules have been tested in inverted solar cells with the architecture ITO/ZnO/P3HT:acceptor (40 nm)/MoO3 (7 nm)/Ag (100 nm), with a maximum PCE of 1.51% obtained with the molecules containing cyclopentadithiophene.

In 2017, Zhan et al. [45] firstly reported the synthesis of a new acceptor small molecule based on a diketopyrrole core and two BODIPY lateral moieties (**2**, Scheme 1). The molecule has a LUMO of −3.79 eV and is present in a solid-state, an intense plateau-like absorption with two maxima at 681 nm and 756 nm, which can be attributed to the delocalization of the LUMO along all the molecule.

The small molecule has been tested as an acceptor in direct OSC with architecture glass/ITO/PEDOT:PSS/active layer/Ca/Al, using as donor P3HT, p-DTS-(FBTTh$_2$)$_2$, and PTB7-Th. The best results have been obtained with P3HT in D/A ratio 1:1.5 with a PCE of 0.888%. To improve the efficiency, the small molecule has been tested also in ternary blend solar cells with p-DTS-(FBTTh$_2$)$_2$ and P3HT in D1/D2/A ratio 0.5:0.5:1 and a PCE of 2.836% has been obtained.

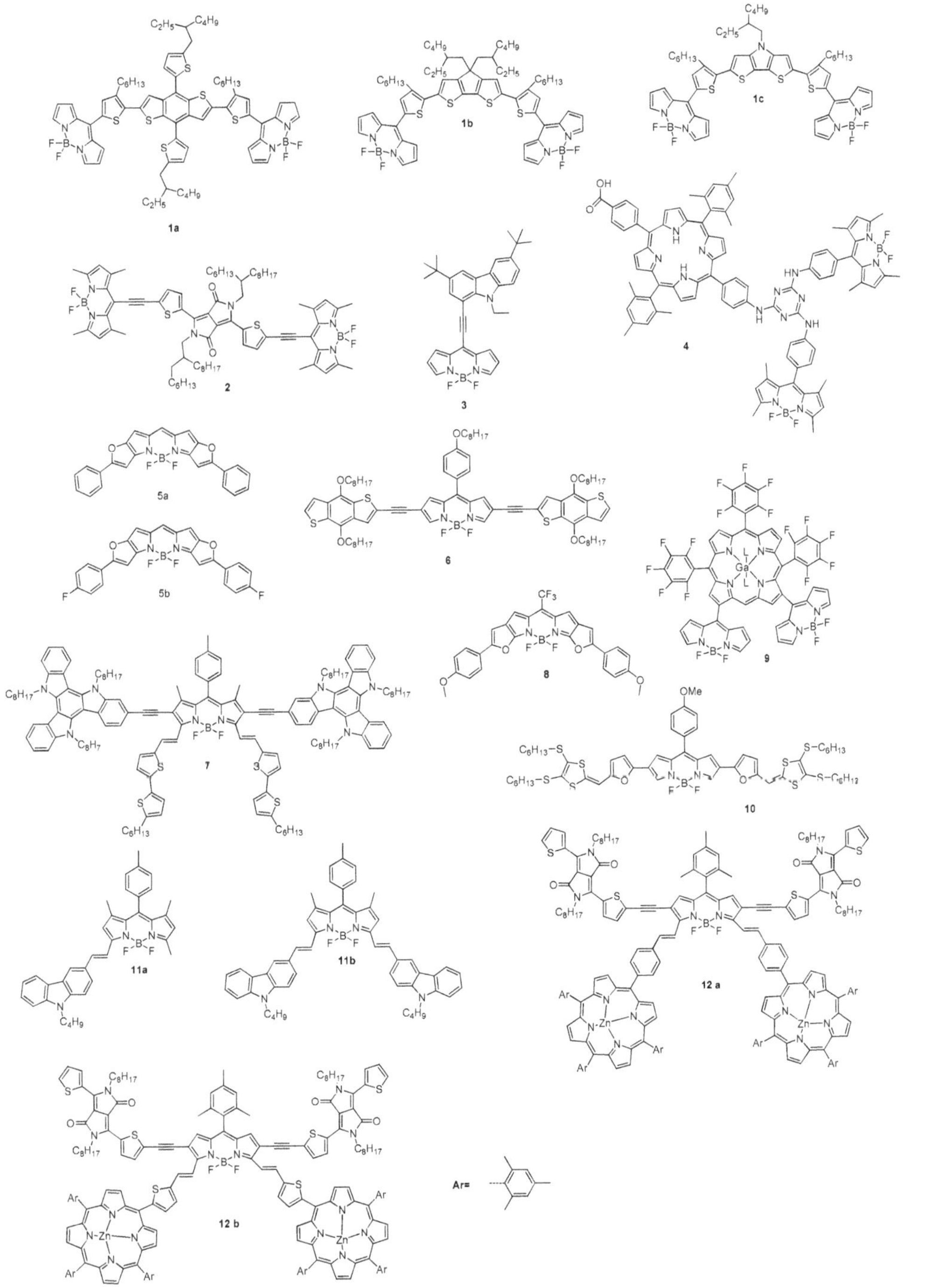

Scheme 1. Chemical structure of BODIPY-based molecules used as an acceptor in organic photovoltaic (OPV) devices (**1** and **2**) and as donor **3**–**12**.

2.2. BODIPY Molecules as Donor Materials in Organic Solar Cells

In Table 1, the main works on BODIPY-based molecules as a donor material in the OPV devices are reported. For each research work, the year of the research publication, the peak of the absorption and the HOMO and the LUMO position of the new molecule are reported, together with the main parameters of an organic photovoltaic cell. The works are listed in chronological order and we observe that the first solar cell with a BODIPY-based active layer dates back to 2009 with a PCE of 1.7% and that just 6 years later, in 2015, a PCE of around 5% was achieved.

Table 1. Photophysical and photovoltaic characteristics of bulk heterojunction organic solar cells based on BODIPY small molecules.

Year	Abs^{Film}_{Max} Film [nm]	HOMO [eV]	LUMO [eV]	J_{sc} [mA/cm^2]	V_{oc} [V]	FF [%]	PCE [%]	Reference
2009	572	−5.69	−3.66	4.7	0.866	42	1.7	[27]
2009	646	−5.56	−3.75	4.14	0.753	44	1.34	[26]
2010	649	-	-	7.00	0.75	38	2.17	[46]
2011	733	−4.71	−2.57	6.3	-	67	3.7	[47]
2012	580	−5.47	−3.48	8.25	0.988	39.5	3.22	[48]
2012	673	−4.26	−3.75	2.9	0.51	35	0.52	[49]
2012	714	−5.32	−3.86	14.3	0.7	47	4.70	[50]
2014	748	−5.00	−3.59	7	0.68	31	1.50	[51]
2014	733	−5.02	−3.64	6.8	0.67	34.3	1.56	[52]
2014	760	−5.02	3.37	8.9	0.51	34	1.5	[53]
2014	643	−5.31	−3.50	3.39	0.71	27	0.65	[54]
2015	680	−5.3	−2.75	8.42	0.82	55	3.76	[55]
2015	652 *	−5.62	−3.52	10.48	0.9	56	5.29	[56]
2015	774	−5.23	−3.72	10.32	0.97	46.5	4.75	[57]
2015	614	−5.48	−3.44	10.20	0.90	55	5.05	[58]
2015	514 *	−5.33	−3.86	3.03	0.81	24	0.58	[59]
2015	655–792	−5.01	−3.73	13.39	0.73	37.3	3.6	[60]
2015	761	−5.34	−3.66	8.17	00.85	39	2.70	[61]
2015	696	−5.40	−3.81	7.64	0.73	38	2.12	[62]
2016	748	−5.19	−3.60	6.77	0.78	41	2.15	[63]
2017	627	−4.93	−3.28	13.79	0.768	66.5	7.2	[64]
2017	550	−5.11	−3.65	11.84	0.73	53.8	4.61	[65]
2017	765	−5.26	−3.91	13.9	0.64	65	5.8	[66]
2017	800	−5.23	−3.87	13.3	0.73	63	6.1	[42]
2018	668	−5.06	−3.60	12.98	0.7	62	5.61	[67]
2018	720 *	−5.39	−3.74	12.43 **	0.88	61	6.67	[68]
2018	752 *	−5.36	−3.79	14.32 **	0.95	67	8.98	[68]
2018	550–640	−5.37	−3.46	11.46 **	0.915	63	6.60	[39]
2019	717	−5.00	−3.42	5.17	0.672	40.8	1.62	[69]
2019	580	−5.28	−3.61	10.9	0.83	60	5.5	[70]
2019	725	−5.16	−3.43	7.72	1	31	2.79	[71]
2019	716	−5.47	−3.76	10.58	0.769	56.4	4.58	[72]
2020	538	−5.35	−3.08	2.27	0.67	27	0.37	[73]
2020	586	−5.16	−3.17	13.56	0.78	61	6.45	[29]
2020	672 *	−4.99	−3.27	16.24	0.71	66	7.61	[29]
2020	586*	−5.91	−4.09	0.87	0.45	21	1.36	[74]

* In solution: ** After Solvent Vapor Annealing (SVA).

In Table 2, only devices with PCE above 5% are analyzed in detail. The table ranges from the least to the most efficient device. The parameters reported are the absorption window, the optical energy gap and the holes mobility of the molecule, and the geometry of the device with any thermal or solvent annealing treatments.

Table 2. Photophysical and photovoltaic characteristics of bulk heterojunction organic solar cells based on BODIPY small molecules with PCE above 5%.

Molecule [Ref.]	Optical Energy GAP [eV]	Treatment	PCE [%]	FF [%]	Hole Mobility [cm^2/Vs]	Electron Mobility [cm^2/Vs]	Abs. Band [nm]
3 [58]	1.72	-	2.71	38	7.84×10^{-6}	-	450–650
		TA	3.99	48	5.34×10^{-5}	-	
		TA + SVA	5.05	55	8.45×10^{-5}	-	
4 [56]	1.84	NO	3.48	46	8.5×10^{-6}	2.34×10^{-4}	350–700
		Processed with pyridine 4% v/v	5.29	56	8.15×10^{-5}	2.29×10^{-4}	
5b [70]	1.59	Vacuum processed	5.5	60	9.2×10^{-6}	-	350–1000
6 [67]	1.74	-	2.74	34	-	-	300–850
		TA	4.72	57	2.09×10^{-4}	-	
		TA +SVA	5.61	62	4.37×10^{-4}	-	
7 [66]	1.52	TA	5.8	65	0.8×10^{-3}	-	400–800
8 [42]	1.32	Vacuum processed	6.1	63	-	-	500–900
9 [39]	1.79	-	2.54	39	5.34×10^{-5}	2.35×10^{-4}	400–700
		SVA	6.6	63	1.13×10^{-4}	2.45×10^{-4}	
10 [63]	1.65	-	5.3	52	0.9×10^{-4}	3.9×10^{-4}	300–800
		TA	7.2	66.5	2.1×10^{-4}	2.8×10^{-4}	
11a [29]	1.78	-	3.21	46	-	-	300–600
		SVA	6.45	61	8.89×10^{-5}	2.41×10^{-4}	
11b [29]	1.58	-	3.76	48	/	/	300–700
		SVA	7.61	66	1.07×10^{-4}	2.47×10^{-4}	
12a [68]	1.52	-	3.32	41	3.95×10^{-5}	2.31×10^{-4}	400–800
		SVA	6.67	61	1.10×10^{-4}	2.36×10^{-4}	
12b [68]	1.44	-	4.73	45	4.85×10^{-5}	2.38×10^{-4}	400–800
		SVA	8.98	67	1.59×10^{-4}	2.43×10^{-4}	

One of the first BODIPY small molecules based OSC with a PCE higher than 5% has been reported by Jadhav et al. [58] in 2015. The group proposes a new D-A type carbazole meso-substituted BODIPY small molecule (3, Scheme 1) as donor along with PC71BM as an electron acceptor in solution-processed BHJ solar cells. The molecule showed an absorption between 500 and 700 nm, with maxima at 568 and 620 nm and E_g^{opt} of 1.75 eV. The best BHJ solar cell, with architecture ITO/PEDOT:PSS/active layer:PC71BM/Al (Figure 2b) shows a PCE of 2.71% which increases to 3.99% after thermal annealing (TA) and 5.05% after thermal and solvent vapor annealing (TSVA). The improvement in the efficiency, clearly induced by the TA and TSVA treatment, seems to be due to the increase in Jsc and the FF which originated from the enhanced absorption, better active layer nanoscale morphology (as shown in TEM images reported in Figure 3) and more balanced charge transport.

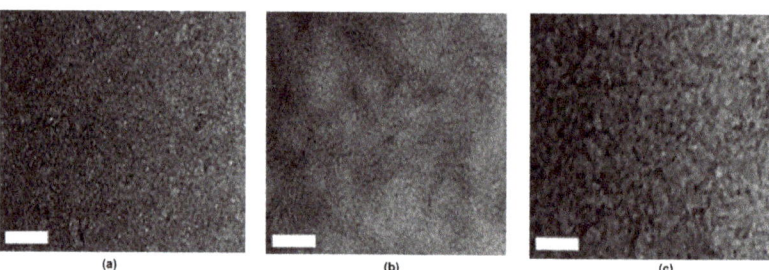

Figure 3. TEM images of the active layer **3**:PC71BM (**a**) as cast, (**b**) with thermal annealing (TA) and (**c**) thermal and solvent vapor annealing (TSVA). The scale bar is 200 nm. Reproduced from Ref. [58] with permission from the PCCP Owner Societies.

In the same year, Coutsolelos et al. [56] proposed a porphyrin-BODIPY-based small molecule as light-harvesting antenna system in BHJ solar cells (**4**, Scheme 1). The molecule consisted of a meso-aryl-substituted free base monocarboxy-porphyrin unit bridged by a central 1,3,5-triazine moiety to 4-aminophenyl-boron dipyrrin units through their arylamino groups. The absorption spectrum of the antenna system was the sum of the BODIPY and porphyrin absorption moiety, indicating that their connection through the triazine bridge does not cause significant electronic interactions between the two parts. In fact, the spectra showed in solution two main absorption maxima at 422 nm, attributed to the Soret band of porphyrin and 502 nm, attributed to the BODIPY unit, and other three less intense maxima (552 nm, 596 nm, and 652 nm) attributed to the Q-bands of the porphyrin unit.

This small molecule was tested in direct BHJ solar cells in blend with PC71BM, with architecture ITO/PEDOT:PSS/BODIPY:PC71BM/Al (Figure 2a). The optimized device, with donor/acceptor ratio 1:1 and processed with THF showed a PCE of 3.48% and an FF of 46%. To improve the performance of the devices, the active layer was processed with a solvent mixture of 4% v/v of pyridine in THF. The efficiency increased to 5.29%, with an FF of 56%, with an increase of an order of magnitude of the hole mobility. In this case, the enhancement can also be attributed to a better morphology and crystallinity of the active layer as evidenced in AFM images reported in Figure 4, leading to a better charge generation/separation/mobility/transport.

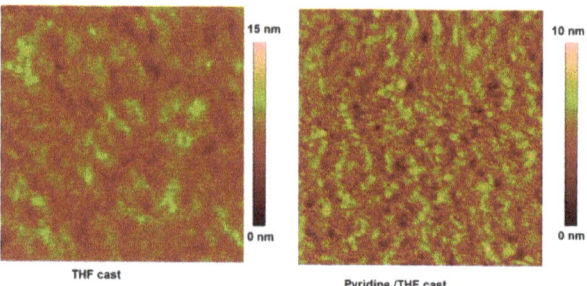

Figure 4. AFM images of **4**: PC71BM 1:1 blend thin films processed with and without pyridine additives. Image size: 3 mm. Reproduced from Ref. [56] with permission from The Royal Society of Chemistry.

The state of aggregation in the materials is another key parameter for understanding the performance of an OPV cell. Recently Leo et al. [70] have studied the effect of H- and J-aggregation [75] in NIR BODIPY-based small molecules BHJ solar cells. In the work, two BODIPYs with furan-fused pyrrole core structures, **5a** and **5b** in Scheme 1, possessing respectively H- and J-aggregation in neat film, have been studied. The two molecules showed similar absorption spectra in solution, while in solid-state **5a** preferred a cofacial H-aggregation and **5b** adopted a J-aggregation, as a result, the solid-state spectra are significantly different. Both BODIPYs presented broadened bands with absorption from 450 to 900 nm. Compared to the solution, the absorption maximum of **5a** is hypsochromically shifted at 580 nm, with a shoulder peak at 417 nm, while **5b** showed bathochromically shifted absorption maximum at 777 nm with a shoulder peak at 630 nm. The two molecules have been tested in vacuum-deposited BHJ solar cells, with architecture ITO/MH250:$W_2(hpp)_4$ (5 nm, 7 wt%)/C70 (15 nm)/BODIPY: C60 (1:2 v/v, 30 nm, 95 °C)/BPAPF (5 nm)/BPAPF:NDP9 (40 nm, 10 wt%)/NDP9 (1 nm)/Al (100 nm); where $W_2(hpp)_4$ (tetrakis(1,3,4,6,7,8-hexahydro-2H-pyrimido[1,2-a]pyrimidinato)ditungsten (II)) and MH250 (*N*,*N*-bis(fluoren-2-yl)-naphthalenetetracarboxylic diimide(bis-Hfl-NTCDI)) are used as electron transporting layer (ETL), and NDP9 (Novaled AG) doped BPAPF (9,9-bis[4-(*N*,*N*-bis-biphenyl-4-yl-amino)phenyl]-9H-fluorene) served as hole transporting layer (HTL) (Figure 2e). The intrinsic layers adjacent to the n-/p-doped layers worked as exciton reflection and hole/electron blocking layer. The C70 intrinsic layer also contributed to the photon absorption and photocurrent generation. The two molecules showed good performances, in particular, **5b** showed an efficiency of 5.5% with an FF of 60%, while **5a** showed a lower efficiency of 4.2% and an FF of 55%. The better performance of 5b may be attributed to bathochromic shifted absorption, a small driving force of exciton dissociation, and lower nonradiative voltage loss, compared to 5a.

Fang et al [67] in 2018 reported a novel donor–acceptor–donor (D–A–D) type small molecules consisting of a BODIPY linked through alkynyl with two benzo[1,2-b:4,5-b']dithiophene (BDT) terminal donors for solution-processed BHJ solar cell (**6**, Scheme 1). The molecule showed an absorption spectrum, in thin-film, from 300 to 800 nm with a maximum at 668 nm and an optical energy gap of 1.46 eV. The small molecule has been tested as an electron donor in solution-processed BHJ with architecture ITO/PEDOT:PSS/active layer/Ca/Al. The optimized device with thermal annealing treatment, showed a PCE of 4.72%, with an FF of 57%. In this case, the efficiency can also be increased up to 5.61% with an FF of 62% when used both thermal annealing and solvent vapor annealing treatment. The improved efficiency can be attributed to a better morphology and reduced roughness from 1.136 to 0.659 nm as evidenced by TEM and AFM reported in Figure 5.

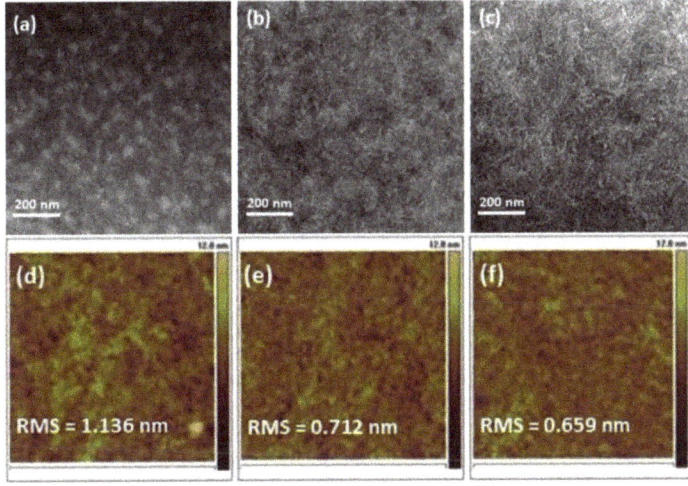

Figure 5. TEM images and AFM height images of an active layer based on **6**:PC71BM: (**a**,**d**) as cast; (**b**,**e**) with TA; (**c**,**f**) with TA and solvent vapor annealing (SVA) treatments. Reproduced from Ref. [67] with permission from 2018 Elsevier B.V.

To increase the molecular packing Bulut et al. [66], inspired by Leclerc and his research group, proposed a BODIPY linked to two triazatruxene (TAT) derivatives as planar end groups (**7**, Scheme 1). This kind of architecture (TAT–dye–TAT) would lead to strengthened molecular stacking behavior, while the optical properties were still essentially determined by the central dye. Moreover, the addition of a TAT platform was able to favor the out-of-plane hole mobility. This small molecule was one of the rare examples of inverted architecture, in fact, it has been tested with the following geometry ITO/PEIE(5 nm)/BODIPY:PC71(61)BM/MoO$_3$(7 nm)/Ag(120 nm) (Figure 2c) reaching an efficiency of 5.8% and FF of 65%. A hole mobility of 0.8×10^{-3} cm^2 V^{-1} s^{-1} was measured using hole-only space-charge limited-current (SCLC) devices.

Another successful strategy to increase the efficiency of solar cells has been to increase the absorption window by developing dyes capable of absorbing in the IR region. A BODIPY with intense and long-wavelength absorption can be achieved by an extension of the π-system and an electron-withdrawing group on the meso-position. Li et al. [42] have developed a BODIPY dye able to combine the absorption in a wide spectral window with an excellent packing reaching an efficiency of 6.1%, with an FF of 63%. The synthesized material was a NIR furan fused BODIPY with a CF$_3$ group in meso-position (**8**, Scheme 1). The small molecule showed an absorption spanning from 500 nm to 900 nm and absorption maxima at 712 nm and 800 nm in the solid-state. The choice of methoxyl group on the peripheral phenyl rings influenced the packing behavior and a brickwork-type arrangement in the neat film is observed, comprising an alternating arrangement of a unit consisting of two antiparallel molecules. Consequently, **8** had a J-aggregation character, moreover, charge transport in OSCs benefits from aggregation, leading to higher overall PCE through improved Jsc and FF, even though Voc (open circuit Voltage) may decrease slightly. The new BODIPY derivatives have been tested as an electron donor in vacuum processed n-i-p BHJ solar cells, with the architecture reported in Figure 2e ITO/MH250:W$_2$(hpp)$_4$ 7 wt% (5 nm)/C70 (15 nm)/**8**:C60 (40 nm)/BPAPF (5 nm)/BPAPF:NDP9 10 wt% (40 nm)/NDP9 (1 nm)/Al (100 nm). W$_2$(hpp)$_4$ and NDP9 were n- and p-dopants and MH250/BPAPF was an electron/hole transporting material. The maximum PCE of 6.1% has been obtained after thermal annealing at 100 °C. The high EQE and Jsc (Short circuit current density) of the device indicate that the photogenerated excitons are efficiently separated into free charge carriers. AFM analysis confirmed that **8** and C60 were well intermixed, providing enough D/A interfaces for exciton dissociation. The molecule has been tested also in tandem with solar cells, in combination with the green absorber DCV5T-Me. In this case, the tandem solar cell has achieved an efficiency of 9.9% and FF of 59%.

Sankar et al. [39] have proposed a corrole-BODIPY derivative as an electron donor in solution-processed BHJ. The molecule consisted of a central Ga(III) corrole with two peripheral BODIPY units (**9**, Scheme 1). The two strong absorption bands of the molecule in thin films centered at 432 and 516 nm are attributed to the Soret bands of corrole and BODIPY and the broadband between 550–640 nm, is attributed to Q-bands typical of corrole-based systems [76]. The molecule has been tested as an electron donor in BHJ with structure ITO/PEDOT:PSS/BODIPY:PC71BM/PFN/Al, where poly[9,9-bis(3'-(*N,N*-dimethylamino)propyl)-2,7-fluorene] (PFN) is used as the cathode interlayer (Figure 2b). The optimized device, with active layer as cast, had an efficiency of 2.54% and FF of 39%, while the optimized device a, with solvent vapor annealing treatment, showed an efficiency of 6.60 and FF of 63%. As shown in Figure 6, better phase separation was found for the SVA-treated active layer compared to the as-cast active layer, forming more bicontinuous interpenetrating networks, which was beneficial for charge-transport efficiency, leading to improved J$_{sc}$ and FF.

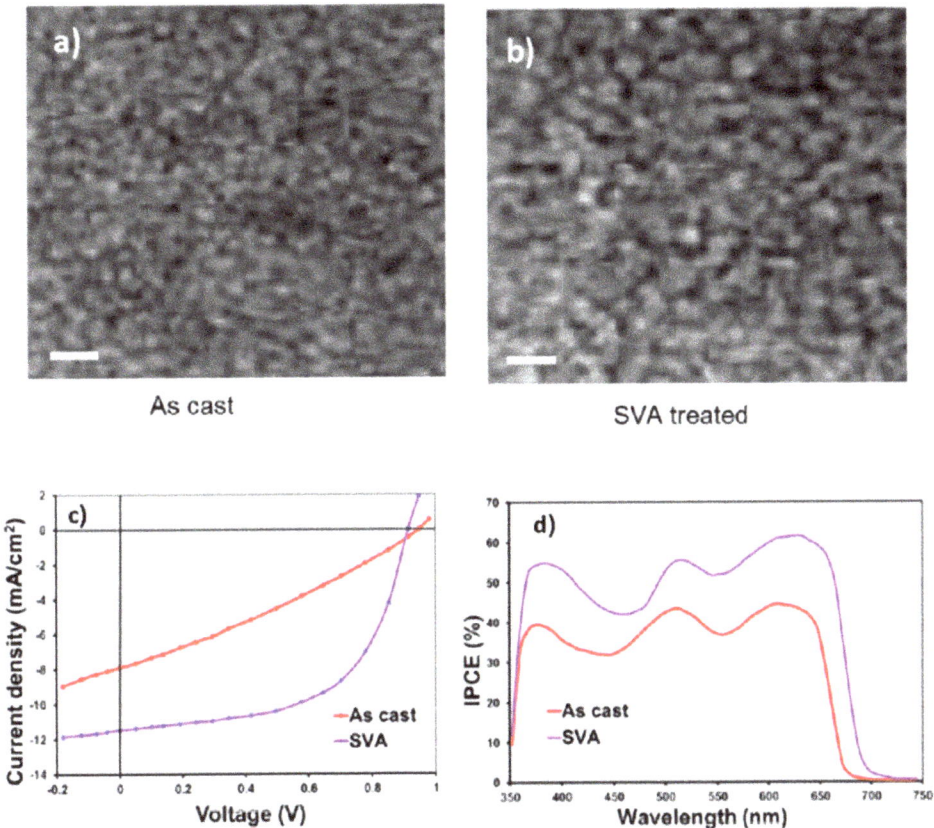

Figure 6. TEM image of as-cast 88 (**a**) and SVA-treated (**b**) 9:PC71BM thin films (Scale bar is 100 nm). (**c**) Current−voltage characteristics under illumination and (**d**) IPCE spectra of the OSCs base on as-cast and SVA-treated 9:PC71BM active layers. Reproduced from Ref. [39] with permission from 2018 American Chemical Society.

In 2017, R. Srinivasa Rao et al. [63] reached one of the highest efficiencies in BODIPY-based solar cells. The groups synthesized a BODIPY molecule decorated with dithiafulvalene wings with a broad absorption profile from 350 to 780 nm and extending up to the near IR region (**10**, Scheme 1). The solution process BHJ device, with architecture ITO/PEDOT:PSS/active layer/Ca/Al (Figure 2b) was prepared with the active layer obtained as cast and reached, after thermal annealing 7.2% of PCE with a FF of 66.5% stressing once again the importance of controlling the nano-scale aggregation. The authors have reached this excellent result for a series of positive factors such as the large absorption window, suitable HOMO and LUMO energy levels, excellent affinity with the acceptor, and an optimal nanomorphology (Figure 7).

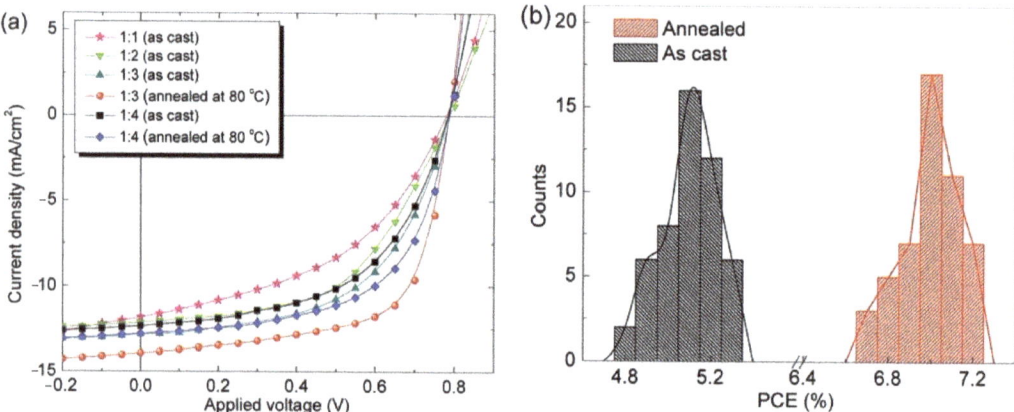

Figure 7. (a) The light J–V characteristics of bulk hetero junction (BHJ) organic solar cells fabricated with "as cast" and "annealed" **10**:PC71BM blend photoactive layers and (b) the efficiency histogram of both categories of devices. Reproduced from Ref. [63] with permission from The Royal Society of Chemistry.

Recently, Yang et al. [29] have designed and synthesized two carbazole–BODIPY-based dyes by Knoevenagel reaction, the first bearing one carbazole unit in position 3, the second bearing two carbazole unit, in position 3 and 5 respectively (**11a** and **11b**, Scheme 1). The introduction of carbazole donating unit in D–A organic semiconductors, lowering the HOMO energy level, resulted in a high open-circuit voltage in OSCs in addition, remarkably red-shifted absorption spectra and better hole transport are reported. The molecules showed classic absorption spectra for BODIPY, with the maxima at 586 nm and 672 nm in DCM solution and broadening and red shift in thin films due to aggregation. The molecules have been tested in solution-processed BHJ solar cells with architecture ITO/PEDOT:PSS/active layer:PC71BM/PFN/Al (Figure 2b), using a conjugated polar polymer for charge regulation at the active layer/electrode interface as previously reported by Sankar and co-worker. In this way the BHJ solar cell reached after solvent vapor annealing, the efficiency of 6.45% and 7.61% and FF of 61% and 66% respectively for the molecules with one carbazole and two carbazole units. Once again, an enhanced crystalline nature and reduced π–π stacking distance for the **11b**:PC71BM active layer may induce better nanoscale phase separation and may be beneficial for charge transport and collection. These observations may be the reason for the higher FF and greater PCE of the OSCs based on the latter.

The highest efficiency for BODIPY small molecule-based OSC was obtained by Bucher and co-worker [68] in 2018. The group designed and synthesized two small molecules comprised of a central BODIPY core, surrounded with two diketopyrrolopyrrole (DPP) and two porphyrin side units, differing only from each other in the aromatic bridge between BODIPY and porphyrins: in one case phenylene (**12a**, Scheme 2) was used, in the other case thienyl (**12b**, Scheme 1). In both cases, complementary absorption properties of the three units allowed a wide solar photons harvest and, as matter of fact, the small molecules showed a panchromatic absorption spectrum from 400 nm to 800 nm in the thin film and an optical E_g of 1.52 and 1.44 eV respectively for **12a** and **12b**. The OSC, with architecture ITO/PEDOT:PSS/BODIPY:PC71BM/PFN/Al (Figure 2b), where the BODIPY is the **12a** molecule, after solvent vapor annealing, presents a PCE of 6.67% and FF of 61%, while with the molecule **12b** has a record efficiency, after SVA, of 8.98% and a FF of 67%. The different performances in OSC might be attributed to the different morphology in the blend, as reported in Figure 8 where the presence of nanofibril and phase separation has increased the charge transport in the layer and subsequently the PCE of the device.

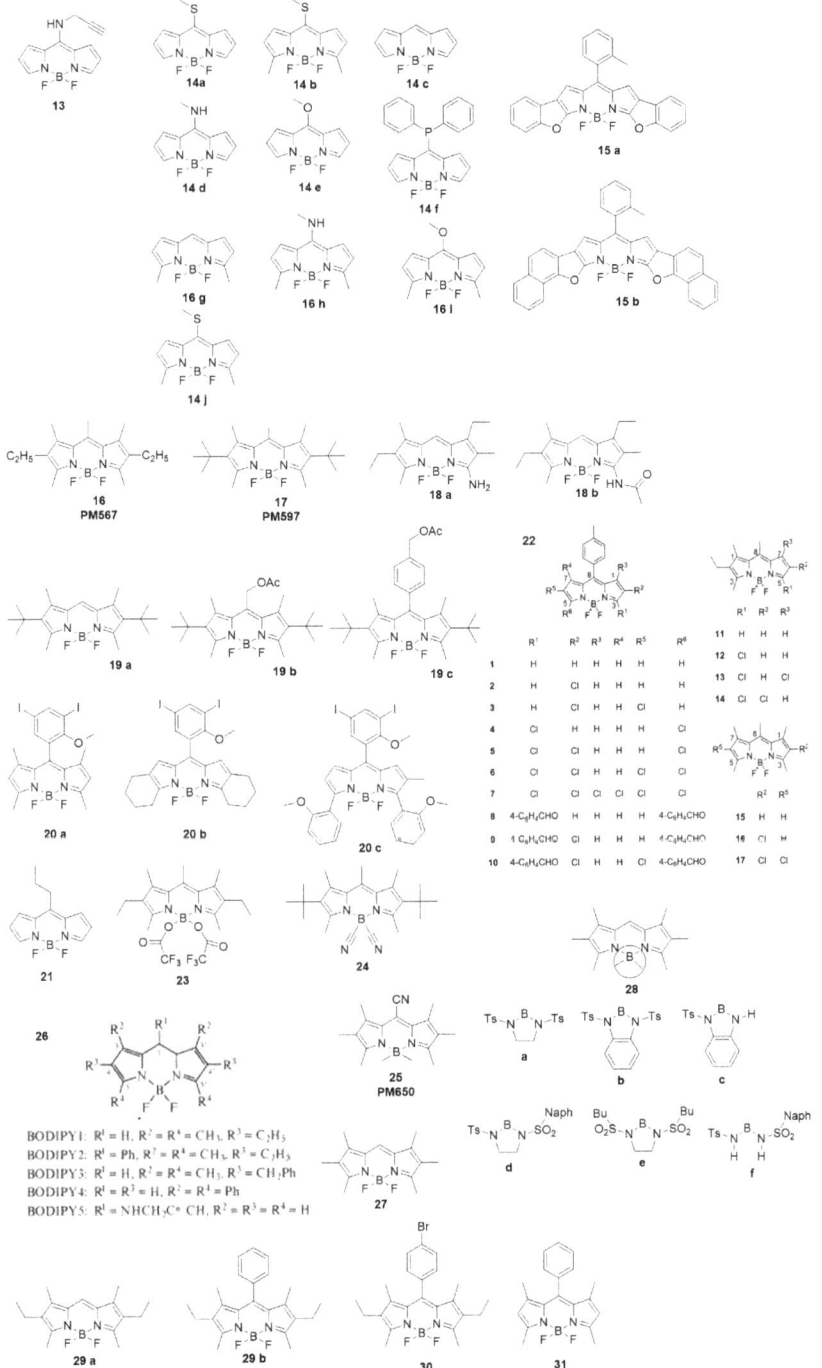

Scheme 2. Chemical structure of BODIPY-based molecules as active materials in solid-state photonic devices. Molecule **22** reproduced from Ref. [77] with permission from 2012 WILEY-VCH Verlag GmbH & Co. KGaA, Weinheim, Germany. Molecule **26** reproduced from Ref. [78] with permission from 2014 Turpion Ltd.

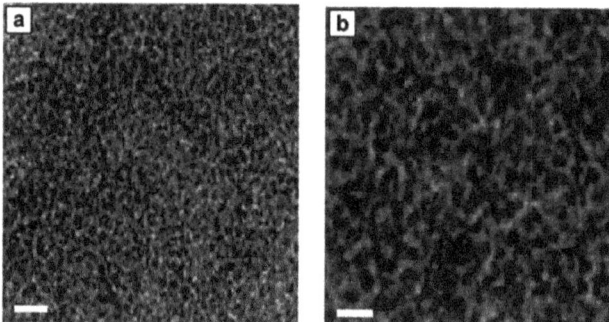

Figure 8. Transmission electron microscopy (TEM) images of optimized (**a**) **12a**:PC71BM and (**b**) **12b**:PC71BM thin films. The scale bar is 100 nm. Reproduced from Ref. [68] with permission from The Royal Society of Chemistry.

From the results reported in Table 2, it is evident that in a solar cell the parameters to be taken into consideration are manifold and all closely connected to each other. Among them, two proved to be particularly important for the development of this class of materials, the absorption spectral window that must clearly be as wide as possible and the active layer morphology. The latter strongly affects the efficiency of exciton dissociation, the charges mobility, and their recombination. For this reason, it is worth to stress the thermal treatments importance and the solvent annealing. We have shown that for the same material, once TA and SVA have been made, the solar cell efficiency can often be doubled.

3. Photonic Applications

Photonics play an important role in driving innovation in an increasing number of fields. The application of photonics spreads across several sectors: from optical data communications to imaging, lighting, and displays; from the manufacturing sector to life sciences, health care, security, and safety. In this context, organic semiconductors are promising candidates to fulfill the capacity of photonics and deliver on their promises. Among all the organic molecules BODIPY derivatives seem to have all the requisites to be effectively used as active materials in photonics. As previously mentioned BODIPY is characterized by a high photoluminescence quantum yield and, it can be easily functionalized on several different sites allowing a fine modulation of its photophysical properties.

BODIPY derivatives have been found to be suitable materials for the development of tunable organic lasers. Diluted solutions of those organic dyes have been exploited as lasing media with excellent results, comparable and even better than most common and largely used dyes such as coumarin, rhodamine, and polymethine [9,79]. Despite these results, there have been only a few studies centered on the solid-state; this was probably related to two main reasons: in the solid-state the quantum yield is usually lower than in solutions and BODIPY complexes seem to be particularly affected by the composition and structure of the polymer matrices in which they generally need to be dispersed to maintain the most of their properties.

For this reason, in view of a possible optimization of the use of this material in this review, we focus on their performances as solid-state depending on different groups or atoms substitutions (the chemical structures of the BODIPY-based molecules have been reported in Scheme 2).

3.1. Fluorescence Emission Optimization in BODIPY Derivatives

In solid-state BODIPY derivatives, the fluorescence is normally strongly quenched mainly due to intermolecular interactions. This is strongly detrimental for photonics applications. The suppression of intermolecular interactions proved to be a useful and valid strategy for decreasing quenching and increasing quantum yield [10]. In the next section, the main researches performed to control interactions between BODIPY molecules and to tune materials emission properties have been reported.

The first method to reduce intermolecular close packing (π-π interactions) is to introduce a bulky group playing as a "spacer" [80,81]. Following this strategy, BODIPY-based emissive compounds have been synthesized by substitution with bulky spacers such as bulky tert-butyl substituents on the meso-phenyl groups [82], bulky arylsily groups [83], or adamantyl group in the 3 and 5 positions of the BODIPY core [84].

A second, different approach is to attach the planar BODIPY core to a twisted or nonplanar scaffold. As an example, a Λ-shaped Tröger's base (Figure 9) skeleton was introduced in the BODIPY core [85]. The new-synthesized dye displayed an intense fluorescence, indicating that quenching caused by aggregation, was suppressed.

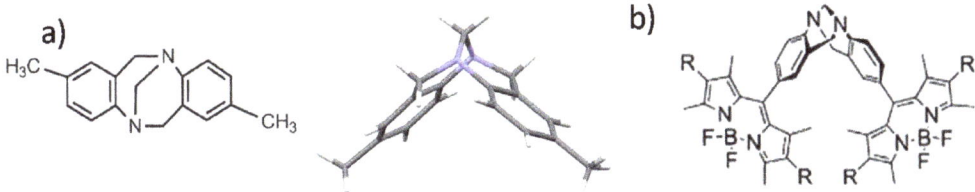

Figure 9. Structure of Tröger's base (**a**) and structure of the new BODIPY (**b**). Reproduced from Ref. [85] with permission from The Royal Society of Chemistry.

A further factor that directly influences the intensity of the fluorescence is the intersystem crossing process between the singlet and the triplet states. This process can be eliminated by making the dyes skeleton more rigid [86] leading, however, to a tighter packing and so to detrimental interchain interactions. A delicate balance between these 2 effects is required. Swamy et al. [87] showed that the molecular free rotation and aggregation-induced fluorescence quenching of BODIPYs can be successfully suppressed by lowering the flexibility of the molecules with a series of appropriate substitutions in the dye.

Aggregation is not always detrimental to the fluorescence quantum yield. In particular, J-aggregation can have a fundamental role and be responsible for red-shifted emission. Taking advantage of this feature, Manzano et al. [88] designed new O-BODIPY derivatives with spontaneous formation of stable J-aggregates and characterized by a conformational rigidity due to the presence of the B-spiranic structure and the disposition of the B-diacyloxyl substituent and the meso-aryl group. So, by controlling the J-aggregation process, it was possible to tune dyes properties, allowing the excitation of efficient and tunable laser emission of both the monomeric and J-aggregated forms. Moreover, by playing with the conformational rigidity of the system, the fluorescence quantum yield was improved, by reducing quenching due to H-aggregates. The authors observed highly efficient and stable red-shifted laser emission from J-aggregates in pure organic solvents and solid-state crystals. Figure 10 shows the emission properties of two representative dyes: 1b with J-aggregation capability and 2e with monomeric units. We observe that, after UV light excitation, both crystals display fluorescence emission, but efficiencies and spectral profiles are very different. This is related to the different molecular arrangement: 2e exhibits a single emission corresponding to the monomer, while in 1b we have a second long-wavelength emission, assigned to the corresponding J-aggregated form.

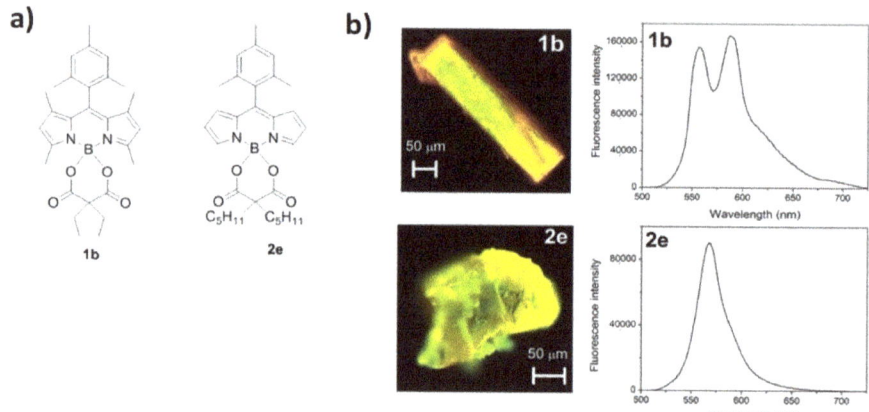

Figure 10. (**a**) Structure of compounds 1b and 2e. (**b**) Fluorescence images and spectra of solid crystals of representative compounds 1b and 2e were recorded under excitation at 450–490 nm with a fluorescence microscope. Reproduced from Ref. [88] with permission from 2016 WILEY-VCH Verlag GmbH & Co. KGaA, Weinheim, Germany.

The impact of aggregation on luminescence was further investigated by Bozdemir et al. [89] studying the emission variation related to different distributions of crystalline and amorphous regions while the range of states produced by system self-assembly (from extended aggregates to other different structures) and influenced by side-group structure was investigated by Musser et al. [90]. In another work, the effect of spiro-structures in J-aggregate formation was studied. It was found that spiro-bicyclic structure has an efficient role in preventing face-to-face π-stacking interactions, which is frequently observed in solid-state BODIPYs, and as a consequence, it can be used as a strategy to enhance the emission intensity [91].

3.2. BODIPY Derivatives as Lasing Dyes

Fluorescent BODIPY derivatives emitted from the green spectral region, above 500 nm, to the near infra-red (NIR) range between 650 and 900 nm; furthermore, functionalization on the 8 meso-position shifted the emission wavelength in the blue spectral region. Therefore, this rather young class of dye materials presents a broad and tunable emission spanning from the blue to the NIR. Here below we describe three research studies where the BODIPY derivative was in solution because we consider them as important landmarks to understand the following studies in solid-state.

In 2010, Gómez-Durán and co-workers [92] synthesized a new dye by incorporating a propargylamine group at the meso-position of the BODIPY core (**13**, Scheme 2). This incorporation leads to a hypsochromic shift of the absorption and fluorescence bands to the blue part of the visible region. The new BODIPY structure exhibits a fluorescence quantum yield up to 0.94 and a laser emission tunable between 455 and 515 nm depending on the solvent. Even if the new dye photostability is still quite-low, this work represents an important benchmark for the capability of tuning BODIPY laser emission.

In 2013, Esnal et al. [93] published a work where they presented BODIPY derivatives with unprecedented laser efficiencies and good photostabilities. In their study, they developed a series of meso-substituted BODIPY dyes. Starting from the core structure A (Figure 11a), they increased efficiency by adding methyl groups at positions 3 and 5 and they changed the laser emission region by changing the heteroatom at position 8 (N, S, O, P, and H) (**14**, Scheme 2). As shown in Figure 11b, the laser efficiency and the emission wavelength depend on the heteroatom incorporated at position 8. Moreover, the laser emission efficiency is greatly improved by incorporation of methyl groups at positions 3 and 5 and in some dyes can reach also a value higher than 70%.

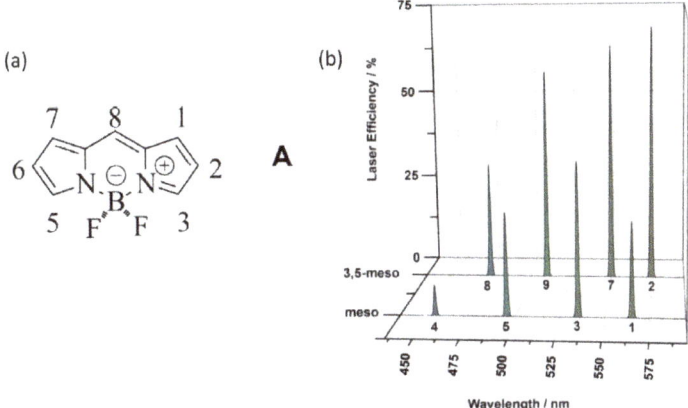

Figure 11. (a) BODIPY core structure. (b) The efficiency of the laser emission of the different dyes meso-substituted and 3,5-meso-substituted in ethyl acetate solution. The numbers under the peaks identify the corresponding dye. Reproduced from Ref. [93] with permission from 2013 WILEY-VCH Verlag GmbH & Co. KGaA, Weinheim, Germany.

The last important step was made by the work performed by Belmonte-Vázquez et al. [94], who developed a new family of benzofuran-fused BODIPY dyes (**15**, Scheme 2) reaching very high fluorescence and laser efficiencies (almost 100% and >40%, respectively); moreover, they obtained emission with a deep shift to the red edge related to the promotion of excited state aggregates at high concentrations (Figure 12). The efficiency and photostability of these new BODIPYs made it possible to overcome the limits of most of the commercially available dyes (such as cyanines or oxazines) in the same spectral region.

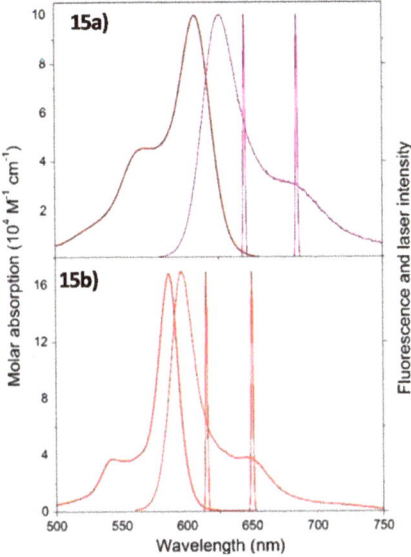

Figure 12. Absorption, normalized fluorescence, and laser (dual emission depending on the dye concentration) spectra of dye **15a** and its analog **15b**, with additional fused phenyl in ethyl acetate. Reproduced from Ref. [94] with permission from 2019 American Chemical Society.

The transition from measuring lasing emission in solutions to have the same in solid-state was not straightforward. In principle, many are the advantages of using polymeric materials as a solid "environment" for lasing dyes: they are optically homogeneous, characterized by good chemical compatibility with organic dyes, and, as it is for solvents, they can be used to control medium polarity and viscoelasticity. On the other hand, these materials are limited by photodegradation processes. To optimize lasing properties and photostability, many new materials based on organic compounds have been synthesized and characterized by the photophysical point of view. In general, BODIPY-based dyes used for solid-state lasing must be dispersed in matrices. This allows for the avoidance of quenching linked to oxidation and in some cases to protect them or slow down the oxidation phenomena.

A key factor in the choice of matrices is also the presence of polar groups (their polarity) which has been shown to have a strong impact [95] on both the stability and efficiency of the films. Studies have been carried out to analyze the dependence of laser properties on several experimental parameters as a dye, pumping repetition rate, the composition of the matrix hosting the dye. Lasing performances have been evaluated by measuring the energy conversion efficiency, given by the ratio between the energy of the laser output and the energy of the pump laser impinging on the sample surface, the laser emission spectrum, and the photostability, which gives us the evolution of the laser output as a function of the number of pump pulses in the same position of the sample at a selected repetition rate.

In the beginning, measurements for laser characterization were performed on a "rod" sample of the dye matrix. As described by Costela et al. [96], the solid laser samples were constituted by cylindrical rods of around 10 mm diameter and 5–10 mm length (shown in Figure 13). In their work, Costela et al. studied the commercial BODIPY derivatives pyrromethene 567 (PM567, **16** in Scheme 2) dissolved in copolymers of MMA with different monomer compositions. After transversely pumping at 534 nm at 5.5 mJ, the lasing efficiency was measured up to 30% with a good photostability.

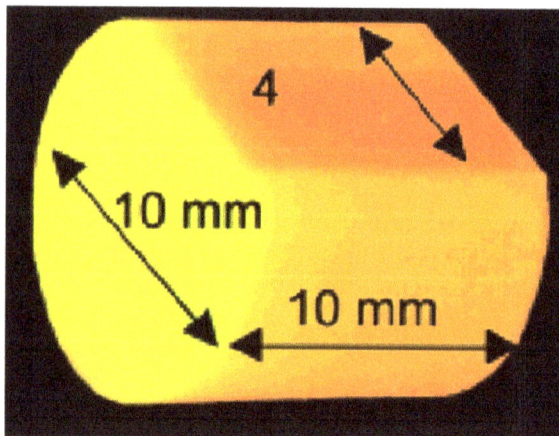

Figure 13. Polymeric solid laser sample. Reproduced from Ref. [96] with permission from 2007 American Chemical Society.

In the following decade, many studies were performed by changing the BODIPY derivative used as a laser dye (summarized in Table 3).

Table 3. Summary of laser parameters and dyes in the reported publications on bulk solid-state samples [1].

Molecule	Efficiency	λ_{LASER} [nm]	λ_{EXC} [nm]	Power [mJ]	Reference
16	up to 30%	559–564	534	5.5	[96]
16	up to 36%	561–564	532	5.5	[95]
17	up to 42%	576–579	532	5.5	[95]
18	25%	563	532	5.5	[97]
19	up to 48%	559–609	532	5	[98]
20	up to 45%	530–615	515 or 532	5	[99]
21	29%	522	355	5.5	[100]
22	up to 30%	565–615	532	5.5	[101]
23	56%	568	532	5	[102]
24	53%	588	532	5	[103]

[1] Efficiency: energy conversion efficiency. λ_{LASER}: peak of the laser emission. λ_{EXC}: pump excitation (transversal). Power: pump power/pulse.

In 2007, the work of García et al. [95] revealed that laser efficiency of commercial BODIPY derivatives PM567 and PM597 (**16** and **17**, Scheme 2) in organic polymeric matrixes was deeply increased by the incorporation of fluorine atoms in the structure of the organic matrixes increasing both photostability and efficiency providing useful information on the potential of the choice of different materials as photostable laser media. They reached up to a 36% and 42% laser efficiency for PM567 and PM597, respectively and they obtained a high laser photostability; thus, paving the way for the development of industrial oriented laser applications of organic solid-state materials for telecommunication, instrumentation, and display technologies.

In the same year, Amat-Guerri's team [97] established an easy synthetic method for the synthesis of asymmetric 3-amino- and 3-acetamido-BODIPY dyes from 2,3,4-trialkyl-substituted pyrroles (**18**, Scheme 2). They obtained materials with good laser emission properties mostly in ethanol solutions but also in solid-state, with an efficiency of up to 48% and 25%, respectively.

A few years later, Costela et al. [98], worked on a simple protocol to synthesized BODIPY with tunable absorption and emission bands in the visible and NIR spectral range. Starting from the commercial PM597 dye, the novel dyes were built by changing the group present at the meso-position: the 8-methyl group was substituted with 8-hydrogen (**19a**, Scheme 2), 8-acetoxymethyl **19b**, Scheme 2), or 8-*p*-acetoxymethylphenyl (**19c**, Scheme 2) groups. The differences induced by those changes were studied; a tunable laser action was observed and the novel analogs showed a high (up to 48%) and photostable laser emission: these results have been so good to consider the new dyes as benchmarks in the extended spectral region (from the green-yellow to red range).

Tunability of laser emission in solid-state was extended even more by Pérez-Ojeda et al., who in 2011 published a work with the synthesis and study of diiodinated BODYPY derivatives [99] (**20**, Scheme 2) both in solution and incorporated into polymeric matrices. They modulated the position of the emission band by choosing the type of substituent attached to the BODIPY core. They found high laser efficiency and photostability of the dyes exhibiting laser emission from 530 nm to 625 nm (see Figure 14) with notable efficiencies, up to 55% in solutions and 45% in PMMA. Moreover, the laser action of these novel BODIPY derivatives were improved in comparison with the lasing properties of commercially available BODIPY dyes (such as PM567, PM597, and PM650), which when incorporated into PMMA, showed lasing efficiencies of 28%, 40%, and 13%, respectively.

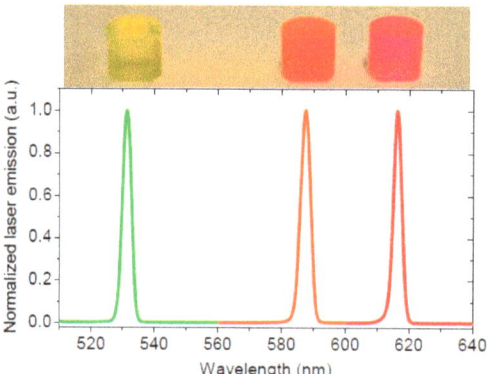

Figure 14. New photosensitized materials based on the new dyes incorporated into PMMA: lasing emission spectra of dyes **20a** (green), **20b** (orange), and **20c** (red). Reproduced from Ref. [99] with permission from 2011 Optical Society of America.

The same group combining the results obtained on fluorinated matrices [95] and on the development of BODIPY dies reported efficient and stable laser emission from BODIPY dye incorporated into polymeric matrixes after high UV pumping excitations [100]. In their study, they synthesized 8-propyl-4,4-difluoro-4-bora-3a,4a-diaza-s-indacene (**21**, Scheme 2), which was then incorporated into PMM), which mimics ethyl acetate solvent, and copolymers with various volumetric proportions of monomers MMA and 2-trifluoromethyl methacrylate (TFMA). Under pumping at 355 nm, lasing efficiencies up to 29% were measured (similar to those obtained in solutions). Moreover, **21** showed photostability higher than what was usually achieved with commercial dyes emitting in the same spectral region, for example, Coumarin 540A. The results pointed out the potential of these class of novel dyes as they are characterized by a relatively easy synthetic buildup with a large variety of possible substituents that can be used to tune the range of solid-state dye lasers to the blue region of the visible range.

Regarding the enhancement of laser properties of BODIPY derivatives with respect to their commercially available analogs, a second work was published in 2012 by Durán-Sampedro et al. [77]. In the article, the novel chlorinated BODIPYs (**22**, Scheme 2) exhibited enhanced laser action in solution and the solid phase. Pumped at 532 nm, the chlorinated BODIPY derivatives showed high efficiency (up to 30%) and photostable laser emission centered between 565 and 615 nm. Fluorescence QY was increased by introducing electron acceptor chlorine substituents in the 3- and 5-positions. As a consequence, lasing properties are enhanced both in a liquid solution and in the solid phase. This is related to the improvement in the photophysical properties of the molecules and in particular for their dipole moments. In fact, the dipole orientations are modified to match pump laser polarization, leading to significant enhancement of system efficiencies.

Up to now, we have reported studies performed in the conventional two-mirror lasers with a thick (bulky) sample. Nevertheless, from 2000, there has been significant work to develop organic thin film (TF) lasers based on dye-doped polymers. It was clear that TF lasers have great potential applications as coherent light sources to be integrated into optoelectronic devices, spectroscopy, and sensors.

In 2013, Durán-Sampedro et al. [78], implemented newly synthesized BODIPY in distributed feedback (DFB) lasers, the most common resonator type for organic thin-film lasers. They developed O-BODIPYs from commercial dyes by replacing fluorines with electron acceptor carboxylate groups (**23**, Scheme 2). They found that in solid-state samples the O-BODIPY derivatives with electron acceptor carboxylate groups with PMMA as matrices, lasing efficiencies were higher than what measured in the corresponding commercial dyes, with the best result up to 56%. A similar outcome was reported also in

DFB lasers: in O-BODIPY dyes, laser thresholds were lower than what was found in the parent dyes, and moreover, lasing intensities were up to one order of magnitude higher.

One year after, the same group synthesized a new library of BODIPY derivatives [101] with different Boron-substituent groups (alkynyl, cyano, vinyl, aryl, and alkyl) (**24**, Scheme 2). They prepared both bulk solid-state samples (Table 3) and TFs (Table 4) of the new dyes incorporated in organic matrices. Even in this case, the new classes of BODIPY showed better performances than the parent dyes. In bulk solid-state, good lasing efficiencies (up to 53%) were gained; while in TFs, the authors found lower laser thresholds and greater lasing intensities, up to three orders of magnitude higher as shown in Figure 15.

Table 4. Summary of laser parameters and dyes in the reported publications on thin films [1].

Molecule	λ_{LASER} [nm]	λ_{EXC} [nm]	Threshold [MW/cm^2]	Reference
23	570	532	1.7×10^{-3}	[78]
24	588	532	20×10^{-3}	[101]
25	680	580	0.34	[102]
26	562–566	532	3	[103]
27	557	532	10–20	[104]
28	569	532	17×10^{-3}	[105]
29	562–566	532	3	[106]

[1] λ_{LASER}: peak of the laser emission; λ_{EXC}: pump excitation.

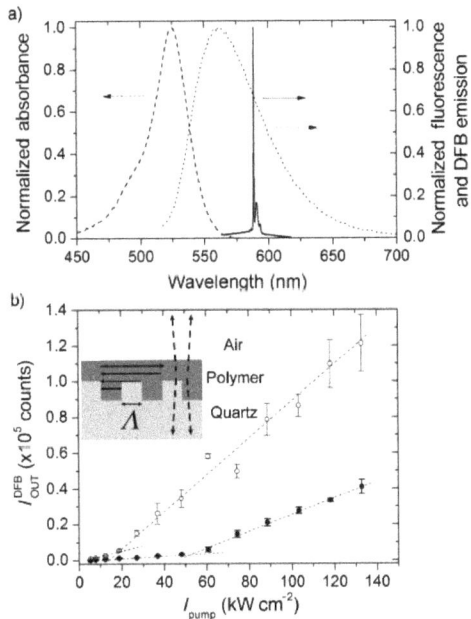

Figure 15. (**a**) Absorption (dashed line), fluorescence (λ_{exc} = 500 nm, dotted line), and laser spectra (solid line) of derivative **24** (23 mm) doped in PMMA. (**b**) DFB output intensity as a function of pump intensity from thin films of commercial PM597 (**17**, Scheme 2) 19 mm (filled circles) and derivative **24** (23 mm; hollow circles) doped in PMMA. Inset: sketch of DFB laser. The arrows represent in-plane feedback owing to second-order diffraction (solid arrows) and outcoupled laser emission owing to first-order diffraction (dashed arrow). Reproduced from Ref. [101] with permission from 2014 WILEY-VCH Verlag GmbH & Co. KGaA, Weinheim, Germany.

In 2014, an interesting work was published on the relationship between small modifications in the material used as a matrix in DFB laser and laser performances of different dyes [102]. The authors slightly modified the commercially available PAZO-Na azobenzene and they used the obtained new materials (carboxylic acid and different salts) as matrices for several common laser dyes: rhodamine, DCM, kiton red, pyrromethene, pyridine, LDS, and exalite. They found that even small modifications in PAZO-Na deeply affect laser stability and efficiency, which are strongly improved due to changes in polarity, refractive index, and morphology of the polymer matrix used. In particular, for PAZO-H, laser generation with the commercial BODIPY dye PM650 (**25** Scheme 2), has been detected, which was attributed to the lower polarity of the matrix while with other PAZO-Na derivatives no lasing was measured with the same dyes. This indicates that in BODIPY dyes, lasing can be easily optimized by changing matrix parameters such as polarity.

In 2014, also Kuznetsova et al. [103] exploited how modifications of the matrix impacted dye lasing efficiency and other parameters. In their research, they incorporated different BODIPY complexes (**26**, Scheme 2) into bulk PMMA matrix and matrix modified with the addition of polyhedral silsesquioxane particles. The interesting result was that modification of the matrices affected only slightly the spectral features but instead, it significantly improved the lasing efficiency and threshold. As an example, it has been reported that in a modified matrix a simple BODIPY derivative (3,3′,5,5′-tetramethyl-4,4′-diethyl-2,2-dipyrromethene difluoroborate, BODIPY1) shows a maximum efficiency up to 90% while 70% in the unmodified matrix.

Further encouraging results obtained by implementing the laser media have been reported in 2016 by Kuznetsova et al. [104] by incorporating alkyl BODIPY derivatives (**27**, Scheme 2) into three component silicate demonstrated the possibility of developing solid-state laser media working in the visible range between 550 and 564 nm and with a laser efficiency of 12%.

Ray et al. [77] finely tuned photophysical properties of stabilized N-BODIPYs (**28**, Scheme 2) thanks to the proper substitution of the fluorine linked to boron atom with electron-poor diamine moieties. Specifically, they demonstrated stabilized N-BODIPYs based on electron-poor amine moieties based on sulfonylated amines with both low flexibility and low steric hindrance, and electron-rich BODIPY cores were able to give rise to bright fluorophores even in the solid crystalline state, with notable lasing capacities in the liquid phase (surpassing 60% laser efficiencies) and when doped into polymers, improving the laser performance of their commercial counterpart PM567.

A series of BODIPYs with different substituted compounds was studied by Kuznetsova et al. In their work [78], they investigated how ligand structures and optical properties were related. They founded that alkyl-, cycloalkyl-, phenyl-, benzyl-, and meso-propargylamino substituted difluoroborates BODIPYs were photostable and laser active in liquid and solid-state in the spectral range between 475 and 687 nm. In particular, in solid-state, two derivatives, **29a** and **29b** (Scheme 2) with ethanyl and methyl groups (with and without phenyl group, respectively) showed very good performances: in PMMA they gained a laser efficiency up to 38% in PMMA and up to 58% in 8MMA-PMMA with a quite low threshold.

3.3. BODIPY as Active Layer in Optical Microcavity

In 2017, Cookson et al. [105] used BODIPY-Br (**30**, Scheme 2) to demonstrate polariton condensation in optical cavities. As shown in Figure 16, the dye was dispersed in a transparent polymeric matrix and then incorporated into a microcavity.

BODIPY-Br is characterized by relatively high photoluminescence (PL) quantum yield and optical amplification at 589 nm after pumping at 500 nm (see Figure 17). The work showed strong evidence of nonlinear PL with increasing excitation density, associated with a high linewidth narrowing and a continuous blueshift. This shift was assigned to polariton interactions with other polaritons and the exciton reservoir.

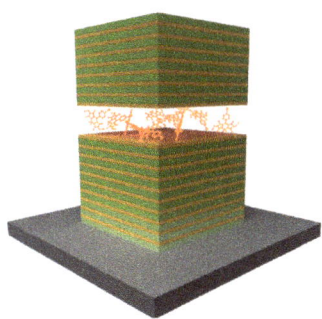

Figure 16. Schematic of the dye-filled microcavity. Reproduced from Ref. [105] with permission from 2017 WILEY-VCH Verlag GmbH & Co. KGaA, Weinheim, Germany.

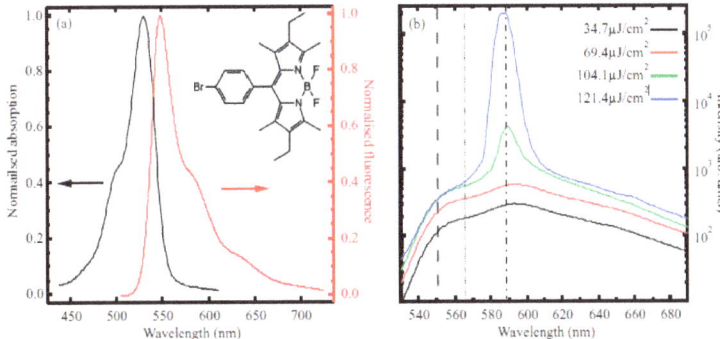

Figure 17. (**a**) The normalized absorption (black) and fluorescence spectrum (red) of BODIPY-Br dispersed in a transparent polystyrene matrix. The chemical structure is shown in the insert. (**b**) Amplified spontaneous emission from a 186 nm thick film of the BODIPY-Br (**39**, Scheme 2) dispersed in a polystyrene matrix and deposited on a quartz substrate. A threshold is observed at a pump fluence of 104 µJ cm^{-2} with a peak forming at 589 nm (dashed–dotted line). The dotted line indicates the polariton emission from the microcavity, while the large dashed line indicates the fluorescence emission peak shown in (**a**). Reproduced from Ref. [105] with permission from 2017 WILEY-VCH Verlag GmbH & Co. KGaA, Weinheim, Germany.

This work demonstrated the capability of achieving condensation at room temperature, in a dilute molecular system. This result opens the possibility to select polariton condensation wavelengths for the development of future optoelectronic devices, as hybrid organic-inorganic polariton laser diodes. There are many different molecular dyes that can be embedded in a microcavity to have polariton condensation at different wavelengths covering the visible and the NIR range.

In 2019, Sannikov et al. [106] developed a material system with polariton lasing at room temperature over a broad spectral range. The system is based on molecule BODIPY-G1 (Scheme 2), which is a derivative presenting high extinction coefficients and high PL quantum yield. The dye is dispersed in a polystyrene matrix and used as the active layer in a strongly coupled microcavity. The authors found that by engineering a thickness gradient across the microcavities (see Figure 18a), it was possible to gain a broad range of exciton–photon detuning conditions. As shown in Figure 18b, in this configuration, BODIPY-G1 microcavity can undergo polariton lasing over a broad range (around 33 nm) of wavelengths in the green-yellow spectral range, with a highly monochromatic emission line of 0.1 nm.

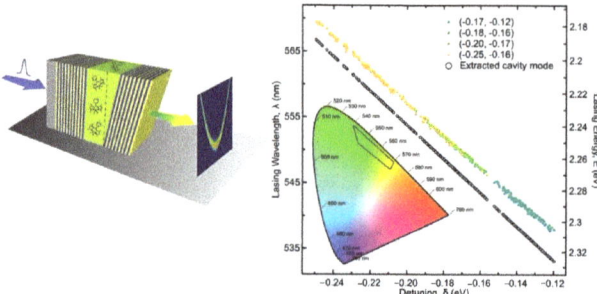

Figure 18. (**left**) Schematic representation of the wedged dye-filled microcavity and the excitation and detection configuration. (**right**) The colored solid-square data points indicate the wavelength (photon energy, right axis) where polariton lasing was observed versus the corresponding exciton–photon detuning. The black open circles indicate the corresponding bare cavity mode for each realization. Polariton lasing spans the green-yellow part of the visible spectrum as indicated with a black solid line on the CIE 1931 chromaticity diagram, bottom left inset. Reproduced from Ref. [106] with permission from 2019 WILEY-VCH Verlag GmbH & Co. KGaA, Weinheim, Germany.

In conclusion, we have shown that a lot of work has been done on the achievement of lasing from BODIPY solutions or from "bulky" solid-state rods, but just in 2013 Durán-Sampedro [78] reported the use of a BODIPY thin film as an active layer in a photonic device as a Distributed Feedback Laser. Few other groups, since then, have implemented the use of thin films in photonic devices like DFB or microcavities lasers demonstrating the good properties of this material and its versatility.

4. Conclusions and Perspectives

The development of new organic semiconductor materials is certainly one of the pillars on which the possibility of bringing advanced technologies to the market and consequently making them accessible to all rests.

Since its discovery BODIPY molecule has shown great versatility thanks to the possibility of easy functionalization in different positions, fine modulation of optoelectronic properties, and good stability moreover it was possible its application in biochemical labeling chemosensors and fluorescent switches. Only recently BODIPY has emerged as effective building blocks for OPV and photonic applications and both these areas are indisputably at the basis of sustainable development.

In the last 5 years, essential improvements have been obtained in both applications thanks to the study of the effect of substituents and matrices for its dispersion.

This review explores the recent progress (2009–2020) in the field of solid-state BODIPY-based semiconducting molecules for photonics and OPVs applications. The review is divided into sections and provides a concise and precise description of the experimental results, their interpretation as well as the conclusions that can be drawn. After the first introductory section, in Section 2, we report the best results obtained in solar cells by focusing on devices with an efficiency greater than 5% and taking into consideration their use as an acceptor and as a donor molecule. In Section 3, we highlight the best results obtained in photonic applications such as lasing in DFB structures or microcavities. The chemical structure and the main optoelectronic properties, as well as their device performances and, where possible, their morphological properties are deeply investigated and summarized in Tables 1–3 evidencing that the most promising results were mostly achieved in the past 2–3 years.

Despite all advantages of using BODIPY-based chromophores, there are several challenges, which require more attention. For example, the effect of substituents on the Boron positions on the BODIPY properties has been only marginally investigated. These substitutions can allow to improve the stability of the material and to modulate the absorption and emission spectra towards the NIR spectral window. The development of hybrid systems with 2D materials could give a valuable contribution to the development of new multifunctional materials [107,108].

Finally, it is necessary to combine the sustainable aspect of the material, whose biocompatibility has already been demonstrated, with sustainable synthetic strategy and solubility in green solvents.

We hope that this review will provide a useful tool for the design and understanding of the structure–property relationship of BODIPY chromophores to facilitate their use in the optoelectronic research area.

Funding: This work was carried out with the financial support of two Regione Lombardia Projects "Piattaforma tecnologica per lo sviluppo di sonde innovative in ambito biomedicale" (ID 244356) and "I-ZEB Verso edifici Intelligenti a Energia Zero per la crescita della città intelligente".

Conflicts of Interest: The authors declare no conflict of interest.

Abbreviations

BDT	Benzo[1,2-b:4,5-b']dithiophene
BHJ	Bulk-HeteroJunction solar cells
BPAPF	9,9-bis[4-(N,N-bis-biphenyl-4-yl-amino)phenyl]-9H-fluorene
D–A	Donor–Acceptor
DCM	Dichloromethane
DFB	Distributed Feedback laser
DPP	Diketopyrrolopyrrole
ETL	Electron Transporting Layer (ETL)
FF	Fill Factor
HOMO	Highest Occupied Molecular Orbital
HTL	Hole Transporting Layer
ITO	Indium Tin Oxide
J_{sc}	Short-circuit current Density
LUMO	Lowest Unoccupied Molecular Orbital
MH250	N,N-bis(fluoren-2-yl)-naphthalenetetracarboxylic diimide(bis-Hfl-NTCDI)
MMA	Methyl Methacrylate
MoO_3	Molibdenum Oxide
NDP9	Organic p-type dopant of *Novaled* GmbH
NIR	Near InfraRed
OPV	Organic PhotoVoltaic
OSC	Organic Solar Cell
P3HT	Poly(3-hexylthiophene-2,5-diyl)
PAZO-Na	Poly[1-[4-(3-carboxy-4-hydroxyphenylazo)benzenesulfonamido]-1,2-Ethanediyl, sodium salt]
PCE	Power Conversion Efficiency
p-DTS-(FBTTh$_2$)$_2$	7,7'-[4,4-Bis(2-ethylhexyl)-4H-silolo[3,2-b:4,5-b']dithiophene-2,6-diyl]bis[6-Fluoro-4-(5'-hexyl-[2,2'-bithiophen]-5-yl)benzo[c][1,2,5]thiadiazole]
PEDOT:PSS	Poly(3,4-ethylenedioxythiophene) polystyrene sulfonate
PEIE	Polyethylenimine ethoxylated
PFN	Poly[9,9-bis(3'-(N,N-dimethylamino)propyl)-2,7-fluorene
PL	Photoluminescence
PMMA	Poly(methyl methacrylate)
PTB7-Th	Poly([2,6'-4,8-di(5-ethylhexylthienyl)benzo[1,2-b;3,3-b]dithiophene][3-fluoro-2[(2-ethylhexyl)carbonyl]thieno[3,4-b]thiophenediyl])
SCLC	Space-Charge Limited-Current
SVA	Solvent Vapor Annealing
TA	Thermal Annealing
TAT	Triazatruxene
TF	Thin film
TFMA	2-trifluoromethyl methacrylate
TSVA	Thermal and Solvent Annealing
V_{oc}	Open Circuit Voltage
W$_2$(hpp)$_4$	Tetrakis(1,3,4,6,7,8-hexahydro-2H-pyrimido[1,2-a]pyrimidinato)ditungsten (II)
ZnO	Zinc Oxide

References

1. Semiconducting Polymers: Chemistry, Physics and Engineering, 2nd Edition, Two-Volume Set | Wiley. Available online: https://www.wiley.com/en-us/Semiconducting+Polymers%3A+Chemistry%2C+Physics+and+Engineering%2C+2nd+Edition%2C+Two+Volume+Set-p-9783527312719 (accessed on 24 November 2020).
2. Ostroverkhova, O. Organic Optoelectronic Materials: Mechanisms and Applications. *Chem. Rev.* **2016**, *116*, 13279–13412. [CrossRef] [PubMed]
3. Forrest, S.R.; Thompson, M.E. Introduction: Organic Electronics and Optoelectronics. *Chem. Rev.* **2007**, *107*, 923–925. [CrossRef]
4. Zvezdin, A.; Mauro, E.D.; Rho, D.; Santato, C.; Khalil, M. En Route toward Sustainable Organic Electronics. *MRS Energy Sustain.* **2020**, *7*. [CrossRef]
5. Giovanella, U.; Betti, P.; Bolognesi, A.; Destri, S.; Melucci, M.; Pasini, M.; Porzio, W.; Botta, C. Core-Type Polyfluorene-Based Copolymers for Low-Cost Light-Emitting Technologies. *Org. Electron.* **2010**, *11*, 2012–2018. [CrossRef]
6. Vohra, V.; Galeotti, F.; Giovanella, U.; Mróz, W.; Pasini, M.; Botta, C. Nanostructured Light-Emitting Polymer Thin Films and Devices Fabricated by the Environment-Friendly Push-Coating Technique. *ACS Appl. Mater. Interfaces* **2018**, *10*, 11794–11800. [CrossRef] [PubMed]
7. Irimia-Vladu, M. "Green" Electronics: Biodegradable and Biocompatible Materials and Devices for Sustainable Future. *Chem. Soc. Rev.* **2014**, *43*, 588–610. [CrossRef] [PubMed]
8. Treibs, A.; Kreuzer, F.-H. Difluorboryl-Komplexe von Di- Und Tripyrrylmethenen. *Justus Liebigs Annalen Chem.* **1968**, *718*, 208–223. [CrossRef]
9. Banuelos, J. BODIPY dye, the most versatile fluorophore ever? *Chem. Rec.* **2016**, *16*, 335–348. [CrossRef]
10. Loudet, A.; Burgess, K. BODIPY Dyes and Their Derivatives: Syntheses and Spectroscopic Properties. *Chem. Rev.* **2007**, *107*, 4891–4932. [CrossRef]
11. Squeo, B.M.; Gregoriou, V.G.; Avgeropoulos, A.; Baysec, S.; Allard, S.; Scherf, U.; Chochos, C.L. BODIPY-Based Polymeric Dyes as Emerging Horizon Materials for Biological Sensing and Organic Electronic Applications. *Prog. Polym. Sci.* **2017**, *71*, 26–52. [CrossRef]
12. Squeo, B.M.; Pasini, M. BODIPY Platform: A Tunable Tool for Green to NIR OLEDs. *Supramol. Chem.* **2020**, *32*, 56–70. [CrossRef]
13. Menges, N. Computational Study on Aromaticity and Resonance Structures of Substituted BODIPY Derivatives. *Comput. Theor. Chem.* **2015**, *1068*, 117–122. [CrossRef]
14. Ulrich, G.; Ziessel, R.; Harriman, A. The Chemistry of Fluorescent Bodipy Dyes: Versatility Unsurpassed. *Angew. Chem. Int. Ed.* **2008**, *47*, 1184–1201. [CrossRef] [PubMed]
15. Poddar, M.; Misra, R. Recent Advances of BODIPY Based Derivatives for Optoelectronic Applications. *Coord. Chem. Rev.* **2020**, *421*, 213462. [CrossRef]
16. Singh, A.; Yip, W.-T.; Halterman, R.L. Fluorescence-On Response via CB7 Binding to Viologen–Dye Pseudorotaxanes. *Org. Lett.* **2012**, *14*, 4046–4049. [CrossRef]
17. Frath, D.; Yarnell, J.E.; Ulrich, G.; Castellano, F.N.; Ziessel, R. Ultrafast Photoinduced Electron Transfer in Viologen-Linked BODIPY Dyes. *ChemPhysChem* **2013**, *14*, 3348–3354. [CrossRef]
18. Tao, J.; Sun, D.; Sun, L.; Li, Z.; Fu, B.; Liu, J.; Zhang, L.; Wang, S.; Fang, Y.; Xu, H. Tuning the Photo-Physical Properties of BODIPY Dyes: Effects of 1, 3, 5, 7- Substitution on Their Optical and Electrochemical Behaviours. *Dyes Pigment.* **2019**, *168*, 166–174. [CrossRef]
19. Llano, R.S.; Zaballa, E.A.; Bañuelos, J.; Durán, C.F.A.G.; Vázquez, J.L.B.; Cabrera, E.P.; Arbeloa, I.L. Tailoring the Photophysical Signatures of BODIPY Dyes: Toward Fluorescence Standards across the Visible Spectral Region. *Photochem. Photophys. Fundam. Appl.* **2018**. [CrossRef]
20. Littler, B.J.; Miller, M.A.; Hung, C.-H.; Wagner, R.W.; O'Shea, D.F.; Boyle, P.D.; Lindsey, J.S. Refined Synthesis of 5-Substituted Dipyrromethanes. *J. Org. Chem.* **1999**, *64*, 1391–1396. [CrossRef]
21. Shimizu, S.; Iino, T.; Araki, Y.; Kobayashi, N. Pyrrolopyrrole Aza-BODIPY Analogues: A Facile Synthesis and Intense Fluorescence. *Chem. Commun.* **2013**, *49*, 1621–1623. [CrossRef]
22. Adarsh, N.; Shanmugasundaram, M.; Avirah, R.R.; Ramaiah, D. Aza-BODIPY Derivatives: Enhanced Quantum Yields of Triplet Excited States and the Generation of Singlet Oxygen and Their Role as Facile Sustainable Photooxygenation Catalysts. *Chem. A Eur. J.* **2012**, *18*, 12655–12662. [CrossRef] [PubMed]
23. Ho, D.; Ozdemir, R.; Kim, H.; Earmme, T.; Usta, H.; Kim, C. BODIPY-Based Semiconducting Materials for Organic Bulk Heterojunction Photovoltaics and Thin-Film Transistors. *ChemPlusChem* **2019**, *84*, 18–37. [CrossRef] [PubMed]
24. Wanwong, S.; Sangkhun, W.; Kumnorkaew, P.; Wootthikanokkhan, J. Improved Performance of Ternary Solar Cells by Using BODIPY Triads. *Materials* **2020**, *13*, 2723. [CrossRef]
25. Boens, N.; Verbelen, B.; Dehaen, W. Postfunctionalization of the BODIPY Core: Synthesis and Spectroscopy. *Eur. J. Org. Chem.* **2015**, *2015*, 6577–6595. [CrossRef]
26. Rousseau, T.; Cravino, A.; Bura, T.; Ulrich, G.; Ziessel, R.; Roncali, J. BODIPY Derivatives as Donor Materials for Bulk Heterojunction Solar Cells. *Chem. Commun.* **2009**, 1673. [CrossRef] [PubMed]
27. Rousseau, T.; Cravino, A.; Bura, T.; Ulrich, G.; Ziessel, R.; Roncali, J. Multi-Donor Molecular Bulk Heterojunction Solar Cells: Improving Conversion Efficiency by Synergistic Dye Combinations. *J. Mater. Chem.* **2009**, *19*, 2298–2300. [CrossRef]

28. Liu, B.; Ma, Z.; Xu, Y.; Guo, Y.; Yang, F.; Xia, D.; Li, C.; Tang, Z.; Li, W. Non-Fullerene Organic Solar Cells Based on a BODIPY-Polymer as Electron Donor with High Photocurrent. *J. Mater. Chem. C* **2020**, *8*, 2232–2237. [CrossRef]
29. Yang, J.; Devillers, C.H.; Fleurat-Lessard, P.; Jiang, H.; Wang, S.; Gros, C.P.; Gupta, G.; Sharma, G.D.; Xu, H. Carbazole-Based Green and Blue-BODIPY Dyads and Triads as Donors for Bulk Heterojunction Organic Solar Cells. *Dalton Trans.* **2020**, *49*, 5606–5617. [CrossRef]
30. Liu, Q.; Jiang, Y.; Jin, K.; Qin, J.; Xu, J.; Li, W.; Xiong, J.; Liu, J.; Xiao, Z.; Sun, K.; et al. 18% Efficiency Organic Solar Cells. *Sci. Bull.* **2020**, *65*, 272–275. [CrossRef]
31. Clarke, T.M.; Durrant, J.R. Charge Photogeneration in Organic Solar Cells. *Chem. Rev.* **2010**, *110*, 6736–6767. [CrossRef]
32. Achieving a High Fill Factor for Organic Solar Cells—Journal of Materials Chemistry A (RSC Publishing). Available online: https://pubs.rsc.org/en/content/articlelanding/2016/ta/c6ta00126b#!divAbstract (accessed on 24 November 2020).
33. Squeo, B.M.; Gregoriou, V.G.; Han, Y.; Palma-Cando, A.; Allard, S.; Serpetzoglou, E.; Konidakis, I.; Stratakis, E.; Avgeropoulos, A.; Heeney, M.; et al. α,β-Unsubstituted Meso -Positioning Thienyl BODIPY: A Promising Electron Deficient Building Block for the Development of near Infrared (NIR) p-Type Donor–Acceptor (D–A) Conjugated Polymers. *J. Mater. Chem. C* **2018**, *6*, 4030–4040. [CrossRef]
34. Zampetti, A.; Minotto, A.; Squeo, B.M.; Gregoriou, V.G.; Allard, S.; Scherf, U.; Chochos, C.L.; Cacialli, F. Highly Efficient Solid-State Near-Infrared Organic Light-Emitting Diodes Incorporating A-D-A Dyes Based on α,β -Unsubstituted "BODIPY" Moieties. *Sci. Rep.* **2017**, *7*, 1611. [CrossRef] [PubMed]
35. Chochos, C.L.; Drakopoulou, S.; Katsouras, A.; Squeo, B.M.; Sprau, C.; Colsmann, A.; Gregoriou, V.G.; Cando, A.-P.; Allard, S.; Scherf, U.; et al. Beyond Donor–Acceptor (D–A) Approach: Structure–Optoelectronic Properties—Organic Photovoltaic Performance Correlation in New D–A1–D–A2 Low-Bandgap Conjugated Polymers. *Macromol. Rapid Commun.* **2017**, *38*, 1600720. [CrossRef]
36. Mahesh, K.; Karpagam, S.; Pandian, K. How to Design Donor–Acceptor Based Heterocyclic Conjugated Polymers for Applications from Organic Electronics to Sensors. *Top. Curr. Chem.* **2019**, *377*, 12. [CrossRef]
37. Salzner, U. Effect of Donor–Acceptor Substitution on Optoelectronic Properties of Conducting Organic Polymers. *J. Chem. Theory Comput.* **2014**, *10*, 4921–4937. [CrossRef]
38. Porzio, W.; Destri, S.; Pasini, M.; Giovanella, U.; Ragazzi, M.; Scavia, G.; Kotowski, D.; Zotti, G.; Vercelli, B. Synthesis and Characterisation of Fluorenone–Thiophene-Based Donor–Acceptor Oligomers: Role of Moiety Sequence upon Packing and Electronic Properties. *New J. Chem.* **2010**, *34*, 1961–1973. [CrossRef]
39. Mishra, R.; Basumatary, B.; Singhal, R.; Sharma, G.D.; Sankar, J. Corrole-BODIPY Dyad as Small-Molecule Donor for Bulk Heterojunction Solar Cells. *ACS Appl. Mater. Interfaces* **2018**, *10*, 31462–31471. [CrossRef]
40. Squeo, B.M.; Carulli, F.; Lassi, E.; Galeotti, F.; Giovanella, U.; Luzzati, S.; Pasini, M. Benzothiadiazole-Based Conjugated Polyelectrolytes for Interfacial Engineering in Optoelectronic Devices. *Pure Appl. Chem.* **2019**, *91*, 477–488. [CrossRef]
41. Carulli, F.; Scavia, G.; Lassi, E.; Pasini, M.; Galeotti, F.; Brovelli, S.; Giovanella, U.; Luzzati, S. A Bifunctional Conjugated Polyelectrolyte for the Interfacial Engineering of Polymer Solar Cells. *J. Colloid Interface Sci.* **2019**, *538*, 611–619. [CrossRef]
42. Li, T.; Meyer, T.; Ma, Z.; Benduhn, J.; Körner, C.; Zeika, O.; Vandewal, K.; Leo, K. Small Molecule Near-Infrared Boron Dipyrromethene Donors for Organic Tandem Solar Cells. *J. Am. Chem. Soc.* **2017**, *139*, 13636–13647. [CrossRef]
43. Lin, Y.; Li, Y.; Zhan, X. Small Molecule Semiconductors for High-Efficiency Organic Photovoltaics. *Chem. Soc. Rev.* **2012**, *41*, 4245–4272. [CrossRef]
44. Poe, A.M.; Pelle, A.M.D.; Subrahmanyam, A.V.; White, W.; Wantz, G.; Thayumanavan, S. Small Molecule BODIPY Dyes as Non-Fullerene Acceptors in Bulk Heterojunction Organic Photovoltaics. *Chem. Commun.* **2014**, *50*, 2913–2915. [CrossRef]
45. Liu, W.; Yao, J.; Zhan, C. A Novel BODIPY-Based Low-Band-Gap Small-Molecule Acceptor for Efficient Non-fullerene Polymer Solar Cells. *Chin. J. Chem.* **2017**, *35*, 1813–1823. [CrossRef]
46. Rousseau, T.; Cravino, A.; Ripaud, E.; Leriche, P.; Rihn, S.; Nicola, A.D.; Ziessel, R.; Roncali, J. A Tailored Hybrid BODIPY–Oligothiophene Donor for Molecular Bulk Heterojunction Solar Cells with Improved Performances. *Chem. Commun.* **2010**, *46*, 5082–5084. [CrossRef]
47. Kubo, Y.; Watanabe, K.; Nishiyabu, R.; Hata, R.; Murakami, A.; Shoda, T.; Ota, H. Near-Infrared Absorbing Boron-Dibenzopyrromethenes That Serve As Light-Harvesting Sensitizers for Polymeric Solar Cells. *Org. Lett.* **2011**, *13*, 4574–4577. [CrossRef] [PubMed]
48. Lin, H.-Y.; Huang, W.-C.; Chen, Y.-C.; Chou, H.-H.; Hsu, C.-Y.; Lin, J.T.; Lin, H.-W. BODIPY Dyes with β-Conjugation and Their Applications for High-Efficiency Inverted Small Molecule Solar Cells. *Chem. Commun.* **2012**, *48*, 8913. [CrossRef] [PubMed]
49. Hayashi, Y.; Obata, N.; Tamaru, M.; Yamaguchi, S.; Matsuo, Y.; Saeki, A.; Seki, S.; Kureishi, Y.; Saito, S.; Yamaguchi, S.; et al. Facile Synthesis of Biphenyl-Fused BODIPY and Its Property. *Org. Lett.* **2012**, *14*, 866–869. [CrossRef] [PubMed]
50. Bura, T.; Leclerc, N.; Fall, S.; Lévêque, P.; Heiser, T.; Retailleau, P.; Rihn, S.; Mirloup, A.; Ziessel, R. High-Performance Solution-Processed Solar Cells and Ambipolar Behavior in Organic Field-Effect Transistors with Thienyl-BODIPY Scaffoldings. *J. Am. Chem. Soc.* **2012**, *134*, 17404–17407. [CrossRef] [PubMed]
51. Kolemen, S.; Cakmak, Y.; Ozdemir, T.; Erten-Ela, S.; Buyuktemiz, M.; Dede, Y.; Akkaya, E.U. Design and Characterization of Bodipy Derivatives for Bulk Heterojunction Solar Cells. *Tetrahedron* **2014**, *70*, 6229–6234. [CrossRef]
52. Liu, W.; Tang, A.; Chen, J.; Wu, Y.; Zhan, C.; Yao, J. Photocurrent Enhancement of BODIPY-Based Solution-Processed Small-Molecule Solar Cells by Dimerization via the Meso Position. *ACS Appl. Mater. Interfaces* **2014**, *6*, 22496–22505. [CrossRef]

53. Sutter, A.; Retailleau, P.; Huang, W.-C.; Lin, H.-W.; Ziessel, R. Photovoltaic Performance of Novel Push–Pull–Push Thienyl–Bodipy Dyes in Solution-Processed BHJ-Solar Cells. *New J. Chem.* **2014**, *38*, 1701–1710. [CrossRef]
54. Cortizo-Lacalle, D.; Howells, C.T.; Pandey, U.K.; Cameron, J.; Findlay, N.J.; Inigo, A.R.; Tuttle, T.; Skabara, P.J.; Samuel, I.D.W. Solution Processable Diketopyrrolopyrrole (DPP) Cored Small Molecules with BODIPY End Groups as Novel Donors for Organic Solar Cells. *Beilstein J. Org. Chem.* **2014**, *10*, 2683–2695. [CrossRef] [PubMed]
55. Chen, J.J.; Conron, S.M.; Erwin, P.; Dimitriou, M.; McAlahney, K.; Thompson, M.E. High-Efficiency BODIPY-Based Organic Photovoltaics. *ACS Appl. Mater. Interfaces* **2015**, *7*, 662–669. [CrossRef] [PubMed]
56. Sharma, G.D.; Siddiqui, S.A.; Nikiforou, A.; Zervaki, G.E.; Georgakaki, I.; Ladomenou, K.; Coutsolelos, A.G. A Mono(Carboxy)Porphyrin Triazine-(Bodipy) 2 Triad as a Donor for Bulk Heterojunction Organic Solar Cells. *J. Mater. Chem. C* **2015**, *3*, 6209–6217. [CrossRef]
57. Xiao, L.; Wang, H.; Gao, K.; Li, L.; Liu, C.; Peng, X.; Wong, W.-Y.; Wong, W.-K.; Zhu, X. A-D-A Type Small Molecules Based on Boron Dipyrromethene for Solution-Processed Organic Solar Cells. *Chem. Asian J.* **2015**, *10*, 1513–1518. [CrossRef] [PubMed]
58. Jadhav, T.; Misra, R.; Biswas, S.; Sharma, G.D. Bulk Heterojunction Organic Solar Cells Based on Carbazole–BODIPY Conjugate Small Molecules as Donors with High Open Circuit Voltage. *Phys. Chem. Chem. Phys.* **2015**, *17*, 26580–26588. [CrossRef] [PubMed]
59. Zou, L.; Guan, S.; Li, L.; Zhao, L. Dipyrrin-Based Complexes for Solution-Processed Organic Solar Cells. *Chem. Res. Chin. Univ.* **2015**, *31*, 801–808. [CrossRef]
60. Liu, W.; Yao, J.; Zhan, C. Performance Enhancement of BODIPY Dimer-Based Small-Molecule Solar Cells Using a Visible-Photon-Capturing Diketopyrrolopyrrole π-Bridge. *RSC Adv.* **2015**, *5*, 74238–74241. [CrossRef]
61. Zhang, X.; Zhang, Y.; Chen, L.; Xiao, Y. Star-Shaped Carbazole-Based BODIPY Derivatives with Improved Hole Transportation and near-Infrared Absorption for Small-Molecule Organic Solar Cells with High Open-Circuit Voltages. *RSC Adv.* **2015**, *5*, 32283–32289. [CrossRef]
62. Liao, J.; Xu, Y.; Zhao, H.; Wang, Y.; Zhang, W.; Peng, F.; Xie, S.; Yang, X. Synthesis, Optical, Electrochemical Properties and Photovoltaic Performance of Novel Panchromatic Meso-Thiophene BODIPY Dyes with a Variety of Electron-Donating Groups at the 3,5-Positions. *RSC Adv.* **2015**, *5*, 86453–86462. [CrossRef]
63. Liao, J.; Zhao, H.; Xu, Y.; Cai, Z.; Peng, Z.; Zhang, W.; Zhou, W.; Li, B.; Zong, Q.; Yang, X. Novel D–A–D Type Dyes Based on BODIPY Platform for Solution Processed Organic Solar Cells. *Dyes Pigment.* **2016**, *128*, 131–140. [CrossRef]
64. Rao, R.S.; Bagui, A.; Rao, G.H.; Gupta, V.; Singh, S.P. Achieving the Highest Efficiency Using a BODIPY Core Decorated with Dithiafulvalene Wings for Small Molecule Based Solution-Processed Organic Solar Cells. *Chem. Commun.* **2017**, *53*, 6953–6956. [CrossRef]
65. Liao, J.; Xu, Y.; Zhao, H.; Zong, Q.; Fang, Y. Novel A-D-A Type Small Molecules with β-Alkynylated BODIPY Flanks for Bulk Heterojunction Solar Cells. *Org. Electron.* **2017**, *49*, 321–333. [CrossRef]
66. Bulut, I.; Huaulmé, Q.; Mirloup, A.; Chávez, P.; Fall, S.; Hébraud, A.; Méry, S.; Heinrich, B.; Heiser, T.; Lévêque, P.; et al. Rational Engineering of BODIPY-Bridged Trisindole Derivatives for Solar Cell Applications. *ChemSusChem* **2017**, *10*, 1878–1882. [CrossRef]
67. Liao, J.; Zhao, H.; Cai, Z.; Xu, Y.; Qin, F.G.F.; Zong, Q.; Peng, F.; Fang, Y. BODIPY-Based Panchromatic Chromophore for Efficient Organic Solar Cell. *Org. Electron.* **2018**, *61*, 215–222. [CrossRef]
68. Bucher, L.; Desbois, N.; Koukaras, E.N.; Devillers, C.H.; Biswas, S.; Sharma, G.D.; Gros, C.P. BODIPY–Diketopyrrolopyrrole–Porphyrin Conjugate Small Molecules for Use in Bulk Heterojunction Solar Cells. *J. Mater. Chem. A* **2018**, *6*, 8449–8461. [CrossRef]
69. Marques dos Santos, J.; Jagadamma, L.K.; Latif, N.M.; Ruseckas, A.; Samuel, I.D.W.; Cooke, G. BODIPY Derivatives with near Infra-Red Absorption as Small Molecule Donors for Bulk Heterojunction Solar Cells. *RSC Adv.* **2019**, *9*, 15410–15423. [CrossRef]
70. Li, T.; Benduhn, J.; Qiao, Z.; Liu, Y.; Li, Y.; Shivhare, R.; Jaiser, F.; Wang, P.; Ma, J.; Zeika, O.; et al. Effect of H- and J-Aggregation on the Photophysical and Voltage Loss of Boron Dipyrromethene Small Molecules in Vacuum-Deposited Organic Solar Cells. *J. Phys. Chem. Lett.* **2019**, *10*, 2684–2691. [CrossRef]
71. Aguiar, A.; Farinhas, J.; da Silva, W.; Ghica, M.E.; Brett, C.M.A.; Morgado, J.; Sobral, A.J.F.N. Synthesis, Characterization and Application of Meso-Substituted Fluorinated Boron Dipyrromethenes (BODIPYs) with Different Styryl Groups in Organic Photovoltaic Cells. *Dyes Pigment.* **2019**, *168*, 103–110. [CrossRef]
72. Thumuganti, G.; Gupta, V.; Singh, S.P. New Dithienosilole- and Dithienogermole-Based BODIPY for Solar Cell Applications. *New J. Chem.* **2019**, *43*, 8735–8740. [CrossRef]
73. Aguiar, A.; Farinhas, J.; da Silva, W.; Susano, M.; Silva, M.R.; Alcácer, L.; Kumar, S.; Brett, C.M.A.; Morgado, J.; Sobral, A.J.F.N. Simple BODIPY Dyes as Suitable Electron-Donors for Organic Bulk Heterojunction Photovoltaic Cells. *Dyes Pigment.* **2020**, *172*, 107842. [CrossRef]
74. Ivaniuk, K.; Pidluzhna, A.; Stakhira, P.; Baryshnikov, G.V.; Kovtun, Y.P.; Hotra, Z.; Minaev, B.F.; Ågren, H. BODIPY-Core 1,7-Diphenyl-Substituted Derivatives for Photovoltaics and OLED Applications. *Dyes Pigment.* **2020**, *175*, 108123. [CrossRef]
75. Würthner, F.; Kaiser, T.E.; Saha-Möller, C.R. J-Aggregates: From Serendipitous Discovery to Supramolecular Engineering of Functional Dye Materials. *Angew. Chem. Int. Ed.* **2011**, *50*, 3376–3410. [CrossRef] [PubMed]
76. Ding, T.; Alemán, E.A.; Modarelli, D.A.; Ziegler, C.J. Photophysical Properties of a Series of Free-Base Corroles. *J. Phys. Chem. A* **2005**, *109*, 7411–7417. [CrossRef]
77. Ray, C.; Díaz-Casado, L.; Avellanal-Zaballa, E.; Bañuelos, J.; Cerdán, L.; García-Moreno, I.; Moreno, F.; Maroto, B.L.; López-Arbeloa, Í.; de la Moya, S. N-BODIPYs Come into Play: Smart Dyes for Photonic Materials. *Chem. A Eur. J.* **2017**, *23*, 9383–9390. [CrossRef]

78. Kuznetsova, R.T.; Aksenova, I.V.; Prokopenko, A.A.; Pomogaev, V.A.; Antina, E.V.; Berezin, M.B.; Antina, L.A.; Bumagina, N.A. Photonics of Boron(III) and Zinc(II) Dipyrromethenates as Active Media for Modern Optical Devices. *J. Mol. Liq.* **2019**, *278*, 5–11. [CrossRef]
79. Kuehne, A.J.C.; Gather, M.C. Organic Lasers: Recent Developments on Materials, Device Geometries, and Fabrication Techniques. *Chem. Rev.* **2016**, *116*, 12823–12864. [CrossRef]
80. Pasini, M.; Giovanella, U.; Betti, P.; Bolognesi, A.; Botta, C.; Destri, S.; Porzio, W.; Vercelli, B.; Zotti, G. The Role of Triphenylamine in the Stabilization of Highly Efficient Polyfluorene-Based OLEDs: A Model Oligomers Study. *ChemPhysChem* **2009**, *10*, 2143–2149. [CrossRef]
81. Jakubiak, R.; Bao, Z.; Rothberg, L. Dendritic Sidegroups as Three-Dimensional Barriers to Aggregation Quenching of Conjugated Polymer Fluorescence. *Synth. Metals* **2000**, *114*, 61–64. [CrossRef]
82. Ozdemir, T.; Atilgan, S.; Kutuk, I.; Yildirim, L.T.; Tulek, A.; Bayindir, M.; Akkaya, E.U. Solid-State Emissive BODIPY Dyes with Bulky Substituents As Spacers. *Org. Lett.* **2009**, *11*, 2105–2107. [CrossRef]
83. Lu, H.; Wang, Q.; Gai, L.; Li, Z.; Deng, Y.; Xiao, X.; Lai, G.; Shen, Z. Tuning the Solid-State Luminescence of BODIPY Derivatives with Bulky Arylsilyl Groups: Synthesis and Spectroscopic Properties. *Chem. A Eur. J.* **2012**, *18*, 7852–7861. [CrossRef] [PubMed]
84. Vu, T.T.; Dvorko, M.; Schmidt, E.Y.; Audibert, J.-F.; Retailleau, P.; Trofimov, B.A.; Pansu, R.B.; Clavier, G.; Méallet-Renault, R. Understanding the Spectroscopic Properties and Aggregation Process of a New Emitting Boron Dipyrromethene (BODIPY). *J. Phys. Chem. C* **2013**, *117*, 5373–5385. [CrossRef]
85. Xi, H.; Yuan, C.-X.; Li, Y.-X.; Liu, Y.; Tao, X.-T. Crystal Structures and Solid-State Fluorescence of BODIPY Dyes Based on Λ-Shaped Tröger's Base. *CrystEngComm* **2012**, *14*, 2087–2093. [CrossRef]
86. Pasini, M.; Destri, S.; Porzio, W.; Botta, C.; Giovanella, U. Electroluminescent Poly(Fluorene-Co-Thiophene-S,S-Dioxide): Synthesis, Characterisation and Structure–Property Relationships. *J. Mater. Chem.* **2003**, *13*, 807–813. [CrossRef]
87. Mukherjee, S.; Thilagar, P. Tuning the Solid State Emission of Meso-Me3SiC6H4 BODIPYs by Tuning Their Solid State Structure. *J. Mater. Chem. C* **2013**, *1*, 4691–4698. [CrossRef]
88. Manzano, H.; Esnal, I.; Marqués-Matesanz, T.; Bañuelos, J.; López-Arbeloa, I.; Ortiz, M.J.; Cerdán, L.; Costela, A.; García-Moreno, I.; Chiara, J.L. Unprecedented j-aggregated dyes in pure organic solvents. *Adv. Funct. Mater.* **2016**, *26*, 2756–2769. [CrossRef]
89. Bozdemir, Ö.A.; Al-Sharif, H.H.T.; McFarlane, W.; Waddell, P.G.; Benniston, A.C.; Harriman, A. Solid-State Emission from Mono- and Bichromophoric Boron Dipyrromethene (BODIPY) Derivatives and Comparison with Fluid Solution. *Chem. A Eur. J.* **2019**, *25*, 15634–15645. [CrossRef]
90. Musser, A.J.; Rajendran, S.K.; Georgiou, K.; Gai, L.; Grant, R.T.; Shen, Z.; Cavazzini, M.; Ruseckas, A.; Turnbull, G.A.; Samuel, I.D.W.; et al. Intermolecular States in Organic Dye Dispersions: Excimers vs. Aggregates. *J. Mater. Chem. C* **2017**, *5*, 8380–8389. [CrossRef]
91. Yuan, K.; Wang, X.; Mellerup, S.K.; Kozin, I.; Wang, S. Spiro-BODIPYs with a Diaryl Chelate: Impact on Aggregation and Luminescence. *J. Org. Chem.* **2017**, *82*, 13481–13487. [CrossRef]
92. Gómez-Durán, C.F.A.; García-Moreno, I.; Costela, A.; Martin, V.; Sastre, R.; Bañuelos, J.; Arbeloa, F.L.; Arbeloa, I.L.; Peña-Cabrera, E. 8-PropargylaminoBODIPY: Unprecedented Blue-Emitting Pyrromethene Dye. Synthesis, Photophysics and Laser Properties. *Chem. Commun.* **2010**, *46*, 5103–5105. [CrossRef]
93. Esnal, I.; Valois-Escamilla, I.; Gómez-Durán, C.F.A.; Urías-Benavides, A.; Betancourt-Mendiola, M.L.; López-Arbeloa, I.; Bañuelos, J.; García-Moreno, I.; Costela, A.; Peña-Cabrera, E. Blue-to-Orange Color-Tunable Laser Emission from Tailored Boron-Dipyrromethene Dyes. *Chemphyschem* **2013**, *14*, 4134–4142. [CrossRef] [PubMed]
94. Belmonte-Vázquez, J.L.; Avellanal-Zaballa, E.; Enríquez-Palacios, E.; Cerdán, L.; Esnal, I.; Bañuelos, J.; Villegas-Gómez, C.; López Arbeloa, I.; Peña-Cabrera, E. Synthetic Approach to Readily Accessible Benzofuran-Fused Borondipyrromethenes as Red-Emitting Laser Dyes. *J. Org. Chem.* **2019**, *84*, 2523–2541. [CrossRef] [PubMed]
95. García, O.; Sastre, R.; del Agua, D.; Costela, A.; García-Moreno, I.; López Arbeloa, F.; Bañuelos Prieto, J.; López Arbeloa, I. Laser and Physical Properties of BODIPY Chromophores in New Fluorinated Polymeric Materials. *J. Phys. Chem. C* **2007**, *111*, 1508–1516. [CrossRef]
96. Costela, A.; García-Moreno, I.; Barroso, J.; Sastre, R. Laser Performance of Pyrromethene 567 Dye in Solid Matrices of Methyl Methacrylate with Different Comonomers. *Appl. Phys. B* **2000**, *70*, 367–373. [CrossRef]
97. Liras, M.; Bañuelos Prieto, J.; Pintado-Sierra, M.; García-Moreno, I.; Costela, Á.; Infantes, L.; Sastre, R.; Amat-Guerri, F. Synthesis, Photophysical Properties, and Laser Behavior of 3-Amino and 3-Acetamido BODIPY Dyes. *Org. Lett.* **2007**, *9*, 4183–4186. [CrossRef]
98. Costela, A.; García-Moreno, I.; Pintado-Sierra, M.; Amat-Guerri, F.; Sastre, R.; Liras, M.; Arbeloa, F.L.; Prieto, J.B.; Arbeloa, I.L. New Analogues of the BODIPY Dye PM597: Photophysical and Lasing Properties in Liquid Solutions and in Solid Polymeric Matrices. *J. Phys. Chem. A* **2009**, *113*, 8118–8124. [CrossRef]
99. Pérez-Ojeda, M.E.; Thivierge, C.; Martín, V.; Costela, Á.; Burgess, K.; García-Moreno, I. Highly Efficient and Photostable Photonic Materials from Diiodinated BODIPY Laser Dyes. *Opt. Mater. Express* **2011**, *1*, 243–251. [CrossRef]
100. Pérez-Ojeda, M.E.; Martín, V.; Costela, A.; García-Moreno, I.; Arroyo Córdoba, I.J.; Peña-Cabrera, E. Unprecedented Solid-State Laser Action from BODIPY Dyes under UV-Pumping Radiation. *Appl. Phys. B* **2012**, *106*, 911–914. [CrossRef]

101. Duran-Sampedro, G.; Esnal, I.; Agarrabeitia, A.R.; Bañuelos Prieto, J.; Cerdán, L.; García-Moreno, I.; Costela, A.; Lopez-Arbeloa, I.; Ortiz, M.J. First Highly Efficient and Photostable E and C Derivatives of 4,4-Difluoro-4-Bora-3a,4a-Diaza-s-Indacene (BODIPY) as Dye Lasers in the Liquid Phase, Thin Films, and Solid-State Rods. *Chem. A Eur. J.* **2014**, *20*, 2646–2653. [CrossRef]
102. Goldenberg, L.M.; Lisinetskii, V.; Ryabchun, A.; Bobrovsky, A.; Schrader, S. Influence of the Cation Type on the DFB Lasing Performance of Dye-Doped Azobenzene-Containing Polyelectrolytes. *J. Mater. Chem. C* **2014**, *2*, 8546–8553. [CrossRef]
103. Kuznetsova, R.T.; Aksenova, Y.V.; Solodova, T.A.; Kopylova, T.N.; Tel'minov, E.N.; Mayer, G.V.; Berezin, M.B.; Antina, E.V.; Burkova, S.L.; Semeikin, A.S. Lasing Characteristics of Difluoroborates of 2,2′-Dipyrromethene Derivatives in Solid Matrices. *Quantum Electron.* **2014**, *44*, 206. [CrossRef]
104. Kuznetsova, R.T.; Aksenova, Y.V.; Prokopenko, A.A.; Bashkirtsev, D.E.; Tel'minov, E.N.; Arabei, S.M.; Pavich, T.A.; Solovyov, K.N.; Antina, E.V. Spectral-Luminescent, Photochemical, and Lasing Characteristics of Boron Dipyrromethene Difluoro (III) Derivatives in Liquid and Solid-State Media. *Russ. Phys. J.* **2016**, *59*, 568–576. [CrossRef]
105. Cookson, T.; Georgiou, K.; Zasedatelev, A.; Grant, R.T.; Virgili, T.; Cavazzini, M.; Galeotti, F.; Clark, C.; Berloff, N.G.; Lidzey, D.G.; et al. A Yellow Polariton Condensate in a Dye Filled Microcavity. *Adv. Opt. Mater.* **2017**, *5*, 1700203. [CrossRef]
106. Sannikov, D.; Yagafarov, T.; Georgiou, K.; Zasedatelev, A.; Baranikov, A.; Gai, L.; Shen, Z.; Lidzey, D.; Lagoudakis, P. Room Temperature Broadband Polariton Lasing from a Dye-Filled Microcavity. *Adv. Opt. Mater.* **2019**, *7*, 1900163. [CrossRef]
107. Nwahara, N.; Nkhahle, R.; Ngoy, B.P.; Mack, J.; Nyokong, T. Synthesis and Photophysical Properties of BODIPY-Decorated Graphene Quantum Dot–Phthalocyanine Conjugates. *New J. Chem.* **2018**, *42*, 6051–6061. [CrossRef]
108. Sun, S.; Zhuang, X.; Wang, L.; Liu, B.; Zhang, B.; Chen, Y. BODIPY-Based Conjugated Polymer Covalently Grafted Reduced Graphene Oxide for Flexible Nonvolatile Memory Devices. *Carbon* **2017**, *116*, 713–721. [CrossRef]

Article

Sulfonate-Conjugated Polyelectrolytes as Anode Interfacial Layers in Inverted Organic Solar Cells

Elisa Lassi [1,†], Benedetta Maria Squeo [1,†], Roberto Sorrentino [1], Guido Scavia [1], Simona Mrakic-Sposta [2], Maristella Gussoni [1], Barbara Vercelli [3], Francesco Galeotti [1], Mariacecilia Pasini [1,*] and Silvia Luzzati [1,*]

[1] Institute of Chemical Sciences and Technologies "G. Natta"-SCITEC, National Research Council, CNR-SCITEC, via Corti 12, 20133 Milan, Italy; lassi@ismac.cnr.it (E.L.); benedetta.squeo@scitec.cnr.it (B.M.S.); roberto.sorrentino@scitec.cnr.it (R.S.); guido.scavia@scitec.cnr.it (G.S.); maristella.gussoni@unimi.it (M.G.); francesco.galeotti@scitec.cnr.it (F.G.)

[2] Institute of Clinical Physiology, National Research Council, CNR-IFC, Piazza Ospedale Maggiore 3, 20162 Milan, Italy; simona.mrakicsposta@cnr.it

[3] Institute of Condensed Matter Chemistry and Technologies for Energy, National Research Council, CNR-ICMATE, Via Roberto Cozzi 53, 20125 Milan, Italy; barbara.vercelli@cnr.it

* Correspondence: mariacecilia.pasini@scitec.cnr.it (M.P.); silvia.luzzati@scitec.cnr.it (S.L.)

† These authors contributed equally to this work.

Abstract: Conjugated polymers with ionic pendant groups (CPEs) are receiving increasing attention as solution-processed interfacial materials for organic solar cells (OSCs). Various anionic CPEs have been successfully used, on top of ITO (Indium Tin Oxide) electrodes, as solution-processed anode interlayers (AILs) for conventional devices with direct geometry. However, the development of CPE AILs for OSC devices with inverted geometry is an important topic that still needs to be addressed. Here, we have designed three anionic CPEs bearing alkyl-potassium-sulfonate side chains. Their functional behavior as anode interlayers has been investigated in P3HT:PC$_{61}$BM (poly(3-hexylthiophene): [6,6]-phenyl C61 butyric acid methyl ester) devices with an inverted geometry, using a hole collecting silver electrode evaporated on top. Our results reveal that to obtain effective anode modification, the CPEs' conjugated backbone has to be tailored to grant self-doping and to have a good energy-level match with the photoactive layer. Furthermore, the sulfonate moieties not only ensure the solubility in polar orthogonal solvents, induce self-doping via a right choice of the conjugated backbone, but also play a role in the gaining of hole selectivity of the top silver electrode.

Keywords: conjugated polyelectrolytes; inverted organic solar cells; anode interfacial layers

Citation: Lassi, E.; Squeo, B.M.; Sorrentino, R.; Scavia, G.; Mrakic-Sposta, S.; Gussoni, M.; Vercelli, B.; Galeotti, F.; Pasini, M.; Luzzati, S. Sulfonate-Conjugated Polyelectrolytes as Anode Interfacial Layers in Inverted Organic Solar Cells. *Molecules* **2021**, *26*, 763. https://doi.org/10.3390/molecules26030763

Academic Editor: Minas M. Stylianakis

Received: 21 December 2020
Accepted: 27 January 2021
Published: 2 February 2021

Publisher's Note: MDPI stays neutral with regard to jurisdictional claims in published maps and institutional affiliations.

Copyright: © 2021 by the authors. Licensee MDPI, Basel, Switzerland. This article is an open access article distributed under the terms and conditions of the Creative Commons Attribution (CC BY) license (https://creativecommons.org/licenses/by/4.0/).

1. Introduction

The use of interfacial layer materials to improve the charge selectivity and minimize the energy barrier of electrodes plays a central role in promoting their performance and stability in organic electronic devices, such as organic solar cells (OSCs) [1–3]. In view of the technological need to attain fully solution-processed devices, most of the attention has been addressed in the search of efficient interfacial materials with good solubility in polar solvents, included water, which allows deposition from orthogonal solvents to the active layers [4–6]. Conjugated polyelectrolytes (CPEs), composed of a conjugated backbone with side-chains bearing ionic functional groups, have emerged as a promising class of interfacial materials with their proven ability to improving photovoltaic (PV) performances through solution processing [7–10]. CPEs combine several advantages including: solubility in aqueous/alcoholic solvents, orthogonality to organic solvents used for the deposition of the active layers, robust film formation, and chemical flexibility in tailoring both the conjugated backbone as well as the polar/ionic lateral functionalities. Most of the CPEs developed so far are effective in reducing the electrode work function [11–13] and might behave as electron transport layers [14,15]. Therefore there is a huge library of CPEs in the literature that have been investigated as cathode interfacial layers to facilitate the electron collection at the cathode in OSCs [16–21]. There is, however, also a growing number of CPEs that have been reported to behave as promising anode interfacial layers to facilitate hole extraction at the ITO electrode [22–30]. These CPEs were demonstrated to be a valid alternative to the common PEDOT:PSS, not only for the good PV performances, but also because most of these materials are pH-neutral, which is an advantage to avoid possible device instabilities induced by the acidic nature of PEDOT:PSS [31,32].

From the studies so far, some guidelines have been reported for the design of the CPEs where their application as anode modifiers were highlighted. It was shown that upon oxidative p-doping, a fluorene-based CPE with anionic sulfonate side groups deposited on top of an ITO electrode behaved as an efficient AIL in OSCs [23]. The p-doping not only favored the hole transport through the inter-layer, but it was shown to be a viable tool to enhance the ITO anode work function [23]. When combining anionic sulfonate side groups to polymer chains containing more electron rich moieties than fluorene, p-type self-doping effects can occur [33,34]. Self-doping was found to induce similar beneficial effects for the hole transport and ITO anode work function modifications [22,24,35], and it was identified as an important characteristic to achieve efficient AIL materials [24–26,28,36].

In spite of the successful incorporation of various CPEs as anode modifiers into devices, mainly devices with direct geometry were reported. Nevertheless, the devices with inverted geometry are better suited to envisage the scale-up towards industrial compatible fabrication processes. Therefore, the application of CPE materials as anode modifiers in inverted OSC devices is an important topic to be addressed.

The capability of a CPE to engineer the anode interface, established in direct geometry, is not necessarily transposable to inverted devices. For example, most of the efficient CPE anode modifiers reported in the literature are quite hydrophilic and need water to be dissolved. This can be an issue for inverted geometry, where the CPE film is deposited on top of the highly hydrophobic active layer surface. This implies that the tailoring of the chemical structure of CPEs to improve their reaction with the active layer is important when considering the inverted device geometry. Another aspect that may vary with the device geometry is the interaction between the interlayer and the electrode which is not necessarily identical when depositing the interlayer on top of the electrode, as is the case for direct geometry, or when evaporating a metal electrode on top of the interlayer, as is the case for inverted geometry.

In this work, we have designed three anionic conjugated copolymers bearing alkyl-potassium sulfonate side groups, namely **P1**, **P2**, and **P3** (see Figure 1), and their application as anode interfacial layers for OSCS devices with inverted geometry have been investigated. The devices are prepared using standard P3HT:PC$_{61}$BM as an active layer. We have started with a well-known CPE, **P1** [34] in Figure 1, which is reported to have good results when applied in OSCs devices with direct geometry [22,27,31]. **P1** is a pH-neutral CPE that showed self-doping characteristics, in particular in the presence of a proton source, good hole conductivity comparable to PEDOT:PSS, and effective engineering of the ITO anode work function [34,37]. **P1** is hydrophilic owing to the high number of polar groups per monomeric unit and needs water to be dissolved. In order to grant self-doping characteristics, but obtain an alcohol-processable material which guarantees better wettability of the active layer, we have designed a novel copolymer, **P2**, Figure 1, with the same backbone and type of alkyl-sulfonate substituents of **P1** but with a reduced number of side chains per monomeric unit. Finally, for comparison, we have prepared and tested **P3**, in Figure 1, a CPE bearing similar potassium sulfonate alkyl groups as **P1** and **P2** but with a different conjugated backbone based on a fluorenic unit, known to be un-fitted for self-doping [34,38–40]. In the **P3** copolymer, the side chains have been tailored to favor the interaction with the active layer by alternating alkyl-potassium sulfonate and non-polar alkyl side chains. By comparing the results obtained with the three CPEs we tried to elucidate the role that pendant sulfonate groups play in the AIL functionality of this class of polymers, demonstrating that sulfonate anionic CPEs could be a valid approach to obtain solution-processable anode interlayers for inverted OSCs devices.

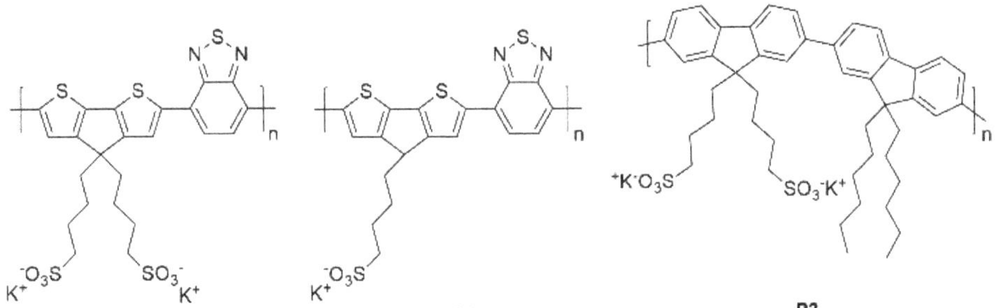

Figure 1. Chemical structure of the synthesized CPEs.

2. Experimental
2.1. Synthetic Methods

General information for synthesis: All glassware was oven-dried. Unless specifically mentioned, all chemicals are commercially available and were used as received. The dialysis membrane (MWCO: 3500–5000 Da) was purchased from Membrane Filtration Products Inc. ^{1}H-NMR spectra were recorded at 600 MHz in D_2O.

4-Bis-potassium butanylsulfonate-4H-cyclopenta-[2,1-b;3,4-b′]-dithiophene (**1**): 4H-cyclopenta-[2,1-b;3,4-b′]-dithiophene (CPDT, 670 mg, 3.76 mmol, 1.0 equivalent), and tetrabutylammonium bromide (60.6 mg, 0.188 mmol, 0.05 equivalent), were dissolved in anhydrous DMSO (18.4 mL), and the solution was degassed by bubbling Ar for 5 min. 50% KOH in H_2O (4.2 g) was added via syringe, followed by the addition of 1,4-butanesultone (924 μL, 9.02 mmol, 2.4 equivalent). After stirring at room temperature for 3 h, the reaction mixture was poured into acetone (100 mL) and the yellowish precipitate was collected by filtration and washed with acetone. The crude was used in the next step without further purification. ^{1}H-NMR (D_2O) δ: 7.35 (d, J = 4.8 Hz, 2H), 7.14 (d, J = 4.8 Hz, 2H), 2.72 (t, J = 8.1 Hz, 4H), 1.98 (t, J = 7.8 Hz, 4H), 1.55 (m, broad, 4H), 0.98 (m, broad, 4H).

2,6-Dibromo-4-bis-potassium butanylsulfonate-4H-cyclopenta-[2,1-b;3,4-b′]-dithiophene (**2**): The crude product **1** was suspended in DMF (15 mL), and H_2O (~2 mL) was added while stirring until dissolved. NBS (1.67 g, 9.4 mmol, 2.5 equivalent) was added in dark conditions by shielding the flask with aluminum foil. The brown solution was stirred at room temperature for 1 h, and poured into acetone. The yellowish precipitate was collected by filtration, and washed with acetone (2 g, 80% yield). ^{1}H-NMR (D_2O) δ: 7.03 (s, 2H), 2.6 (m, broad, 4H), 1.69 (m, broad, 4H), 1.43 (m, broad, 4H), 0.77 (m, broad, 4H).

Polymer **P1**: A mixture of compound **2** (79 mg, 0.115 mol, 1 equivalent), 2,1,3-Benzothiadiazole-4,7-bis(boronic acid pinacol ester) (45 mg, 0.115 mmol, 1 equivalent), and tetrakis(triphenylphosphine)palladium(0) (Pd(PPh$_3$)$_4$) (2.6 mg, 2% mol) was added in a pre-degassed Schlenk flask, followed by three vacuum/nitrogen cycles. Then degassed DMF (1 mL) and degassed potassium carbonate aqueous solution (0.25 mL) were added. The mixture was stirred at 110 °C for 3 h. The reaction mixture was poured in acetone and the dark blue precipitate was collected by filtration and washed with copious amounts of acetone. The precipitate was all dissolved in deionized H_2O and transferred into a dialysis tube (MWCO: 3500–5000). The dialysis tube was placed in a large beaker with H_2O stirring for 3 days, and the H_2O was changed every 12 h. Evaporation of the H_2O provided the title product, a dark blue solid (55 mg, 72%), after drying under vacuum overnight. The NMR of the polymer in D_2O showed only non-informative broad peaks, due to the presence of paramagnetic radical cations [34] (see Scheme S1 in Supplementary Materials).

Synthesis of 4-Potassium butanylsulfonate-4H-cyclopenta-[2,1-b;3,4-b']-dithiophene (**3**): 4H-cyclopenta-[2,1-b;3,4-b']-dithiophene (CPDT, 300 mg, 1.68 mmol, 1.0 equivalent) and tetrabutylammonium bromide (27 mg, 0.084 mmol, 0.05 equivalent) were dissolved in anhydrous DMSO (8.2 mL), and the solution was degassed by bubbling with Ar for 5 min. 50% KOH in H_2O (1.8 g) was added via syringe, followed by the addition of 1,4-butanesultone (924 µL, 9.02 mmol, 1.2 equivalent). After stirring at room temperature for 3 h, the reaction mixture was poured into acetone (50 mL) and the yellowish precipitate was collected by filtration and washed with acetone. The crude was used in the next step without further purification. ^{1}H-NMR (D_2O) δ: 7.05 (d, *J* = 4.8 Hz, 2H), 6.95 (d, *J* = 2.9 Hz, 2H), 3.42(t, *J* = 6.9 Hz, 1H) 2.73 (t, *J* = 7.8 Hz, 2H), 1.63 (m, broad, 2H), 1.56 (m, broad, 2H), 1.3 (m, broad, 2H).

Synthesis of 2,6-Dibromo-4-potassium butanylsulfonate-4H-cyclopenta-[2,1-b;3,4-b']-dithiophene (**4**): The crude product **3** was suspended in DMF (6.7 mL), and H_2O (~1 mL) was added while stirring until dissolved. NBS (747 g, 4.2 mmol, 2.5 equivalent) was added in dark conditions by shielding the flask with aluminum foil. The brown solution was stirred at room temperature for 1 h and poured into acetone. The yellowish precipitate was collected by filtration, and washed with acetone (640 mg, 75% yield). ^{1}H-NMR (D_2O) δ: 7.23 (s, 2H), 3.74 (t, broad, 1H), 2.80 (m, broad, 2H), 2.60–2.50 (m, broad, 2H), 1.85–1.80 (m, broad, 2H), 1.55–1.45 (m, broad, 2H).

Synthesis of polymer **P2**: A mixture of compound **3** (117 mg, 0.230 mol, 1 equivalent), 2,1,3-Benzothiadiazole-4,7-bis(boronic acid pinacol ester) (89 mg, 0.230 mmol, 1 equivalent), and tetrakis(triphenylphosphine)palladium(0) ($Pd(PPh_3)_4$) (5.2 mg, 2% mol) was added in a pre-degassed Schlenk flask, followed by three vacuum/nitrogen cycles. Then degassed DMF (2.2 mL) and degassed potassium carbonate aqueous solution (0.55 mL) were added. The mixture was stirred at 110 °C for 3 h. The reaction mixture was poured in acetone and the dark blue precipitate was collected by filtration and washed with copious amounts of acetone. The precipitate was all dissolved in deionized H_2O and transferred into a dialysis tube (MWCO: 3500–5000). The dialysis tube was placed in a large beaker with H_2O stirring for 3 days, and the H_2O was changed every 12 h. Evaporation of H_2O provided the title product, a dark blue solid (65 mg, 60%), after drying under vacuum overnight. The NMR of the polymer in D_2O showed only non-informative broad peaks, due to the presence of paramagnetic radical cations (see Scheme S2 in Supplementary Materials) [34].

2.2. Device Fabrication and Photovoltaic Characterization

Direct geometry devices fabrication: Solar cells were assembled with the conventional structure Glass/ITO/PEDOT:PSS or AIL/P3HT:$PC_{61}BM$/Ag. Glass ITO (Kintec, Hong Kong) 15 Ω/sq substrates were mechanically cleaned with peeling tape and paper with acetone and then were washed in a sonic bath at 50 °C for 10 min sequentially with water, acetone, and isopropanol. After drying with compressed nitrogen flow, 10 min plasma treatment in the air was used to enhance the ITO wettability for the next deposition. PEDOT:PSS (Al VP 8030 from Heraus, Hanau, Germany) was filtered on a 0.45 µm nylon filter and spin-coated in the air at 2500 rpm for 50 s. **P1** (5 mg/mL in H_2O:MeOH 1:1), **P2**, and **P3** (1 mg/mL in EtOH) were spin-coated at 2000 rpm and 4000 rpm for 60 s. Finally, the substrates were stored in a glovebox and annealed at 110 °C for 10 min. The device assembly was then performed in the glovebox. The active layer was composed of a blend dissolved at 1:0.8 wt/wt of P3HT:PC61BM solution in 1,2-dichlorobenzene at a total concentration of 27 mg/mL. The P3HT was purchased from Plextronics (Pittsburgh, PA, USA, Plexcore OS2100, Mn: 62602; Mw: 119010, 99% regioregularity) and $PC_{61}BM$ (99.5 % purity) was purchased from Solenne BV. The solution was stirred for 12 h on a hotplate in a glovebox at 60 °C. The active layer was spin-coated from the warm solution at 1000 rpm for 60 s, which resulted in a thickness of 130 nm; then, the active layer film was slow dried under a glass petri dish for 1 h. Finally, a 100 nm-thick aluminum electrode was evaporated on the top of the device through a shadow mask under a pressure of 2×10^{-6} mbar.

The deposition rate was 0.5 nm/s. There were six devices on a single substrate, each with an active area of 6.1 mm^2.

Inverted geometry devices fabrication: Solar cells were assembled with the conventional structure Glass/ITO/PEIE/active layer/AIL/Ag. Glass ITO (Kintec, Hong Kong) 15 Ω/sq substrates were mechanically cleaned with peeling tape and paper with acetone and then were washed in a sonic bath at 50 °C for 10 min sequentially with water, acetone, and isopropanol. After drying with compressed nitrogen flow, ITO was treated under UV light from a solar simulator for 20 s to remove O_2 species adsorbed on its surface. Polyethylenimine ethoxylated (PEIE) 0.4% *w/w* in 2-Metoxyethanole solution was spin-coated in N_2 at 5000 rpm for 60 s. PEIE film was then washed with H_2O to remove excess polymers (200 µL were deposited for 10 s on substrate and then removed at 4000 rpm for 60 s). The P3HT:PC$_{61}$BM active layer was prepared upon blending the two components at a 1:0.8 weight ratio in 1,2-dichlorobenzene, at a total concentration of 27 mg/mL. The solution was stirred for 12 h on a hotplate in glovebox at 60 °C; the active layer was spin-coated from the warm solution at 1000 rpm for 60 s, which results in a thickness of 130 nm; the films were then covered with a glass petri dish to perform slow drying for an hour and then treated at 120 °C for 10 min. The PTB7-Th was purchased from Cal-Os (batch number 6) and PC$_{71}$BM from Solenne BV (99.5% purity). The PTB7-Th:PC71BM active layer was prepared by dissolving the two components at a 1:1.5 weight ratio in chlorobenzene, with a solute concentration of 25 mg/mL. This solution was stirred overnight at 65 °C. Next, 10 min after adding 2% *v/v* of 4-methoxybenzaldehyde to the solution, the blend was spin coated at 1200 rpm, permitting the obtaining of films with thickness of 100 nm. The samples were left at room temperature for 10 min and then annealed for 20 min at 60 °C. The AILs were deposited on top of the active layers by dissolving **P2** or **P3** in ethanol at a concentration of 1 mg/mL, and 20 µL of this solution was dropped on a device rotating at 4000 rpm for 60 s. **P1** was dissolved in H_2O:Methanol (1:1) at a concentration of 5 mg/mL, and 20µL of this solution was dropped on a device rotating at 4000 rpm for 60 s. Finally, a 100 nm-thick silver electrode was evaporated on the top of the device through a shadow mask under a pressure of 2×10^{-6} mbar. The deposition rate was 0.3 nm/s. A device without any AIL and a device with an evaporate MoOx AIL were prepared for comparison. The thickness of the MoOx layer was 10 nm, achieved with a deposition rate of 0.1 nm/s. There were six devices on a single substrate, each with an active area of 6.1 mm^2.

Device characterization and measurements: The devices were characterized through current density–voltage and external quantum efficiency characterization. Current density–voltage measurements were performed directly in the glovebox where the solar cells were assembled, with a Keithley 2602 source meter, under an AM 1.5G solar simulator (ABET 2000). The incident power, measured with a calibrated photodiode (Si cell + KG5 filter), was 100 mW/cm^2. The EQE spectral responses were recorded by dispersing an Xe lamp through a monochromator, using an Si solar cell with a calibrated spectral response to measure the incident light power intensity at each wavelength. The devices were taken outside the glovebox for the EQE measurements, after mounting them on a sealed cell to avoid moisture and oxygen exposure. For the aging measurement, the devices were stored in air and measured in glovebox.

3. Results and Discussion

The chemical structures of the three copolymers **P1**, **P2**, and **P3**, are shown in Figure 1. The polymers were synthesized via Suzuki cross-coupling and their purifications were performed according to the literature. The selected experimental procedures were chosen according to the literature [41,42]; **P1** was prepared according to the procedure described by Mai et al. [34] starting from the commercially available 2,1,3-Benzothiadiazole-4,7-bis(boronic acid pinacol ester) and 2,6-dibromo-4,4-bis-potassium butanylsulfonate-4H-cyclopenta-[2,1-b;3,4-b′]-dithiophene. The same procedure was applied to **P2**, starting from the newly synthesized 2,6-dibromo-4-potassium butanylsulfonate-4H-cyclopenta-[2,1-b;3,4-b′]-dithiophene, while **P3** was synthetized according to the literature [39,40] (synthesis details and characterizations are provided in the Supplementary Materials). GPC measurements confirmed the structure of the synthesized polymers, however they only provided an indicative estimation of the real value of the molecular weight due to solubility problems and structural difference with the polystyrene used as standard [43].

Cyclic voltammetry and optical spectroscopy were used to assess the electronic properties of the copolymers and to estimate their HOMO and LUMO energy levels. See Table 1 and the Supplementary Materials.

Table 1. Optical properties and HOMO and LUMO energy levels of **P1**, **P2**, and **P3**.

Polymer	λ_{max} (nm)	λ_{onset} (nm)	E_g [a] (eV)	HOMO (eV)	LUMO (eV)	E_g [b] (eV)
P1	630	898	1.38	−4.87	−3.31	1.56
P2	719	947	1.31	−4.83	−3.13	1.50
P3	389	426	2.91	−5.50	−2.2	3.23

[a] Estimated from the onset wavelength of the optical absorption in the solid state film (Figure 3a). [b] Calculated from the HOMO and LUMO level.

The HOMO and LUMO levels of **P1**, **P2**, and **P3** were obtained from the onsets of the oxidation and reduction potential [44], respectively, and are reported in Table 1 together with UV–vis spectroscopic data. The deviation between the optical and electrochemical results are in quite satisfactory agreement, as previously described in literature [45].

The energy levels of **P2** are similar to **P1**, consistently with the same conjugated backbone and lateral substituents. **P3** shows a variation of the HOMO and LUMO which is coherent to its polyfluorene backbone [39,40]. Importantly, as previously reported [46–49] the functionalization with the polar side chains does not interfere with the conjugated π-orbitals of the polyfluorene backbone and thus leads to no modification in the position of the HOMO and LUMO levels.

The different concentration in the copolymers of the sulfonated groups and alkyl chains is a tool to obtain different hydrophilic/hydrophobic characteristics: **P1** is the most hydrophilic owing to its two sulfonate groups per monomeric unit. The presence of only one polar side chain in **P2** and the presence of two long alkyl chains in **P3** are two approaches to reduce the CPEs hydrophilicity/polarity. This trend is confirmed by contact angle measurements on top of the **P1**, **P2**, and **P3** films using glycerol as a polar solvent. As shown in Figure S10 in the Supplementary Materials, the contact angles were found to be respectively 36° for **P1**, 49° for **P2**, and 68° for **P3**. Such reduction in the wettability to polar solvent going from **P1** to **P3** suggests that thanks to our design we were able to finely modulate the different hydrophilic/hydrophobic characteristics of the polymers and consequently the interaction with the a-polar active layer.

As a consequence, **P1** has a good solubility in water but it is not soluble in pristine methanol or ethanol. **P2** shows an overall poor solubility in water/alcohol solvents but its dispersion in ethanol at low concentrations, 1 mg/mL, leads to stable suspensions which are suitable for film deposition by spin casting. **P3** can be also dissolved and deposited from ethanol at low concentrations (1 mg/mL). A good solubility in alcohol is particularly important with a view to preparing devices in an inverted configuration where a thin CPE film should be deposited on top of the hydrophobic active layer surface.

3.1. Self-Doping Behaviour

The self-doping behavior of **P1**, **P2**, and **P3** was investigated by electron paramagnetic resonance (EPR) measurements in aqueous solutions, see Figure 2b. A strong EPR symmetric signal consistent with the presence of unpaired electrons is observed for both **P1** and **P2**, with peak to peak linewidths and calculated g values respectively of 0.28 mT and 1.9955 for **P1** and 0.25 mT and 1.99314 for **P2**, which are the typical signatures of polaron formation in conjugated polymers [34,36]. It is important to underline that the values of the integrals are comparable and therefore the signal of the **P2** polymer is more intense than that of the **P1** polymer. Moreover, as it is well known, EPR is an intrinsically quantitative technique, since the EPR signals are due to the number of the excited spins [50]. As a consequence, at the same temperature and experimental conditions, EPR signal heights and, even more so double integrals, are comparable.

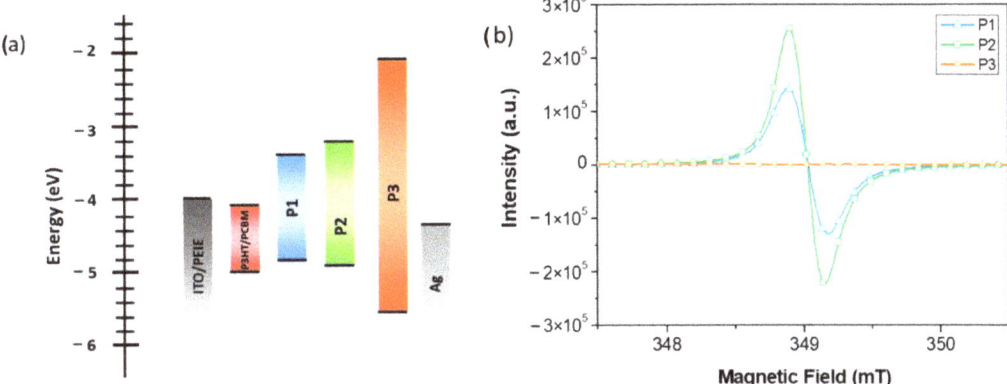

Figure 2. (a) Energy level of the materials used to fabricate inverted solar cells in this contribution; and (b) electron paramagnetic resonance (EPR) spectra of **P1**, **P2**, and **P3** in aqueous solution.

With regard to **P3**, no evidence of polaronic features is detected in the EPR spectrum. Polarons formation was previously reported for a **P1**-type CPE in aqueous solution and it was explained by a self-doping mechanism [33,34] in presence of a proton source like water [51]. Similarly, the polaronic features in the **P2** EPR spectrum indicate that self-doping in solution is occurring also when the PCPDTBT backbone is bearing only one alkylsulfonate lateral group per monomeric unit. On the other side, the absence of polarons in the **P3** EPR spectrum is consistent with the fact that self-doping can only take place in CPEs with a relatively low ionization potential, as is the case for **P1** and **P2**, but not for polyfluorene-based CPEs such as **P3** which have a high ionization potential/deep HOMO energy level. The UV/Vis-NIR absorption spectra of **P1**, **P2**, and **P3** films are reported in Figure 3a. The spectrum of **P1** exhibits two main bands centered at 405 nm and 660 nm that are peculiar for polymers with conjugated ciclopentadithiophene-benzothiadiazole (PCPDTBT) backbones and a weak and broad band peaked at 1140 nm which could be ascribed to the formation of polarons (radical cations) which are stabilized by the pendant sulfonate groups [34]. As shown in Figure 3a, a similar band is observed upon doping a pristine PCPDTBT film through exposure to I_2 vapors. This confirms a polaronic transition assignment for the NIR absorption feature of the **P1** film absorption spectrum. The **P2** spectrum exhibits two main absorption bands, peaking respectively at 410 and 715 nm, which are similarly assigned to the PCPDTBT backbone transitions. The **P2** D-A (donor-acceptor) band at 715 nm has a broad tail in the NIR region which could arise from light scattering effects. A scattering tail is also observed in the absorption spectrum of a **P2** dilute solution, indicating that **P2** is partially aggregated even in diluted solution (0.1 mg/mL in ethanol or water, see Figure S2 in Supplementary Materials). The granular morphology in the solid state, evident from the atomic force microscopy (AFM) images reported in Figure S3, confirms the presence of aggregation and it is consistent to significant light scattering from **P2** films. As a matter of fact, the **P2** polarons spectral features, already evidenced by EPR spectra, are probably covered by the mentioned scattering, in the **P2** film absorption spectrum. Consequently, we can reasonably suppose that, similar to the well-known **P1**, self-doping is maintained also in **P2** films. The **P3** film shows a π-π* transition peak at 365 nm as expected for a polyfluorene backbone. No polaronic features are detected, which is consistent to the previously discussed absence of self-doping in solution for this fluorene-based CPE.

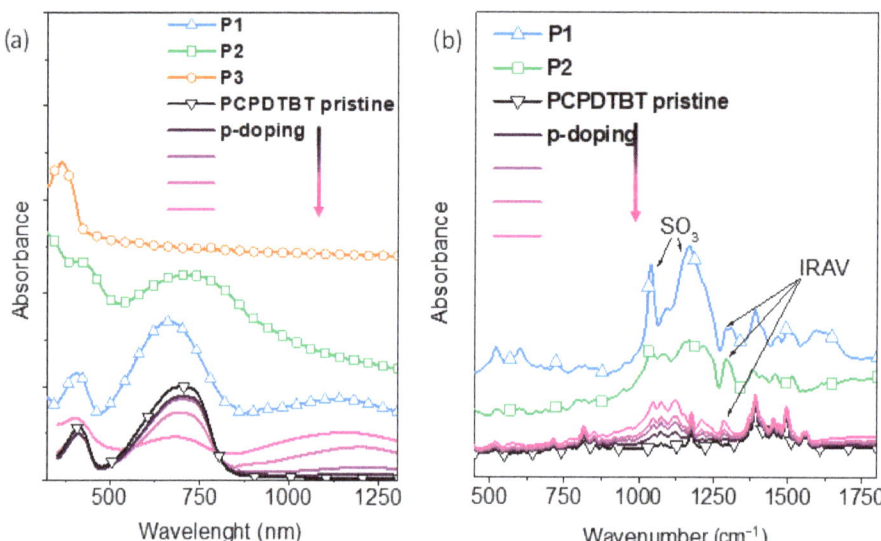

Figure 3. (a) UV/Vis-NIR absorption spectra of **P1**, **P2**, and **P3** films and (b) IR absorption spectra of **P2** and **P3** films with a pristine PCPDTBT film during its p-doping through I2 vapor exposure.

The self-doping of CPEs films can be monitored by IR (Infra-Red) absorption spectroscopy. The absorption spectral signatures of polaronic charged states in conjugated polymers consist of both electronic transitions and IR vibrational modes, the so-called IRAV bands [52,53]. Figure 3b displays the IR absorption spectral pattern of **P1** and **P2**, with a pristine and I$_2$-doped PCPDTBT for comparison. The **P1** and **P2** IR spectra show two intense and broad peaks at 1035 and 1165 cm^{-1}. These bands are attributed to the symmetric and antisymmetric stretching modes of the SO$_3^-$ groups [54]. In **P3**, which has the same alkylsulfonate side chains, similar intense SO$_3^-$ stretching bands are observed (Figure S13, Supplementary Materials). Both **P1** and **P2** polymers show a weak band at 1285 cm^{-1}, that is not observed in the **P3** spectrum of Figure S3. This suggests that an alkylsulfonate vibrational mode cannot account for this feature. Interestingly, this weak band does not match the pristine PCPDTBT IR spectrum but instead has a close similarity to one of the IRAV bands growing up upon p-doping PCPDTBT with I$_2$ vapors. On the basis of these considerations, we suggest that the band observed at 1285 cm^{-1} in the IR spectra of **P1** and **P2** is an IRAV band of the conjugated backbone arising from polaronic charged defects. These features support our previous conjecture that self-doping is maintained also in the **P2** films. To sum up, both **P1** and **P2** display self-doping behavior, whereas **P3** does not. We will show in the following paragraphs that self-doping is one of the important characteristics of anionic CPEs for attaining effective interfacial anode modification in OSCs devices with inverted geometry.

3.2. OSCs Devices Characterization

Direct geometry device: We have characterized the copolymers' interlayers in devices with direct geometry, made with a P3HT:PC$_{61}$BM active layer, taking a standard PEDOT:PSS anode modifier as a reference. The devices architecture and PV characteristics are reported in the Supplementary Materials, Figure S14 and Table S2, while the devices' assembly procedures are given in the experimental section. The **P1** devices exhibited the best PV performances with a power conversion efficiency (PCE) of 2.17, with comparable PV characteristics to the PEDOT:PSS reference device (see Table S1). A similar trend was already reported by the Bazan group using a **P1**-type CPE interlayer in P3HT:PC$_{61}$BM devices [27]. Interestingly, the relatively high fill factor obtained using **P1** (0.58), brings support for an effective hole extraction when using **P1**. This can be ascribed to a combination of factors, including the good matching among the HOMO energy levels of **P1** and P3HT, which is preventing a barrier formation to hole extraction; and the self-doping of **P1**, which is known to induce beneficial effects for the hole transport and for the engineering of the ITO electrode work function [22]. With regard to **P2**, a reduction of the PV performances were observed (PCE 1.39%). **P2** has the same backbone/electronic features than **P1** but lower processability, owing to the previously mentioned low solubility and aggregates formation in solution, that affects its film forming properties as confirmed by the presence of thick aggregates with granular morphology in the AFM images (see Figure S3). Such features can account for the lower performances obtained with **P2** instead of **P1**. Using **P3** as anode interlayer, there is a drastic drop of the PV performances, similarly to the previously reported pristine anionic CPE AILs, with a high energy gap and no self-doping, as **P3** [23]. Kelvin probe measurements were used to investigate how the copolymers' interlayers modify the ITO electrode work function (see Figure S15 in the Supplementary Materials). While the effective work function of the ITO electrode decreased from 4.8 eV to 4.7 eV with the **P3** interlayer, an increase to 5.02 eV and 4.98 eV was observed respectively with **P1** and **P2**, as expected for self-doped anionic CPEs [23]. Hence, the hole selectivity of the ITO electrode increases when using **P1** and **P2**, facilitating charge extraction, but it is reduced with **P3**.

Inverted geometry device: In order to investigate the key characteristics that anionic CPEs should have to obtain solution-processable anode interlayers for inverted OSC devices, we have tested **P1**, **P2**, and **P3** interlayers in P3HT:PC$_{61}$BM-based devices, with a device architecture displayed in Figure 4. PEIE thin film was used as cathode interlayer on top of the ITO electrode. This non-conjugated amino-containing polymer interlayer affords good stability and efficient performance when applied in fullerene-based polymeric solar cells [55,56]. For this study, PEIE offers the advantage of being scarcely affected by the air soaking treatments herein reported. In inverted geometry the interlayers are deposited on top of the hydrophobic active-layer surface and to increase the wettability we used ethanol for the AILs processing; this was possible for **P2** and **P3**, but not for **P1** which needs a water/ethanol mixture to be dissolved and processed. The details of the devices' assembly procedures are given in the experimental section. The PV characteristics are reported in Figure 4, Figure S16 in the Supplementary Materials, and Table 2.

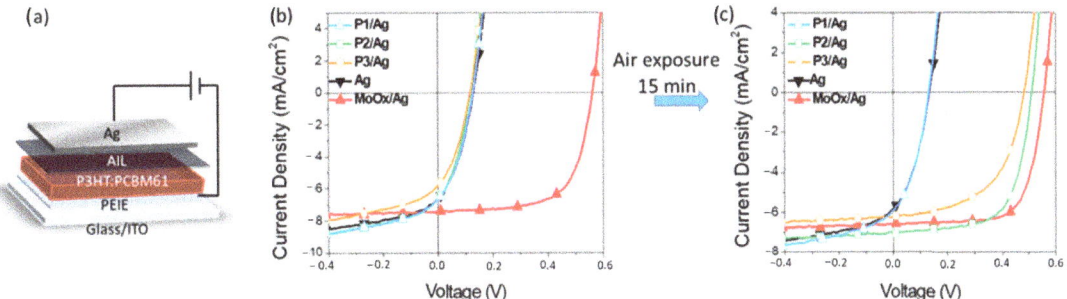

Figure 4. (**a**) Inverted devices architecture; (**b**,**c**) current density–voltage (J–V) curves under AM1.5 G irradiation at 100 mW/cm^2. (**b**) before air exposure; and (**c**) after 15 min of air exposure.

The typical current density–voltage (J–V) curves of the devices featuring the polymeric interlayers **P1**, **P2**, and **P3** are depicted in Figure 4b; we have taken for comparison a device with a state-of-the-art evaporated MoOx interlayer (MoOx/Ag) and a device prepared without any AIL (Ag). The PV characteristics are summarized in Table 2. The devices with the copolymer interlayers showed poor performances, with PCEs around 0.3–0.25%, similar to the Ag devices and about one order of magnitude lower than the reference MoOx/Ag devices, with a PCE around 2.5%. In one of our previous works we have highlighted that the air exposure of devices with polar polymers interfacial layers improved their performances [57]. Interestingly, after a short air soaking treatment of 15 min, a significant enhancement of the PV performances was observed in the **P2** and **P3** devices, see Figure 4c and Table 2. Namely, the devices made with **P2** reached the best performances, comparable to the reference MoOx/Ag device, with a PCE of 2.62%; in **P3** devices, the PCE was 1.9%. In contrast to the **P2** and **P3** devices and irrespective to air exposure, **P1**-based devices showed the same poor performances of the Ag devices, prepared without any AIL. Such poor **P1** functionality in inverted devices might arise from the strong hydrophilic character of **P1**, leading to scarce adhesion and film formation on top of the hydrophobic P3HT:PC$_{61}$BM active layer.

Table 2. Summary of the photovoltaic parameters [a] using pristine Ag or MoOx, **P1**, **P2**, and **P3** AILs before and after an air exposure treatment of 15 min.

Device	V_{oc} (V) [a]	FF [a]	J_{sc} (mA/cm^2) [a]	PCE (%) [a]	R_s [b] (Ωcm^2)	R_{sh} [c] (kΩcm^2)
Before air exposure treatment						
Ag	0.13	0.365	6.91	0.33 ± 0.01	9.19	1.70
MoOx	0.56	0.645	6.86	2.48 ± 0.2	6.85	123.0
P1	0.13	0.356	6.67	0.31 ± 0.01	9.17	2.26
P2	0.12	0.355	6.30	0.28 ± 0.01	8.21	1.37
P3	0.12	0.349	6.20	0.26 ± 0.01	9.99	0.79
After air exposure treatment (15 min)						
Ag	0.13	0.382	6.54	0.33 ± 0.01	9.86	1.71
MoOx	0.56	0.693	6.83	2.63 ± 0.12	6.87	174.9
P1	0.13	0.382	6.64	0.33 ± 0.01	9.94	3.49
P2	0.51	0.653	7.86	2.62 ± 0.15	7.20	31.03
P3	0.48	0.527	7.54	1.90 ± 0.2	9.42	64.46

[a] average values across 12 devices; [b] R_s are calculated from the light J–V curve inverse slope at voltages around the V_{OC}; [c] R_{sh} are calculated from the dark J–V curves at voltages around $V_{OC} = 0$.

The active layer (AL) coverage by the copolymer interlayers was analyzed by AFM. As shown in Figure 5, the surface morphology after **P1** deposition is similar to the AL substrate, thus suggesting a bad AL coverage by the **P1** interlayer. On the other hand, after **P2** and **P3** deposition, the morphology and roughness are significantly different. RMS values are reported in Figure 5 and are consistent with the presence of **P2** and **P3** on top of the active layer.

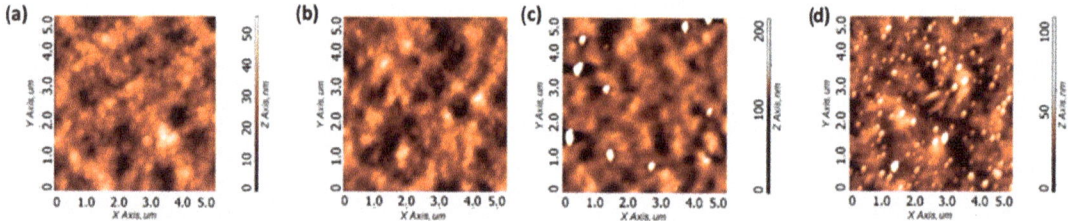

Figure 5. Surface topographic AFM images of (**a**) active layer; (**b**) active layer/**P1**; (**c**) active layer/**P2**; and (**d**) active layer/**P3** deposition. RMS values: 8.5 nm, 8.2 nm, 10.9 nm, 10.6 nm for (**a**–**d**), respectively. The active layer and CPE depositions were carried out in the devices assembly conditions.

To confirm the adhesion of our polar CPEs to the active layer, we measured the water contact angles prior to and after the interlayers deposition. As reported in the Supplementary Materials, Figure S12, the AL water contact angle reduces respectively from 107° to 96° and 102° after the **P2** and **P3** interlayers deposition. Such increase of hydrophilicity arises from the adhesion of the **P2** and **P3** polar polymers on top of the AL hydrophobic surface. Interestingly, by spin casting **P1** the water contact angle remained the same (at 107° as for the active layer) revealing a scarce/absent adhesion of **P1** on the active layer. This explains the identical PV characteristics observed for the **P1** device and the control Ag device.

The above observations highlight an important requirement that an AIL CPE for solution-processed inverted solar cells has to satisfy: a correct balance between the hydrophilic and hydrophobic parts. This ensures a good adhesion to the active layers while maintaining solution processing from orthogonal solvents, which is a key for devices multi-stacking fabrication.

As depicted in Table 2, the pristine Ag devices exhibit quite low V_{OC}, poor FF, relatively high R_s, and low R_p. Such features are the typical fingerprints of the bad hole selectivity of a silver electrode. As a matter of fact, the Ag work function was reported at −4.3 eV, while the P3HT HOMO energy level was around −5 eV. As such, there is a bad electrode-organic band alignment, leading to a barrier to hole extraction.

The insertion of the **P2** or **P3** interlayers in the devices, combined with an air soaking treatment (15 min), induces a significant enhancement of the V_{OC} and FF parameters as compared to the pristine Ag. This testifies the ability of the **P2** and **P3** interlayers to improve the hole collecting character of the top Ag electrode. As depicted, the V_{OC} and FF parameters of the **P2** and **P3** devices are reaching values close to the ones obtained with the state-of-the-art MoOx interlayer. This behavior evidences an overall effective modification of the silver electrode-active layer band alignment by the insertion of these copolymers interlayers. However, **P2** appears to be a more effective anode modifier than **P3**, as self-doping occurs in **P2** but not in **P3**. Self-doping should favor the hole transport within the **P2** interlayer, leading to beneficial effects for hole extraction in the devices. Besides self-doping, **P2** exhibits a better energy level alignment to the P3HT AL component than **P3**, see Figure 2. This reduces the barrier to hole collection at the AL/CPE interface when using **P2** rather than **P3** [25].

The above comparison of **P2** and **P3** AILs clearly highlights that the choice of the conjugated backbone is a key in the development of effective AILs for inverted solar cells. To favor hole collection at the top silver electrode, the conjugated backbone should be designed to grant self-doping and provide a good band energy alignment at the AL–AIL interface.

Interestingly, in direct geometry, **P3** did not function as anode interlayer. In inverted geometry, even if **P3** is a less effective AIL than **P2**, due to the mentioned absence of self-doping and poor band alignment the **P3** interlayer is able to induce a good hole collecting character to the top Ag electrode. Since both **P3** and **P2** have alkylsulfonate side groups, but different conjugated backbones, we deduce that the sulfonate side groups are playing a role in the anode engineering of the inverted solar cells.

It should be noticed that the ability of the sulfonate groups to impart a hole collection character to the top electrode could be not necessarily relate to this peculiar side group, but just to its inherent hydrophilicity. A hydrophilic material close to an Ag electrode upon air exposure may induce a shift from the vacuum of the Ag electrode work function via an oxidation mechanism [58].

In an attempt to clarify this issue, we have monitored, within a time scale of few days, the evolution of the J–V curves and PV characteristics of our devices versus their time of storage under air atmosphere. As depicted in Figure 6, with the **P2** and **P3** interlayers, most of the gain in hole selectivity of the top electrode is obtained rather quickly, at the very beginning of air exposure ("day 0" here corresponds to the previous 15 min air treatment). The Ag control device exhibits a completely different behavior upon aging under ambient atmosphere. At the very beginning of air exposure, the PV characteristics are not substantially affected. For longer times, a steady and continuous gain in hole selectivity of the top Ag electrode is observed. Such effect is ascribed to the formation of a thin oxide layer at the inner Ag surface, shifting the electrode work function to −5 eV [58]. It is known that this is a gradual process, driven by the slow diffusion of oxygen from the edges of the electrode [59]. For this reason, the gain in hole selectivity in the reference Ag devices is not completed in a few days [58,59]. By inserting the **P2** and **P3** interlayers, a faster process could be envisaged, owing to the hydrophilic nature of the CPEs which is attracting moisture at the buried silver. Moisture, in fact, provides a medium for adsorption of gases and the subsequent formation of an oxide layer [60,61]. However, a gradual oxidation process would be anyhow expected [18], which is by far different from the **P2** and **P3** trend depicted in Figure 6. For this reason, the formation of an Ag_2O layer is not enough to explain the functional behavior of the **P2** and **P3** anode interlayers. This was indeed confirmed by monitoring the **P2** device PV performances after 15 min storage under different ambient conditions, see the J–V curves in Figure 7 and corresponding PV parameters in the Supplementary Materials. We have found that a N_2 atmosphere, without oxygen but with a moisture content as in ambient air (50% RH), is identically effective for the PV performances as a standard air soaking treatment. Therefore, it is moisture rather than oxygen which plays a role in the **P2** and **P3** functional behavior. As shown in Figure 7b, by exposing the devices back to dry N_2 atmosphere and/or by drying them up by vacuum treatments, the anode modification is very stable and almost no reversibility is observed.

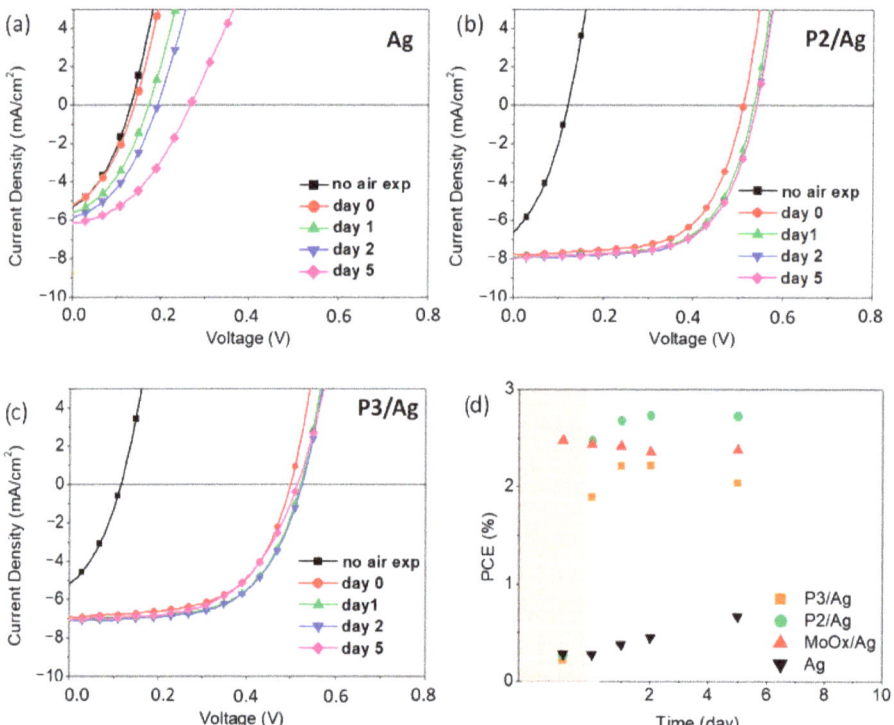

Figure 6. Evolution of the J–V curves and PCEs of devices with inverted geometry as a function of time of storage under ambient atmosphere under AM1.5G irradiation at 100 mW/cm². Top electrode: (**a**) pristine Ag; (**b**) **P2**/Ag; (**c**) **P3**/Ag; and (**d**) PCEs of pristine Ag, MoOx/Ag, **P2**/Ag, and **P3**/Ag; before air exposure: pink background; after air exposure: white background. The data at day 0 refers to the devices exposed to air for 15 min, similar to Figure 4c.

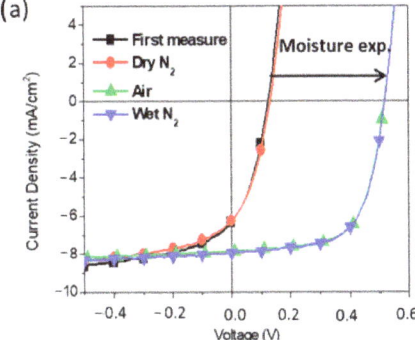 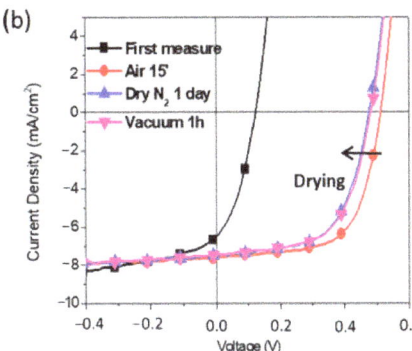

Figure 7. J–V curves of inverted devices stored for 15 min in different atmosphere conditions; (**a**) moisture exposure in air or in N_2 atmosphere with a similar moisture content to air (50 % RH): Air and wet N_2; first measure and dry N_2 correspond respectively to a first and a second curve recorded after 15 min, for a device kept inside the glove box. (**b**) Drying of a device exposed to air for 15 min upon storage in a glovebox overnight prior and after 1 h under vacuum (10^{-7} atm.): Air 15 min, dry N_2 overnight, 1 h vacuum; first measure is the J–V curve prior to air exposure.

To summarize, the above results indicate that the mechanism that explains the common ability of **P2** and **P3** in modifying the top anode electrode is related to the combined effect of the sulfonate side groups and water molecules. Note that in fact water is a proton source able to dope this class of polymers, as recently reported by Bazan and coworkers [51].

The importance of sulfonate groups in the engineering of the top Ag electrode was also confirmed by using a PTB7-Th-based active layer (see the Supplementary Materials). Here, the non-optimal alignment of the levels between the PTB7-Th and AIL led to PV poor performance, but similarly to P3HT-based devices, a quite relevant gain in hole selectivity was also observed.

According to the above discussion, it has been identified that the choice of the polar group is another extremely important factor in the design of the AILs. We infer that the sulfonate moieties not only assure the solubility in polar orthogonal solvents, inducing self-doping via a right choice of the conjugated backbone, but also play a role for the anode engineering in inverted solar cells.

However, a simple picture that may possibly explain this mechanism is proposed in the following Figure 8. In the presence of moisture/water, SO_3^- anions get partially solvated and therefore the ionic pair with the alkaline metal K^+ is weakened, gaining the freedom to better interact with the silver surface. Moisture may also facilitate the orientation of the sulfonate groups towards the metal interface by sweeping the voltage (electric field) in the device. As a result, dipoles oriented towards the inner silver electrode should be formed, shifting from vacuum the silver electrode work function. As a result, the hole collecting character of the top electrode is improved.

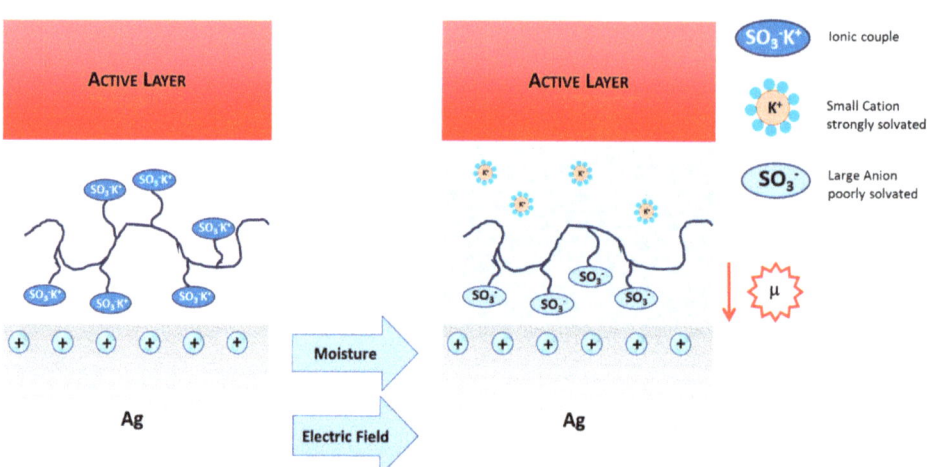

Figure 8. Proposed mechanism for moisture annealing.

4. Conclusions

In this work we have conducted a study about the unexplored application of sulfonate anionic CPE solution-processable anode interlayer materials in inverted organic photovoltaics. Based on the results obtained in our study it is possible to establish a good understanding of AIL structure–property–PV performance relationships for inverted devices applications. In fact, by designing and investigating the functional behavior of three different polymers **P1**, **P2**, and **P3** bearing a different number of sulfonate groups and modifications of their conjugated backbone, we could assess the important material features that should be taken in account for the development of effective AIL CPE materials for inverted OSC devices. First, it is mandatory to develop anionic CPEs with good wettability to the active layer and this can be achieved by the correct balancing of hydrophilic and hydrophobic substituents. Second, similarly to conventional direct geometry devices, the conjugate backbone should be suitably designed to ensure the self-doping of the CPE materials and grant a good energy level match with the photoactive layer. Finally, the choice of the polar groups is another important factor in the design of the AILs. The sulfonate moieties not only assure the solubility in polar orthogonal solvents, induce self-doping via a right choice of the conjugated backbone, but also play, combined with moisture exposure, a role for the anode engineering in inverted solar cells. Moisture annealing is a simple, easily accessible, and low cost procedure and could be a valuable alternative to electrical or thermal annealing. However a more comprehensive understanding of the importance of sulfonate for the self-doping mechanism of CPEs and its effect on charge transport and mobility is required to improve the use and design of this class of polymers. We believe that our insight could give a valuable contribution to the advancement in the development of engineering all polymeric solution-processable inverted solar cells.

Supplementary Materials: The following are available online, H-NMR spectra, GPC spectra, IR and UV-Vis absorption spectra, AFM images, contact angle, Kelvin probe and photovoltaic characterizations.

Author Contributions: E.L. and B.M.S. contributed equally to this work with the OPV preparation and characterization and the synthesis of materials respectively; G.S. performed AFM images; S.M.-S. and M.G. were responsible for EPR measurements; B.V. was responsible for CV; R.S. contributed to PV measurements; F.G. contributed to discussion of the results; M.P. and S.L. designed the idea plan, discussed the data and prepared the manuscript. All authors have read and agreed to the published version of the manuscript.

Funding: This work was carried out with the financial support of Regione Lombardia Project "Piattaforma tecnologica per lo sviluppo di sonde innovative in ambito biomedicale" (ID 244356).

Institutional Review Board Statement: Not applicable.

Informed Consent Statement: Not applicable

Data Availability Statement: Data is contained within the article and supplementary material.

Conflicts of Interest: The authors declare no conflict of interest.

Sample Availability: Samples of the compounds are not available from the authors.

References

1. Steim, R.; Kogler, F.R.; Brabec, C.J. Interface materials for organic solar cells. *J. Mater. Chem.* **2010**, *20*, 2499–2512. [CrossRef]
2. Yip, H.-L.; Jen, A.K.-Y. Recent advances in solution-processed interfacial materials for efficient and stable polymer solar cells. *Energy Environ. Sci.* **2012**, *5*, 5994–6011. [CrossRef]
3. Corzo, D.; Bihar, E.; Alexandre, E.B.; Rosas-Villalva, D.; Baran, D. Ink Engineering of Transport Layers for 9.5% Efficient All-Printed Semitransparent Nonfullerene Solar Cells. *Adv. Funct. Mater.* **2020**, 2005763. [CrossRef]
4. Chueh, C.-C.; Li, C.-Z.; Jen, A.K.-Y. Recent progress and perspective in solution-processed Interfacial materials for efficient and stable polymer and organometal perovskite solar cells. *Energy Environ. Sci.* **2015**, *8*, 1160–1189. [CrossRef]
5. Jiang, Y.; Peng, H.; Mai, R.; Meng, Y.; Rong, Q.; Cabanetos, C.; Nian, L.; Roncali, J.; Zhou, G.; Liu, J.; et al. Alcohol-soluble anode modifier for highly efficient inverted solar cells with oligo-oxyethylene chains. *Org. Electron.* **2019**, *68*, 200–204. [CrossRef]
6. Liu, Z.; Ouyang, X.; Peng, R.; Bai, Y.; Mi, D.; Jiang, W.; Facchetti, A.; Ge, Z. Efficient polymer solar cells based on the synergy effect of a novel non-conjugated small-molecule electrolyte and polar solvent. *J. Mater. Chem. A* **2016**, *4*, 2530–2536. [CrossRef]
7. Seo, J.H.; Gutacker, A.; Sun, Y.; Wu, H.-B.; Huang, F.; Cao, Y.; Scherf, U.; Heeger, A.J.; Bazan, G.C. Improved High-Efficiency Organic Solar Cells via Incorporation of a Conjugated Polyelectrolyte Interlayer. *J. Am. Chem. Soc.* **2011**, *133*, 8416–8419. [CrossRef]
8. He, Z.; Zhong, C.; Su, S.; Xu, M.; Wu, H.; Cao, Y. Enhanced power-conversion efficiency in polymer solar cells using an inverted device structure. *Nat. Photon.* **2012**, *6*, 591–595. [CrossRef]
9. Torimtubun, A.A.A.; Sánchez, J.G.; Pallarès, J.; Marsal, L.F. A cathode interface engineering approach for the comprehensive study of indoor performance enhancement in organic photovoltaics. *Sustain. Energy Fuels* **2020**, *4*, 3378–3387. [CrossRef]
10. Nam, M.; Baek, S.; Ko, D. Unraveling optimal interfacial conditions for highly efficient and reproducible organic photovoltaics under low light levels. *Appl. Surf. Sci.* **2020**, *526*, 146632. [CrossRef]
11. Oh, S.-H.; Na, S.-I.; Jo, J.; Lim, B.; Vak, D.; Kim, D.-Y. Water-Soluble Polyfluorenes as an Interfacial Layer Leading to Cathode-Independent High Performance of Organic Solar Cells. *Adv. Funct. Mater.* **2010**, *20*, 1977–1983. [CrossRef]
12. He, Z.; Zhong, C.; Huang, X.; Wong, W.-Y.; Wu, H.; Chen, L.; Su, S.; Cao, Y. Simultaneous Enhancement of Open-Circuit Voltage, Short-Circuit Current Density, and Fill Factor in Polymer Solar Cells. *Adv. Mater.* **2011**, *23*, 4636–4643. [CrossRef] [PubMed]
13. Lee, B.H.; Jung, I.H.; Woo, H.Y.; Shim, H.-K.; Kim, G.; Lee, K. Multi-Charged Conjugated Polyelectrolytes as a Versatile Work Function Modifier for Organic Electronic Devices. *Adv. Funct. Mater.* **2013**, *24*, 1100–1108. [CrossRef]
14. Hu, L.; Wu, F.; Li, C.; Hu, A.; Hu, X.; Zhang, Y.; Chen, L.; Chen, Y. Alcohol-Soluble n-Type Conjugated Polyelectrolyte as Electron Transport Layer for Polymer Solar Cells. *Macromolecules* **2015**, *48*, 5578–5586. [CrossRef]
15. Hu, Z.; Chen, Z.; Zhang, K.; Zheng, N.; Xie, R.; Liu, X.; Yang, X.; Huang, F.; Cao, Y. Self-Doped N-Type Water/Alcohol Soluble-Conjugated Polymers with Tailored Backbones and Polar Groups for Highly Efficient Polymer Solar Cells. *Sol. RRL* **2017**, *1*, 1700055. [CrossRef]
16. Kesters, J.; Ghoos, T.; Penxten, H.; Drijkoningen, J.; Vangerven, T.; Lyons, D.M.; Verreet, B.; Aernouts, T.; Lutsen, L.; Vanderzande, D.; et al. Imidazolium-Substituted Polythiophenes as Efficient Electron Transport Materials Improving Photovoltaic Performance. *Adv. Energy Mater.* **2013**, *3*, 1180–1185. [CrossRef]
17. Liu, Y.; Page, Z.A.; Russell, T.P.; Emrick, T. Finely Tuned Polymer Interlayers Enhance Solar Cell Efficiency. *Angew. Chem. Int. Ed.* **2015**, *54*, 11485–11489. [CrossRef]
18. Carulli, F.; Scavia, G.; Lassi, E.; Pasini, M.; Galeotti, F.; Brovelli, S.; Giovanella, U.; Luzzati, S. A bifunctional conjugated polyelectrolyte for the interfacial engineering of polymer solar cells. *J. Colloid Interface Sci.* **2019**, *538*, 611–619. [CrossRef]
19. Zhang, W.; Li, Y.; Zhu, L.; Liu, X.; Song, C.; Li, X.; Sun, X.; Zhang, W. A PTB7-based narrow band-gap conjugated polyelectrolyte as an efficient cathode interlayer in PTB7-based polymer solar cells. *Chem. Commun.* **2017**, *53*, 2005–2008. [CrossRef]
20. Carulli, F.; Mróz, W.; Lassi, E.; Sandionigi, C.; Squeo, B.M.; Meazza, L.; Scavia, G.; Luzzati, S.; Pasini, M.; Giovanella, U.; et al. Effect of the introduction of an alcohol-soluble conjugated polyelectrolyte as cathode interlayer in solution-processed organic light-emitting diodes and photovoltaic devices. *Chem. Pap.* **2018**, *72*, 1753–1759. [CrossRef]
21. Squeo, B.M.; Carulli, F.; Lassi, E.; Galeotti, F.; Giovanella, U.; Luzzati, S.; Pasini, M. Benzothiadiazole-based conjugated polyelectrolytes for interfacial engineering in optoelectronic devices. *Pure Appl. Chem.* **2019**, *91*, 477–488. [CrossRef]
22. Zhou, H.; Zhang, Y.; Mai, C.-K.; Collins, S.D.; Nguyen, T.; Bazan, G.C.; Heeger, A.J. Conductive Conjugated Polyelectrolyte as Hole-Transporting Layer for Organic Bulk Heterojunction Solar Cells. *Adv. Mater.* **2013**, *26*, 780–785. [CrossRef] [PubMed]

23. Lee, B.H.; Lee, J.-H.; Jeong, S.Y.; Park, S.B.; Lee, S.H.; Lee, K. Broad Work-Function Tunability of p-Type Conjugated Polyelectrolytes for Efficient Organic Solar Cells. *Adv. Energy Mater.* **2014**, *5*, 1401653. [CrossRef]
24. Cui, Y.; Jia, G.; Zhu, J.; Kang, Q.; Yao, H.; Lu, L.; Xu, B.; Hou, J. The Critical Role of Anode Work Function in Non-Fullerene Organic Solar Cells Unveiled by Counterion-Size-Controlled Self-Doping Conjugated Polymers. *Chem. Mater.* **2018**, *30*, 1078–1084. [CrossRef]
25. Cui, Y.; Xu, B.; Yang, B.; Yao, H.; Li, S.; Hou, J. A Novel pH Neutral Self-Doped Polymer for Anode Interfacial Layer in Efficient Polymer Solar Cells. *Macromolecules* **2016**, *49*, 8126–8133. [CrossRef]
26. Jo, J.W.; Jung, J.W.; Bae, S.; Ko, M.J.; Kim, H.; Jo, W.H.; Jen, A.K.-Y.; Son, H.J. Development of Self-Doped Conjugated Polyelectrolytes with Controlled Work Functions and Application to Hole Transport Layer Materials for High-Performance Organic Solar Cells. *Adv. Mater. Interfaces* **2016**, *3*, 1500703. [CrossRef]
27. Moon, S.; Khadtare, S.; Wong, M.; Han, S.-H.; Bazan, G.C.; Choi, H. Hole transport layer based on conjugated polyelectrolytes for polymer solar cells. *J. Colloid Interface Sci.* **2018**, *518*, 21–26. [CrossRef]
28. Xie, Q.; Zhang, J.; Xu, H.; Liao, X.; Chen, Y.; Li, Y.; Chen, L. Self-doped polymer with fluorinated phenylene as hole transport layer for efficient polymer solar cells. *Org. Electron.* **2018**, *61*, 207–214. [CrossRef]
29. Xu, H.; Zou, H.; Zhou, D.; Zeng, G.; Chen, L.; Liao, X.; Chen, Y. Printable Hole Transport Layer for 1.0 cm2 Organic Solar Cells. *ACS Appl. Mater. Interfaces* **2020**, *12*, 52028–52037. [CrossRef]
30. Xu, H.; Yuan, F.; Zhou, D.; Liao, X.; Chen, L.; Chen, Y. Hole transport layers for organic solar cells: Recent progress and prospects. *J. Mater. Chem. A* **2020**, *8*, 11478–11492. [CrossRef]
31. Zhou, H.; Zhang, Y.; Mai, C.-K.; Seifter, J.; Nguyen, T.; Bazan, G.C.; Heeger, A.J. Solution-Processed pH-Neutral Conjugated Polyelectrolyte Improves Interfacial Contact in Organic Solar Cells. *ACS Nano* **2014**, *9*, 371–377. [CrossRef] [PubMed]
32. Choi, H.; Mai, C.-K.; Kim, H.-B.; Jeong, J.; Song, S.; Bazan, G.C.; Kim, J.Y.; Heeger, A.J. Conjugated polyelectrolyte hole transport layer for inverted-type perovskite solar cells. *Nat. Commun.* **2015**, *6*, 7348. [CrossRef] [PubMed]
33. Patil, A.; Ikenoue, Y.; Basescu, N.; Colaneri, N.; Chen, J.; Wudl, F.; Heeger, A. Self-doped conducting polymers. *Synth. Met.* **1987**, *20*, 151–159. [CrossRef]
34. Mai, C.-K.; Zhou, H.; Zhang, Y.; Henson, Z.B.; Nguyen, T.-Q.; Heeger, A.J.; Bazan, G.C. Facile Doping of Anionic Narrow-Band-Gap Conjugated Polyelectrolytes During Dialysis. *Angew. Chem. Int. Ed.* **2013**, *52*, 12874–12878. [CrossRef] [PubMed]
35. Li, S.; Wan, L.; Chen, L.; Deng, C.; Tao, L.; Lu, Z.; Zhang, W.; Fang, J.; Song, W. Self-Doping a Hole-Transporting Layer Based on a Conjugated Polyelectrolyte Enables Efficient and Stable Inverted Perovskite Solar Cells. *ACS Appl. Energy Mater.* **2020**, *3*, 11724–11731. [CrossRef]
36. Xu, H.; Fu, X.; Cheng, X.; Huang, L.; Zhou, D.; Chen, L.; Chen, Y. Highly and homogeneously conductive conjugated polyelectrolyte hole transport layers for efficient organic solar cells. *J. Mater. Chem. A* **2017**, *5*, 14689–14696. [CrossRef]
37. Cui, Q.; Bazan, G.C. Narrow Band Gap Conjugated Polyelectrolytes. *Accounts Chem. Res.* **2018**, *51*, 202–211. [CrossRef]
38. Pace, G.; Tu, G.; Fratini, E.; Massip, S.; Huck, W.T.; Baglioni, P.; Friend, R.H. Poly(9,9-dioctylfluorene)-Based Conjugated Polyelectrolyte: Extended π-Electron Conjugation Induced by Complexation with a Surfactant Zwitterion. *Adv. Mater.* **2010**, *22*, 2073–2077. [CrossRef]
39. Zhu, X.; Xie, Y.; Li, X.; Qiao, X.; Wang, L.; Tu, G. Anionic conjugated polyelectrolyte–wetting properties with an emission layer and free ion migration when serving as a cathode interface layer in polymer light emitting diodes (PLEDs). *J. Mater. Chem.* **2012**, *22*, 15490. [CrossRef]
40. Stay, D.; Lonergan, M.C. Varying Anionic Functional Group Density in Sulfonate-Functionalized Polyfluorenes by a One-Phase Suzuki Polycondensation. *Macromolecules* **2013**, *46*, 4361–4369. [CrossRef]
41. Murugesan, V.; De Bettignies, R.; Mercier, R.; Guillerez, S.; Perrin, L. Synthesis and characterizations of benzotriazole based donor–acceptor copolymers for organic photovoltaic applications. *Synth. Met.* **2012**, *162*, 1037–1045. [CrossRef]
42. Pasini, M.; Destri, S.; Porzio, W.; Botta, C.; Giovanella, U. Electroluminescent poly(fluorene-co-thiophene-S,S-dioxide): Synthesis, characterisation and structure–property relationships. *J. Mater. Chem.* **2003**, *13*, 807–813. [CrossRef]
43. Tian, Y.; Kuzimenkova, M.V.; Halle, J.; Wojdyr, M.; Mendaza, A.D.D.Z.; Larsson, P.-O.; Müller, C.; Scheblykin, I.G. Molecular Weight Determination by Counting Molecules. *J. Phys. Chem. Lett.* **2015**, *6*, 923–927. [CrossRef] [PubMed]
44. Iosip, M.; Destri, S.; Pasini, M.; Porzio, W.; Pernstich, K.; Batlogg, B. New dithieno [3,2-b:2′,3′-d]thiophene oligomers as promising materials for organic field-effect transistor applications. *Synth. Met.* **2004**, *146*, 251–257. [CrossRef]
45. Vercelli, B.; Pasini, M.; Berlin, A.; Casado, J.; Navarrete, J.T.L.; Ortiz, R.P.; Zotti, G. Phenyl- and Thienyl-Ended Symmetric Azomethines and Azines as Model Compounds for n-Channel Organic Field-Effect Transistors: An Electrochemical and Computational Study. *J. Phys. Chem. C* **2014**, *118*, 3984–3993. [CrossRef]
46. Castelli, A.; Meinardi, F.; Pasini, M.; Galeotti, F.; Pinchetti, V.; Lorenzon, M.; Manna, L.; Moreels, I.; Giovanella, U.; Brovelli, S. High-Efficiency All-Solution-Processed Light-Emitting Diodes Based on Anisotropic Colloidal Heterostructures with Polar Polymer Injecting Layers. *Nano Lett.* **2015**, *15*, 5455–5464. [CrossRef] [PubMed]
47. Zalar, P.; Nguyen, T.-Q. Charge Injection Mechanism in PLEDs and Charge Transport in Conjugated Polyelectrolytes. In *Conjugated Polyelectrolytes*; John Wiley & Sons: Hoboken, NJ, USA, 2013; pp. 315–344.
48. Cho, N.S.; Hwang, D.-H.; Lee, J.I.; Jung, B.J.; Shim, H.-K. Synthesis and Color Tuning of New Fluorene-Based Copolymers. *Macromolecules* **2002**, *35*, 1224–1228. [CrossRef]

49. Prosa, M.; Benvenuti, E.; Pasini, M.; Giovanella, U.; Bolognesi, M.; Meazza, L.; Galeotti, F.; Muccini, M.; Toffanin, S. Organic Light-Emitting Transistors with Simultaneous Enhancement of Optical Power and External Quantum Efficiency via Conjugated Polar Polymer Interlayers. *ACS Appl. Mater. Interfaces* **2018**, *10*, 25580–25588. [CrossRef]
50. Mrakic-Sposta, S.; Gussoni, M.; Montorsi, M.; Porcelli, S.; Vezzoli, A. Assessment of a Standardized ROS Production Profile in Humans by Electron Paramagnetic Resonance. Available online: https://www.hindawi.com/journals/omcl/2012/973927/ (accessed on 18 January 2021).
51. Cao, D.X.; Leifert, D.; Brus, V.V.; Wong, M.S.; Phan, H.; Yurash, B.; Koch, N.; Bazan, G.C.; Nguyen, T.-Q. The importance of sulfonate to the self-doping mechanism of the water-soluble conjugated polyelectrolyte PCPDTBT-SO3K. *Mater. Chem. Front.* **2020**, *4*, 3556–3566. [CrossRef]
52. Etemad, S.; Pron, A.; Heeger, A.J.; MacDiarmid, A.G.; Mele, E.J.; Rice, M.J. Infrared-active vibrational modes of charged solitons in (CH)x and (CD)x. *Phys. Rev. B* **1981**, *23*, 5137–5141. [CrossRef]
53. Anderson, M.; Ramanan, C.; Fontanesi, C.; Frick, A.; Surana, S.; Cheyns, D.; Furno, M.; Keller, T.; Allard, S.; Scherf, U.; et al. Displacement of polarons by vibrational modes in doped conjugated polymers. *Phys. Rev. Mater.* **2017**, *1*, 055604. [CrossRef]
54. Ohno, K.; Mandai, Y.; Matsuura, H. Vibrational spectra and molecular conformation of taurine and its related compounds. *J. Mol. Struct.* **1992**, *268*, 41–50. [CrossRef]
55. Yeo, J.-S.; Kang, M.; Jung, Y.-S.; Kang, R.; Lee, S.-H.; Heo, Y.-J.; Jin, S.-H.; Kim, D.-Y.; Na, S.-I. In-depth considerations for better polyelectrolytes as interfacial materials in polymer solar cells. *Nano Energy* **2016**, *21*, 26–38. [CrossRef]
56. Zhou, Y.; Fuentes-Hernandez, C.; Shim, J.; Meyer, J.; Giordano, A.J.; Li, H.; Winget, P.; Papadopoulos, T.; Cheun, H.; Kim, J.; et al. A Universal Method to Produce Low-Work Function Electrodes for Organic Electronics. *Science* **2012**, *336*, 327–332. [CrossRef]
57. Giovanella, U.; Pasini, M.; Lorenzon, M.; Galeotti, F.; Lucchi, C.; Meinardi, F.; Luzzati, S.; Dubertret, B.; Brovelli, S. Efficient Solution-Processed Nanoplatelet-Based Light-Emitting Diodes with High Operational Stability in Air. *Nano Lett.* **2018**, *18*, 3441–3448. [CrossRef]
58. Lloyd, M.T.; Olson, D.C.; Lu, P.; Fang, E.; Moore, D.L.; White, M.S.; Reese, M.O.; Ginley, D.S.; Hsu, J.W.P. Impact of contact evolution on the shelf life of organic solar cells. *J. Mater. Chem.* **2009**, *19*, 7638–7642. [CrossRef]
59. Savva, A.; Burgués-Ceballos, I.; Papazoglou, G.; Choulis, S.A. High-Performance Inverted Organic Photovoltaics Without Hole-Selective Contact. *ACS Appl. Mater. Interfaces* **2015**, *7*, 24608–24615. [CrossRef]
60. Graedel, T. Corrosion Mechanisms for Silver Exposed to the Atmosphere. *J. Electrochem. Soc.* **1992**, *139*, 1963–1970. [CrossRef]
61. Yoon, Y.; Angel, J.D.; Hansen, D.C. Atmospheric Corrosion of Silver in Outdoor Environments and Modified Accelerated Corrosion Chambers. *Corrosion* **2016**, *72*, 1424–1432. [CrossRef]

Article

Molecular Engineering Enhances the Charge Carriers Transport in Wide Band-Gap Polymer Donors Based Polymer Solar Cells

Siyang Liu [1], Shuwang Yi [2], Peiling Qing [1], Weijun Li [1], Bin Gu [1], Zhicai He [2] and Bin Zhang [1,3,*]

1. School of Materials Science and Engineering, Baise University, Baise 533000, China; ray-lsy@hotmail.com (S.L.); plqing110@163.com (P.Q.); liweijun_bsu@163.com (W.L.); 18276637865@163.com (B.G.)
2. Institute of Polymer Optoelectronic Materials and Devices, State Key Laboratory of Luminescent Materials and Devices, South China University of Technology, Guangzhou 510640, China; shuwangyi399@sina.com (S.Y.); zhicaihe@scut.edu.cn (Z.H.)
3. Jiangsu Engineering Laboratory of Light-Electricity-Heat Energy-Converting Materials and Applications, School of Materials Science and Engineering, Changzhou University, Changzhou 213164, China
* Correspondence: msbinzhang@outlook.com; Tel.: +86-776-284-8131

Academic Editors: Tersilla Virgili and Mariacecilia Pasini
Received: 10 August 2020; Accepted: 4 September 2020; Published: 8 September 2020

Abstract: The novel and appropriate molecular design for polymer donors are playing an important role in realizing high-efficiency and high stable polymer solar cells (PSCs). In this work, four conjugated polymers (PIDT-O, PIDTT-O, PIDT-S and PIDTT-S) with indacenodithiophene (IDT) and indacenodithieno [3,2-b]thiophene (IDTT) as the donor units, and alkoxy-substituted benzoxadiazole and benzothiadiazole derivatives as the acceptor units have been designed and synthesized. Taking advantages of the molecular engineering on polymer backbones, these four polymers showed differently photophysical and photovoltaic properties. They exhibited wide optical bandgaps of 1.88, 1.87, 1.89 and 1.91 eV and quite impressive hole mobilities of 6.01×10^{-4}, 7.72×10^{-4}, 1.83×10^{-3}, and 1.29×10^{-3} cm^2 V^{-1} s^{-1} for PIDT-O, PIDTT-O, PIDT-S and PIDTT-S, respectively. Through the photovoltaic test via using PIDT-O, PIDTT-O, PIDT-S and PIDTT-S as donor materials and [6,6]-phenyl-C-71-butyric acid methyl ester (PC$_{71}$BM) as acceptor materials, all the PSCs presented the high open circuit voltages (V_{oc}s) over 0.85 V, whereas the PIDT-S and PIDTT-S based devices showed higher power conversion efficiencies (PCEs) of 5.09% and 4.43%, respectively. Interestingly, the solvent vapor annealing (SVA) treatment on active layers could improve the fill factors (FFs) extensively for these four polymers. For PIDT-S and PIDTT-S, the SVA process improved the FFs exceeding 71%, and ultimately the PCEs were increased to 6.05%, and 6.12%, respectively. Therefore, this kind of wide band-gap polymers are potentially candidates as efficient electron-donating materials for constructing high-performance PSCs.

Keywords: molecular engineering; polymer donors; high hole mobility; polymer solar cells

1. Introduction

The unique advantages of polymer solar cells (PSCs), such as mechanical flexibility, semi-transparency and large-scale production, mean they are the most promising next-generation photovoltaic technologies in the future [1]. Recently, the highest power conversion efficiency (PCE of >18% has been achieved, which is potentially possible for commercial application [2].

In order to optimize the generation of electricity in PSCs, various methods are used in PSCs, including the development of new active materials and device engineering. Generally, bulk-heterojunction (BHJ) structure is a widely utilized device technique, which comprises at least two types of semiconducting components in photoactive layer, where the electron donors (D) support the electrons, and electron acceptors (A) transport electrons. To obtain high-performance PSCs, the formation of interpenetrating networks between donor and acceptor domains, within nano-scale sizes and the connectivity among these two phases, deposit the significant influences on the charge carrier transport, recombination and collection in BHJ solar cell [3–5]. Furthermore, an electron donor is usually a conjugated polymer, while the acceptor is often a fullerene derivatives, small molecules and polymers. Among these materials, they take part in light absorption, exciton generation and exciton dissociation. Furthermore, they also play a role in transporting charge carriers through respective electrodes to the external circuit. In this regard, the development on the donors and acceptors are essential in adjusting the optical, electrical, and photovoltaic properties in PSCs.

At present, one of the important acceptors, used in PSCs, are fullerene derivatives, such as [6,6]-phenyl-C-61-butyric acid methyl ester ($PC_{61}BM$), [6,6]-phenyl-C-71-butyric acid methyl ester ($PC_{71}BM$), and 1′,1′′,4′,4′′-tetrahydro-di [1,4] methanonaphthaleno [5,6] fullerene-C60 (ICBA) [6]. Fullerenes and their derivatives exhibit ultrafast electron transfer from conjugated polymers by their versatile isotropic electron transport properties. This is due to their large conjugated spheres and electron-deficient centre. This characteristic can facilitate charge separation by delocalizing charges, low internal reorganization energy for electron transfer. The large spherical size of fullerenes increases tolerance to disorder and elimination of disturbances by the presence of donor polymer. In addition, it helps electrons find effective tunnel out of the mixed phase regions, and the entropy effects decrease the Coulomb barrier for charge separation [7,8]. Overall, fullerene and its derivatives are useful electron-accepting materials in achieving high-efficiency PSCs.

With the development of polymer donors, the PSCs combine a wide band-gap polymer donor and a narrow band-gap fullerene acceptor, and have great potential in achieving relatively good performance [9]. The design strategy for polymer donors that match with small-molecule acceptors requires: (1) Suitable energy levels, which could be finely adjusted by introducing heteroatoms or functional groups, and offering a channel for charge carriers transport among the small-molecule acceptors [10]; (2) the side chain or backbone modification provides moderate solubility, together with proper hole mobility, aggregation properties and prior molecular orientation for the optimal phase separation [11–13]; and (3) a broad and complementary absorption spectrum for harvesting much more sunlight. Hence, the proper design on molecular structure would realize the high-performance polymer donors for PSCs.

Molecular conformation has proven that polymer backbone planarity could affect energy levels, BHJ morphology, energetic disorder and charge transport dramatically. To control the molecular structure, noncovalent intramolecular interactions that favor a certain intramolecular conformation could be the suitable way for molecular structure design. In particular, sulphur-oxygen interaction is known to increase conjugated backbone planarity in a D-A copolymer, leading to high charge carrier mobility and device performance [14]. Instead of simply decreasing rotatable single bonds between units, the use of these conformational strategies may provide excellent control over organic semiconducting properties, and allow for more facile approaches to controlling molecular structure. For instance, Kim et al. introduced the ortho-hydrogen to promote a more planar conjugated backbone by sulphur-oxygen interactions, leading to the improvement on morphology and degradation [15]. Ma and co-workers optimized two BHJ polymers' structures to realize the high charge carrier mobility, long lifetime, and great free-carrier diffusion length [16]. Zhang et al. also achieved a high PCE of 9.0% using a conjugated polymer as donor, with a non-fullerene ITIC as the acceptor, with a remarkably low energy loss of 0.53 V, but without any negative impacts on the morphology of the blend films [17]. Liang et al. summarized that the silicon, germanium, sulfur and nitrogen as bridge atoms can change the degree of coplanarity between consecutive backbone units, and more effectively, by flattening the π-conjugated molecular framework to tailor the physical properties of IDT based p-type materials [18]. Currently, PSCs are focused on materials synthesis and device engineering, where both of these efforts are dedicated to further improve the photovoltaic performance. However, few investigations have been performed to explore the fundamental properties of polymer materials, which is extremely important and helpful in understanding the relationship between materials and device performance, as well as potentially providing guidelines to design novel materials and device architecture.

In this work, we designed and synthesized four planar D-A copolymers (PIDT-O, PIDTT-O, PIDT-S, PIDTT-S) based on the indacenodithiophene (IDT), indacenodithieno [3,2-b]thiophene (IDTT), alkoxy-substituted benzoxadiazole and benzothiadiazole derivatives [19–22]. These four polymers displayed wide band-gap properties with optical bandgap around 1.9 eV. Through the photovoltaic characterization via using these polymers as donors and $PC_{71}BM$ as acceptor, it was found that the alkoxy-substituted benzothiadiazole based polymers (PIDT-S and PIDTT-S) showed higher hole mobilities than the alkoxy-substituted benzoxadiazole based polymers (PIDT-O and PIDTT-O). Under the association of solvent annealing in device engineering, both of PIDT-S and PIDTT-S gave the final PCEs exceeding 6.0%, while the other two polymers of PIDT-O and PIDTT-O showed lower PCEs around 4%. We found that such an imbalance between photovoltaic performance and charge carrier transport properties can be attributed toward the hole mobilities and blend film morphologies. Therefore, it is believed that these results will be beneficial for understanding the charge transport-morphology-performance relationship for efficient and stable PSCs.

2. Results and Discussion

2.1. Synthesis

To understand the synthesis distinctly, the detailed synthetic routes of PIDT-O, PIDTT-O, PIDT-S and PIDTT-S are shown in Scheme 1. All polymers were synthesized by the general Stille polycondensation reaction in the presence of active $Pd_2(dba)_3$ as the catalyst and tri-o-tolylphopine as the ligand. All reactions were performed in 120 °C, and then the crude polymers were purified via the Soxhlet extraction through the methanol, acetone, hexane and chloroform, respectively. Finally, the final chloroform solution from Soxhlet extraction was precipitated in dry methanol again to attain the target polymers. All of the polymers displayed the red solid, and showed high solubility in the general organic solvents, such as dichloromethane, chloroform, tetrahydrofuran, toluene and chlorobenzene. In order to get the molecular weights of PIDT-O, PIDTT-O, PIDT-S and PIDTT-S, we used the Gel Permeation Chromatography (GPC) characterization with tetrahydrofuran as the eluent and polystyrene as the internal standards to test number-averaged molecular weights (M_ns) and weight-averaged molecular weights (M_ws). It was found that the M_ns and M_ws were 26,300, 43,700, 26,800, 49,400 and 45,000, 91,200, 53,400, 120,100, with the polydispersity index (PDIs) of 1.71, 2.09, 1.99 and 2.34 for PIDT-O, PIDTT-O, PIDT-S and PIDTT-S, respectively. (Table 1) From the test of molecular weights, we can see that all of the polymers show very high molecular weights, which are significantly beneficial for the solution-processible technique in PSCs.

Table 1. Molecular weights and thermal properties for the polymers.

Polymers	M_n	M_w	PDI	T_d (°C)
PIDT-O	26,300	45,000	1.71	337
PIDTT-O	43,700	91,200	2.09	343
PIDT-S	26,800	53,400	1.99	335
PIDTT-S	49,400	120,100	2.34	362

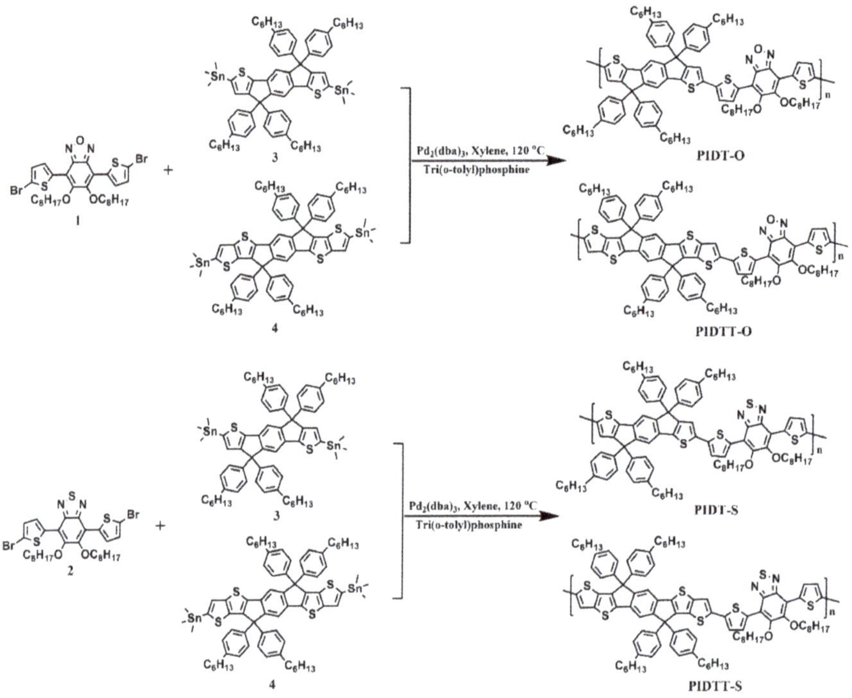

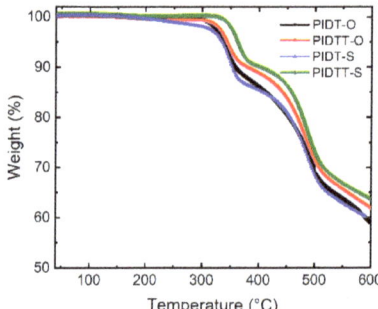

Scheme 1. Synthetic routes of the polymers.

2.2. Thermal Properties

To evaluate thermal stability, the thermal gravimetric analysis (TGA) under N_2 was used to study the thermal properties of PIDT-O, PIDTT-O, PIDT-S and PIDTT-S, and the related TG curves and data were presented in Figure 1 and Table 1, respectively. As shown in Figure 1, the degradation temperature (T_d) at 5% weight loss for PIDT-O, PIDTT-O, PIDT-S and PIDTT-S are 337, 343, 335, and 362 °C, respectively. It is noted that all of these four polymers display the very high thermal stability, which is potentially beneficial for the PSCs.

Figure 1. Thermal gravimetric analysis curves of PIDT-O, PIDTT-O, PIDT-S and PIDTT-S.

2.3. UV-vis Absorption and Electrochemical Properties

The UV-vis absorption spectra of PIDT-O, PIDTT-O, PIDT-S and PIDTT-S in chloroform and in solid film were shown in Figure 2a,b, and the corresponding data were summarized in Table 2. All of them display the similar absorption spectra with two major absorption bands. The absorption spectra in chloroform shows two peaks at 439 and 573 nm for PIDT-O, 453 and 574 nm for PIDTT-O, 447 and 556 nm for PIDT-S, and 453 and 553 nm for PIDTT-S, respectively. The absorption peaks in short wavelength (<450 nm) are assigned to the π-π* transition of the conjugated rigid polymers, whereas the absorption peaks in the range of 530–610 nm are attributed to the intramolecular charge transfer (ICT) transition [23]. In the film state, all the polymers show almost the same features of absorption as that in solution, except PIDT-O and PIDTT-O have one more shoulder peak at 600 and 598 nm, respectively. From the Figure 2b, the absorption spectra in the film state also show two distinct peaks ranged from 350 to 700 nm. Compared with the absorption in the solution, the absorption in the solid state gives the obvious red shift. This may be attributed to the stronger intermolecular interactions between the planar π-conjugated skeletons [24]. Accordingly, this interaction between conjugated backbones could influence solubility and miscibility of bulk-heterojunction blends in solution state, but also affect π-π stacking and crystallization in solid-state films. Finally, the PIDT-O, PIDTT-O, PIDT-S and PIDTT-S show the absorption onsets of 660, 661, 655 and 650 nm with the optical band gaps of 1.88, 1.87, 1.89, and 1.91 eV, respectively, which indicate that they are wide band-gap polymers.

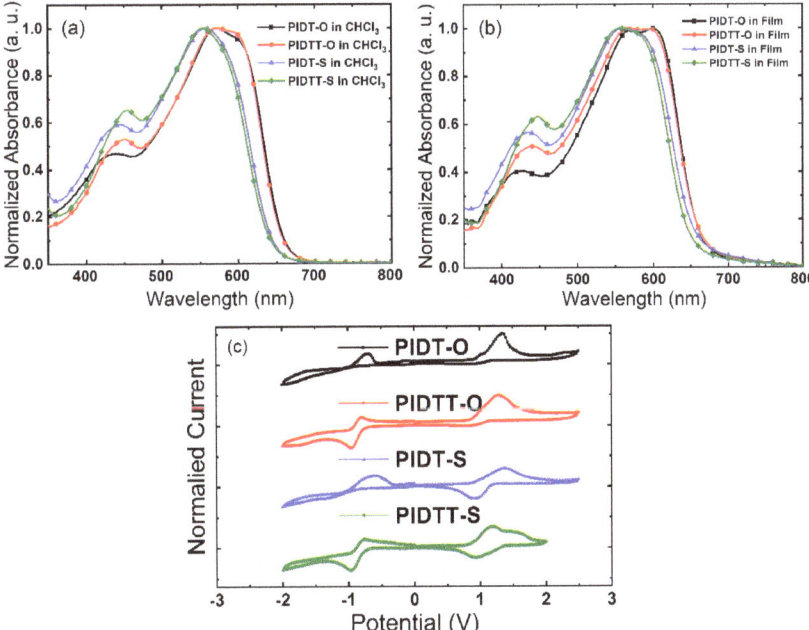

Figure 2. (a) UV-visible spectra in chloroform and (b) in film, (c) Cyclic voltammograms curves of PIDT-O, PIDTT-O, PIDT-S and PIDTT-S.

Table 2. Optical and electrochemical properties of PIDT-O, PIDTT-O, PIDT-S and PIDTT-S.

Polymers	λ_{abs}, Solution (nm)	λ_{abs}, Film (nm)	λ_{abs}, Onset (nm)	E_g, Opt (eV)	E_{ox} (eV)	E_{red} (eV)	E_{HOMO} (eV)	E_{LUMO} (eV)
PIDT-O	439,573	426,570,600	660	1.88	0.92	−0.85	−5.34	−3.57
PIDTT-O	453,574	438,564,598	661	1.87	0.84	−0.79	−5.26	−3.63
PIDT-S	447,556	436,558	655	1.89	0.87	−0.89	−5.29	−3.53
PIDTT-S	453,553	448,556	650	1.91	0.85	−0.78	−5.25	−3.64

The electrochemical properties of PIDT-O, PIDTT-O, PIDT-S and PIDTT-S were characterized by cyclic voltammetry (CV) in dry acetonitrile. The oxiditative/reductive potentials (E_{ox} and E_{red}) were calibrated from CV curves with the ferrocene/ferrocenium (Fc/Fc^+) as the internal standard and the tetra-n-butyl ammonium hexafluorophosphate (TBAF, 0.1 M) as the supporting electrolyte. The corresponding CV curves were shown in Figure 2c and the detailed data were summarized in Table 2. From Figure 2c, it displays the obviously reversible oxidation curves with E_{ox}s of 0.92 V for PIDT-O, 0.84 V for PIDTT-O, 0.87 V for PIDT-S, and 0.85 V for PIDTT-S, respectively. Furthermore, the reversible reduction curves with E_{red}s of −0.85 V for PIDT-O, −0.79 V for PIDTT-O, −0.89 V for PIDT-S, and −0.78 V for PIDTT-S, respectively, are also recorded. The highest occupied molecular orbital (HOMO) and the lowest unoccupied molecular orbital (LUMO) energy levels were calculated according to the empirical formula of $E_{HOMO} = -e(E_{ox} + 4.8 - E_{1/2, (Fc/Fc^+)})$ and $E_{LUMO} = -e(E_{red} + 4.8 - E_{1/2, (Fc/Fc^+)})$, where the $E_{1/2, (Fc/Fc+)}$ was recorded as 0.38 V. Therefore, the HOMOs of PIDT-O, PIDTT-O, PIDT-S and PIDTT-S are −5.34, −5.26, −5.29 and −5.25 eV and the LUMOs are −3.57, −3.63, −3.53 and −3.64 eV, respectively. Interestingly, the PIDTT-O and PIDTT-S have the similar LUMO levels, while the LUMO levels of PIDT-O and PIDT-S are also close to each other, because they contain the same IDTT and IDT units in the polymer backbones, respectively. In addition, the HOMO levels of these polymers have the same features. The PIDT-O and PIDT-S show the lower HOMOs than PIDTT-O and PIDTT-S. This is because that the IDTT unit in PIDT-O and PIDT-S exists the higher conjugation length than IDT unit in the PIDT-O and PIDT-S, which are in good agreement with the published results [19–23]. It is well-known that the open circuit voltage (V_{oc}) is related to the energy level difference between the donor's HOMO and the acceptor's LUMO [22]. Based on the HOMO energy levels of PIDT-O, PIDTT-O, PIDT-S and PIDTT-S here, it can be predicted that the photovoltaic performance in PSCs would lead to the high V_{oc} values.

2.4. Hole Mobilities

To study the charge-transport properties of the PIDT-O, PIDTT-O, PIDT-S and PIDTT-S, the space-charge-limited current (SCLC) method was performed to investigate thoroughly the hole mobility of neat polymers. The hole mobilities (μ_h) from Mott-Gurney equation was measured based on the hole-only devices with the devices structure of ITO/PEDOT: PSS/polymer donor (100 nm)/MoO$_3$/Al, where the current density–voltage (J-V) curves from hole-only devices were presented in Figure 3 and the related data were summarized in Table 3. As shown in Figure 3, it displays the typical J-V curves of the hole-only devices and the corresponding data are summarized in Table 2. It is found that these four polymer donors exhibit the μ_hs of 6.01×10^{-4} for PIDT-O, 7.72×10^{-4} for PIDTT-O, 1.83×10^{-3} for PIDT-S, and 1.29×10^{-3} cm^2 V^{-1} s^{-1} for PIDTT-S, respectively. It is clear that the PIDT-S and PIDTT-S based devices show the relatively higher hole mobilities than PIDT-O and PIDTT-O based counterparts. These higher hole mobility values in PIDT-S and PIDTT-S are possibly resulted from the stronger S-based noncovalent conformational interaction between polymer chains [18,25]. This result implies that the PIDT-S and PIDTT-S based bulk-heterojunction PSCs would support the better hole-transport performance and ultimately lead to the higher photovoltaic properties.

Table 3. Summary of the hole mobilities of the neat polymers.

Polymers	Thickness (nm)	Hole Mobility (cm^2 V^{-1} s^{-1})
PIDT-O	100	6.01×10^{-4}
PIDTT-O	100	7.72×10^{-4}
PIDT-S	100	1.83×10^{-3}
PIDTT-S	100	1.29×10^{-3}

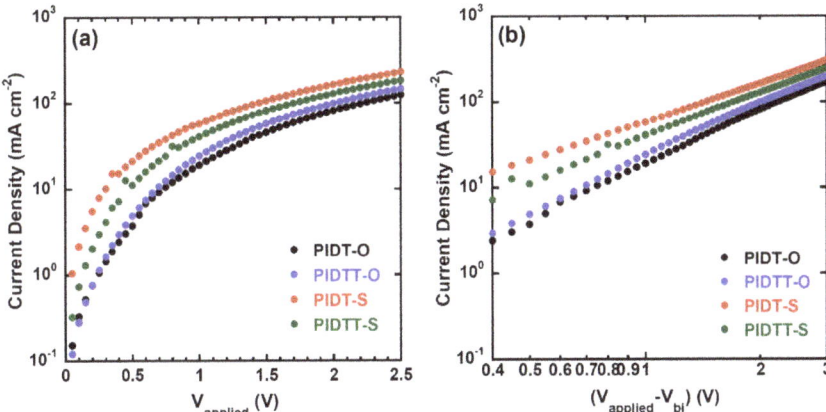

Figure 3. (a) The J-V curves of hole-only devices for neat polymers, and (b) fitting results to (a) from the SCLC model.

2.5. Photovoltaic Properties

To investigate the photovoltaic performance, the BHJ PSCs with PIDT-O, PIDTT-O, PIDT-S and PIDTT-S as electron donors were explored in detail, whereby the fullerene derivative $PC_{71}BM$ was used as the electron acceptor. The relatively conventional devices were assembled with a configuration of ITO/PEDOT: PSS (40 nm)/active layer (80–100 nm)/PFN (5 nm)/Al (100 nm) as presented in Figure 4a. The donor/acceptor (D/A) weight ratio was 1:2 for all the BHJ PSCs. The energy diagrams of the acceptor $PC_{71}CM$ was shown in Figure 4b with HOMO and LUMO levels of −5.9, and −33.9 eV, respectively. The difference of LUMOs between the donor and acceptor materials is near above 0.3 eV, which is sufficient for charge separation [26,27]. Through photovoltaic characterization, the J-V curves were recorded in Figure 4c and the corresponding data were summarized in Table 4. It is noted that the devices without any other post-treatment show lower PCEs of 2.41%, 2.89%, 5.09% and 4.43% for PIDT-O, PIDTT-O, PIDT-S, and PIDTT-S, respectively. Based on these devices, all of them give the high V_{oc}s over 0.85 V ascribing to their lower HOMO energy levels. Compared with the PIDT-O and PIDTT-O based devices, the ones based on PIDT-S and PIDTT-S display the higher PCEs, possibly resulting from their higher hole mobilities than PIDT-O and PIDTT-O. The higher hole mobilities would supply the better hole transport channel and suppress the negative charge carriers recombination in the BHJ systems.

Table 4. Photovoltaic performances of PSCs based on PIDT-O, PIDTT-O, PIDT-S and PIDTT-S under the illumination of AM 1.5 G, 100 mW cm^{-2}.

Polymers	V_{oc} (V)	J_{sc} (mA cm^{-2})	FF (%)	PCE (%)
PIDT-O	0.88	5.22	52.39	2.41
PIDT-O SVA [a]	0.88	7.22	64.83	4.12
PIDTT-O	0.85	7.56	45.02	2.89
PIDTT-O SVA	0.85	8.64	55.22	4.06
PIDT-S	0.88	9.58	60.42	5.09
PIDT-S SVA	0.86	9.76	72.04	6.05
PIDTT-S	0.88	8.89	56.66	4.43
PIDTT-S SVA	0.86	9.92	71.79	6.12

[a]: SVA means THF assisted solvent vapor annealing.

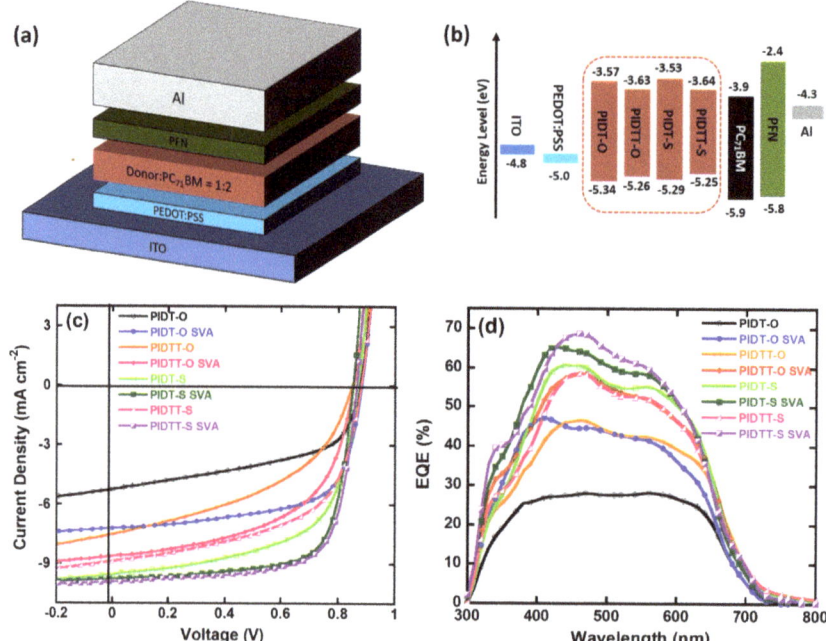

Figure 4. (**a**) The architecture of BHJ devices. (**b**) Energy level diagram of the photovoltaic materials. (**c**) J-V curves; and (**d**) the external quantum efficiency spectra of the BHJ photoactive OSCs.

Besides, in order to improve the photovoltaic performance, we used solvent vapor annealing (SVA, TFH) technique to promote the miscibility and molecular orientation in the active layer. As shown in Figure 4c and Table 4, after THF SVA, all of the PIDT-O, PIDTT-O, PIDT-S and PIDTT-S based devices exhibit the higher PCEs of 4.12%, 4.06%, 6.05%, and 6.12%, respectively. In particular, the fill factors (FFs) realize the greater increase from 52.39%, 45.02%, 60.42% and 56.66% to 64.83%, 55.22%, 72.04% and 71.79% for PIDT-O, PIDTT-O, PIDT-S, and PIDTT-S, respectively. We can see that the PIDT-S and PIDTT-S based devices show the better enhancement in FFs exceeding to 71%, which indicate that the nano-scale phase separation would be improved tremendously after THF SVA [28–30]. Interestingly, when the SVA was used, there is rare change in V_{oc}s, whereas the J_{sc} values realize a little improvement to 7.22, 8.64. 9.76, and 9.92 mA cm^{-2} for PIDT-O, PIDTT-O, PIDT-S and PIDTT-S, respectively.

The external quantum efficiency (EQE) spectra of the conventional and SVA optimized devices for PIDT-O, PIDTT-O, PIDT-S and PIDTT-S as polymer donors were shown in Figure 4d. The devices show the effective photo response in the range from 300 to 700 nm region, which is associated with the UV-vis absorption spectra in the neat films. It is clear that the absorption of donors plays a significant role in the photocurrent spectra. Compared with PIDT-O and PIDTT-O based devices, those of PIDT-S and PIDTT-S based PSCs present the stronger photocurrent responses from 300 to 700 nm, implying the more efficient charge separation and photo harvesting, which are consistent with the higher J_{sc}s and PCEs.

2.6. Atomic Force Microscopy Topographies

In the view of photovoltaic performance, based on PIDT-O, PIDTT-O, PIDT-S and PIDTT-S, it is obvious that the PIDT-S and PIDTT-S based devices exhibit the higher FF values than those based on PIDT-O and PIDTT-O. One of the reasons for these high FFs is possibly because the PIDT-S and PIDTT-S based blend films would form the better nano-scale phase separation than PIDT-O and PIDTT-O, and realize the optimal bi-continuous D/A phases. Hence, the morphology of the D/A phases is crucial to the performance of photovoltaic devices. Here, the atomic force microscopy (AFM) was employed to investigate the surface topographies of the active layers with the polymers as the donors and $PC_{71}BM$ as acceptor [31]. The AFM images of these blend films were shown in Figure 5. It illustrates that the PIDT-O and PIDTT-O based blend films show the very rougher surface with the root mean square (RMS) roughness values of 9.14 and 4.56 nm, respectively. Even though, the RMS roughness also stays very high for PIDT-O and PIDTT-O-based blend films after SVA. Comparably, the PIDT-S and PIDTT-S give the very flat topographies with the 1.22 and 0.45 nm, respectively. Based on the AFM topography characterization, we can see that the PIDT-O and PIDTT-O based active layers would form the vigorous phase separation and lead to lower photovoltaic performance, which is in good agreement with the results from the PSCs characterization.

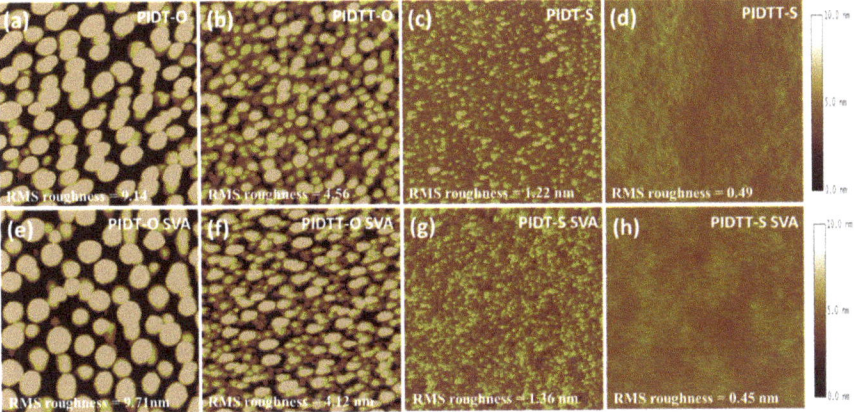

Figure 5. Surface topographic AFM images of polymer: $PC_{71}BM$ blend films (**a**) PIDT-O, (**b**) PIDT-O, (**c**) PIDT-S, (**d**) PIDTT-S without SVA treatment, and (**e**) PIDT-O, (**f**) PIDT-O, (**g**) PIDT-S, (**h**) PIDTT-S with SVA treatment.

3. Experimental Section

3.1. Characterization and Instrumentation

To characterize the polymer structure, the ^1H NMR spectra were tested by using Bruker (500 MHz) DRX spectrometer (Bruker, Karlsruhe, Germany) with tetramethylsilane (TMS) as the internal reference. Through utilizing the linear polystyrene (PS) as internal standards and tetrahydrofuran (THF, J&K Corp., Beijing, China) as eluent, the number-averaged molecular weight (M_n) and weight-averaged molecular weight (M_w) of the final polymers were measured on Waters gel permeation chromatography (GPC). The thermal gravimetric analysis (TGA) was performed on the TG 209 F3 Tarsus (NETZSCHC, Selb, Germany). Cyclic voltammetry (CV) measurement was employed on a PARSTAT2273 electrochemical workstation electrochemical workstation (Princeton Instruments, Trenton, NJ, United States) in the anhydrous acetonitrile under the nitrogen protection with a tetrabutylammonium hexafluorophosphates (Bu_4NPF_6, 0.1 mol L^{-1}) solution as electrolyte, accompanied by using a standard three electrodes cell with a Pt wire counter electrode, a platinum (Pt) working electrode, against saturated calomel electrode (SCE) as reference electrode. In the CV measurement, the ferrocene/ferrocenium (Fc/Fc^+) was utilized as the internal reference. To get the UV-vis absorption, it is performed on a SHIMADZU UV-2700 spectrophotometer (SHIMADZU, Kyoto, Japan). To obtain the topography images of polymer: [6,6]-phenyl-C71-butyric acid methyl ester ($PC_{71}BM$, purchased from ADS Corp., Cambria, CA, United States) based active layers, we used the atomic force microscopy (AFM) under the tapping-mode on a Veeco Nanoscope V scanning probe microscope to test topographies. The film thickness was measured on a Dektak XT step profiler (Bruker, Billerica, MA, United States).

3.2. Device Fabrication

In this work, the indium tin oxide (ITO)-coated glass (the size is 15 mm × 15 mm with the square resistance of 15 Ω) was used as the conductive substrate for fabricating PSCs. Prior to using the ITO substrate, it was treated by UV-ozone process Then, a layer (40 nm) of conductive polymer poly(3,4-ethylenedioxythiophene): poly(4-styrenesulfonate) (PEDOT: PSS) (Clevios 4083) was spin-coated onto ITO within 3000 rpm and then baked at 140 °C for 15 min in air. The active layer was prepared from the mixed solution containing polymers and $PC_{71}BM$ with the weight ratio of 1:2 in chlorobenzene, where the spin-coat speed was 1200 rpm during 40 s inside a glove-box with nitrogen. For the photovoltaic devices, the conventional device structure with ITO/PEDOT-PSS/active layer/PFN/Al was utilized, in which the PFN is a alcohol soluble polymer of poly[(9,9-dioctyl-2,7-fluorene)-alt-(9,9-bis (3′-(N,N-dimethylamino)propyl)-2,7-fluorene)]. Furthermore, a PFN layer (5 nm) was prepared above the active layer by spin-casting a mixed solution (0.2 mg mL^{-1}) in methanol solution with a trace of acetic acid. Finally, the cathode (80 nm) was prepared by thermally evaporating the aluminum under vacuum (~10^{-6} torr) with a shadow mask of 0.16 cm^2. To characterize the photovoltaic performance, the current-voltage (*J-V*) curves were tested on a Keithley 2400 multimeter under standard solar illumination (AM 1.5 G, 100 mW cm^{-2}). The external quantum efficiency (EQE) were measured by a monochromator under calibrating with a silicon photodiode. The hole mobility of neat polymer films was tested by space-charge-limited current (SCLC) method with the device configuration of ITO/PEDOT-PSS/neat polymers/MoO_3/Al.

3.3. Synthesis of Polymers

The starting monomers of 4,7-Bis(5-bromothiophen-2-yl)-5,6-bis(octyloxy)benzoxadiazole 1 and 4,7-bis(5-bromothiophen- 2-yl)-5,6-bis(octyloxy)benzothiadizole 2 were synthesized according to the our published literatures [22,23], where the indacenodithiophene (IDT, 3) and indacenodithieno [3,2-b] thiophene (IDTT, 4) derivatives were purchased commercially (Solarmer Corp., Beijing, China). The final polymers in this work were prepared by the following general Stille reaction.

3.3.1. General Procedure for Preparing Polymers

4,7-Bis(5-bromothiophen-2-yl)-5,6-bis (octyloxy) benzoxadiazole 1 or 4,7-Bis(5-bromothiophen-2-yl)-5,6-bis (octyloxy) benzothiadizole 2 (0.3 mmol), and indacenodithiophene (IDT, 3) or indacenodithieno [3,2-b]thiophene (IDTT, 4) derivatives (0.3 mmol), $Pd_2(dba)_3$ (11.0 mg, 0.012 mmol) (J&K Corp., Beijing, China) and tri-o-tolylphopine (18.3 mg, 0.06 mmol) (J&K Corp., Beijing, China)were dissolved in 6 mL xylene under nitrogen. The mixture was then heated to 120 °C and continued to react for 48 h. After 48 h, the solution was cooled to room temperature, and precipitated in methanol, respectively. The crude red polymers were then purified by using Soxhlet extraction in methanol, acetone, hexane, and chloroform, respectively. At last, the chloroform solution, comprising the target polymers, was precipitated in pure methanol, filtered off under vacuum and then dried at 60 °C in the vacuum overnight.

3.3.2. Poly(indacenodithiophene-alt-4,7-di(thiophen-2-yl)-5,c-bis(octyloxy) benzoxadiazole) (PIDT-O)

A red solid as target polymer was achieved with the yield of 510 mg (88.4%). ^1H NMR ($CDCl_3$, 500 MHz, δ): 8.46–8.38 (m, 2H, Ar-H), 7.43–7.37 (m, 2H, Ar-H), 7.31–7.28 (br, 2H, Ar-H), 7.23–7.08 (br, 18H, Ar-H), 4.18 (br, 4H, CH_2), 2.58–2.57 (m, 8H, CH_2), 2.07–2.01 (br, 4H, CH_2), 1.65–1.60 (br, 4H, CH_2), 1.50–1.30 (br, 48H, CH_2), 0.88–0.86 (m, 18H, CH_3). GPC (THF): M_n = 26,300, M_w = 45,000, PDI = 1.71.

3.3.3. Poly(indacenodithieno[3,2-b]thiophene-alt-4,7-di(thiophen-2-yl)-5,6-bis(octyloxy) benzoxadiazole) (PIDTT-O)

A red solid as target polymer was achieved with the yield of 470 mg (76.6%). ^{1}H NMR (CDCl$_3$, 500 MHz, δ): 8.47–8.40 (m, 2H, Ar-H), 7.53–7.47 (m, 4H, Ar-H), 7.34–7.28 (br, 4H, Ar-H), 7.23–7.05 (br, 14H, Ar-H), 4.20 (br, 4H, CH$_2$), 2.58 (br, 8H, CH$_2$), 2.04 (br, 4H, CH$_2$), 1.64–1.60 (br, 4H, CH$_2$), 1.55–1.29 (br, 48H, CH$_2$), 0.88–0.85 (m, 18H, CH$_3$). GPC (THF): M_n = 43,700, M_w = 91,200, PDI = 2.09.

3.3.4. Poly(indacenodithiophene-alt-4,7-di(thiophen-2-yl)-5,6-bis(octyloxy) benzothiadizole) (PIDT-S)

A red solid as target polymer was achieved with the yield of 450 mg (77.4%). ^{1}H NMR (CDCl$_3$, 500 MHz, δ): 8.56–8.48 (m, 2H, Ar-H), 7.47–7.37 (m, 2H, Ar-H), 7.30–7.28 (br, 2H, Ar-H), 7.23–7.07 (br, 18H, Ar-H), 4.14 (br, 4H, CH$_2$), 2.58–2.57 (m, 8H, CH$_2$), 2.04–1.96 (br, 4H, CH$_2$), 1.66–1.59 (br, 4H, CH$_2$), 1.48–1.29 (br, 48H, CH$_2$), 0.87–0.85 (m, 18H, CH$_3$). GPC (THF): M_n = 26,800, M_w = 53,400, PDI = 1.99.

3.3.5. Poly(indacenodithieno[3,2-b]thiophene-alt-4,7-di(thiophen-2-yl)-5,6-bis(octyloxy) benzothiadizole) (PIDTT-S)

A red solid as target polymer was achieved with the yield of 452 mg (73.1%). ^{1}H NMR (CDCl$_3$, 500 MHz, δ): 8.58–8.52 (m, 2H, Ar-H), 7.52–7.49 (m, 4H, Ar-H), 7.34–7.28 (br, 4H, Ar-H), 7.24–7.08 (br, 14H, Ar-H), 4.1 (br, 4H, CH$_2$), 2.58 (br, 8H, CH$_2$), 2.04–1.98 (br, 4H, CH$_2$), 1.64–1.60 (br, 4H, CH$_2$), 1.50–1.25 (br, 48H, CH$_2$), 0.87–0.85 (m, 18H, CH$_3$). GPC (THF): M_n = 49,400, M_w = 120,100, PDI = 2.43.

4. Conclusions

In summary, four wide band-gap polymers PIDT-O, PIDTT-O, PIDT-S and PIDTT-S were designed, synthesized and used as donor materials for fullerene-based BHJ PSCs. As a result of the introduction of planar IDT and IDTT units to the polymer main chains, the target polymers showed high hole mobilities, which would increase the intramolecular charge transfer from donor to acceptor phases, and thus, the charge transport in PSCs is enhanced effectively. All polymers exhibited the impressive hole mobility as high as 6.01×10^{-4}, 7.72×10^{-4}, 1.83×10^{-3} and 1.29×10^{-3} cm^2 V^{-1} s^{-1} for PIDT-O, PIDTT-O, PIDT-S and PIDTT-S, respectively. The sulfur-substituted (octyloxy)benzothiadizole derivatives (PIDT-S and PIDTT-S) presented the higher hole mobilities than oxygen-substituted (octyloxy)benzoxadiazole derivatives (PIDT-O and PIDTT-O). Through the photovoltaic characterization, the PIDT-S and PIDTT-S based PSCs gave the higher PCEs of 5.09% and 4.43% than PIDT-O and PIDTT-O, due to their higher hole mobilities. Interestingly, the SVA technique can improve the photovoltaic performance dramatically with the PCEs reaching 4.12%, 4.06%, 6.05% and 6.12% for PIDT-O, PIDTT-O, PIDT-S and PIDTT-S, respectively. This obvious enhancement on PCEs were mainly resulted from the distinct improvement on FF values, which suggested that the SVA is an effective methodology for improving photovoltaic properties. These results indicated that these wide band-gap polymers could be promising candidates for the fabrication of high-performance BHJ PSCs.

Author Contributions: B.Z. conceived and designed the experiments; S.L. and B.Z. synthesized and characterized the target polymers; S.Y. performed the experiments of polymer solar cells; P.Q., W.L., B.G., Z.H. and B.Z. analyzed the data; S.L. and B.Z. wrote the paper; All authors have read and agreed to the published version of the manuscript.

Funding: This work was financially supported by the Open Fund of Institute of Metal Powers of Baise University (no. 2020-bsu-03), Funds for the Construction of Master's Degree Granting Units in Guangxi Zhuang Autonomous Region, Guangdong International Science and Technology Cooperation Fund (no. 2020A0505100002) and Guangdong Special Support Program (No. 2017TQ04N559).

Conflicts of Interest: The authors declare no conflict of interest.

References

1. Brabec, C.J. Organic photovoltaics: Technology and market. *Sol. Energy Mater. Sol. Cells.* **2004**, *83*, 273–279. [CrossRef]

2. Liu, Q.; Jiang, Y.; Jin, K.; Qin, J.; Xu, J.; Li, W.; Xiong, J.; Liu, J.; Xiao, Z.; Sun, K.; et al. 18% Efficiency organic solar cells. *Sci. Bull.* **2020**, *65*, 272–275. [CrossRef]
3. Venkatesan, S.; Adhikari, N.; Chen, J.; Ngo, E.C.; Dubey, A.; Galipeau, D.W.; Qiao, Q. Interplay of nanoscale domain purity and size on charge transport and recombination dynamics in polymer solar cells. *Nanoscale* **2014**, *6*, 1011–1019. [CrossRef] [PubMed]
4. McDowell, C.; Abdelsamie, M.; Toney, M.F.; Bazan, G.C. Solvent additives: Key morphology-directing agents for solution-processed organic solar cells. *Adv. Mater.* **2018**, *30*, 1707114. [CrossRef]
5. Zhao, F.; Wang, C.; Zhan, X. Morphology control in organic solar cells. *Adv. Energy Mater.* **2018**, *8*, 1703147. [CrossRef]
6. Sieval, A.B.; Hummelen, J.C. Device physics and manufacturing technologies. In *Organic Photovoltaics: Materials*; Brabec, C., Scherf, U., Dyakonov, V., Eds.; Wiley-Vchweinheim: Weinheim, Germany, 2014; Volume 8, pp. 209–238.
7. Clarke, T.M.; Durrant, J.R. Charge photogeneration in organic solar cells. *Chem. Rev.* **2010**, *110*, 6736–6767. [CrossRef]
8. Gregg, B.A. Entropy of charge separation in organic photovoltaic cells: The benefit of higher dimensionality. *J. Phys. Chem. Lett.* **2011**, *2*, 3013–3015. [CrossRef]
9. Baran, D.; Kirchartz, T.; Wheeler, S.; Dimitrov, S.; Abdelsamie, M.; Gorman, J.; Ashraf, R.S.; Holliday, S.; Wadsworth, A.; Gasparini, N. Restricting the liquid-liquid phase separation of PTB7-Th:PF12TBT:PC71BM by enhanced PTB7-Th solution aggregation to optimize the interpenetrating network. *Adv. Mater.* **2017**, *139*, 17913–17922.
10. Cheng, P.; Li, G.; Zhan, X.; Yang, Y. Next-generation organic photovoltaics based on non-fullerene acceptors. *Nat. Photonics* **2018**, *12*, 131–142. [CrossRef]
11. Huang, B.; Chen, L.; Jin, X.; Chen, D.; An, Y.; Xie, Q.; Tan, Y.; Lei, H.; Chen, Y. Alkylsilyl functionalized copolymer donor for annealing-free high performance solar cells with over 11% efficiency: Crystallinity induced small driving force. *Adv. Funct. Mater.* **2018**, *28*, 1800606. [CrossRef]
12. Lin, Y.; Lu, Y.; Tsao, C.; Saeki, A.; Li, J.; Chen, C.; Wang, H.; Chen, H.; Meng, D.; Wu, K.; et al. Enhancing photovoltaic performance by tuning the domain sizes of a small-molecule acceptor by side-chain-engineered polymer donors. *J. Mater. Chem. A* **2019**, *7*, 3072–3082. [CrossRef]
13. Li, H.; Wu, Q.; Zhou, R.; Shi, Y.; Yang, C.; Zhang, Y.; Zhang, J.; Zou, W.; Deng, D.; Lu, K.; et al. Liquid-crystalline small molecules for nonfullerene solar cells with high fill factors and power conversion efficiencies. *Adv. Energy Mater.* **2019**, *9*, 1803175. [CrossRef]
14. Sariciftci, N.S.; Smilowitz, L.; Heeger, A.J.; Wudl, F. Photoinduced electron transfer from a conducting polymer to buckminsterfullerene. *Science* **1992**, *258*, 1474. [CrossRef] [PubMed]
15. Luke, J.; Speller, E.M.; Kim, J. Twist and degrade—Impact of molecular structure on the photostability of nonfullerene acceptors and their photovoltaic blends. *Adv. Energy Mater.* **2019**, *9*, 1803755. [CrossRef]
16. Ma, W.; Zhang, Q.; Feng, Y.; Larson, W.; Su, M.; Li, Y.; Yuan, J. Understanding the interplay of transport-morphology-performance in PBDB-T-based polymer solar cells. *Sol. RRL* **2020**, *4*, 1900524.
17. Zhang, J.; Jiang, K.; Yang, G.; Ma, T.; Liu, J.; Li, Z.; Lai, J.Y.L.; Ma, W.; Yan, H. Tuning energy levels without negatively affecting morphology: A promising approach to achieving optimal energetic match and efficient nonfullerene polymer solar cells. *Adv. Energy Mater.* **2017**, *7*, 1602119. [CrossRef]
18. Liang, C.; Wang, H. Indacenodithiophene-based D-A conjugated polymers for application in polymer solar cells. *Org. Electron.* **2017**, *50*, 443–457. [CrossRef]
19. Chueh, C.; Yao, K.; Yip, H.; Chang, C.; Xu, Y.; Chen, K.; Li, C.; Liu, P.; Huang, F.; Chen, Y.; et al. Non-halogenated solvents for environmentally friendly processing of high-performance bulk-heterojunction polymer solar cells. *Energy Environ. Sci.* **2013**, *6*, 3241–3248. [CrossRef]
20. Xu, Y.; Chueh, C.; Yip, H.; Chang, C.; Liang, P.; Intemann, J.J.; Chenb, W.; Alex, K.Y. Indacenodithieno[3,2-b]thiophene-based broad bandgap polymers for high efficiency polymer solar cells. *Polym. Chem.* **2013**, *4*, 5220–5223. [CrossRef]
21. Cai, Y.; Zhang, X.; Xue, X.; Wei, D.; Huo, L.; Sun, Y. High-performance wide-bandgap copolymers based on indacenodithiophene and indacenodithieno[3,2-b]thiophene units. *J. Mater. Chem. C* **2017**, *5*, 7777–7783. [CrossRef]
22. Zhang, B.; Hu, X.; Wang, M.; Xiao, H.; Gong, X.; Yang, W.; Cao, Y. Highly efficient polymer solar cells based on poly (carbazole-alt-thiophene-benzofurazan). *New J. Chem.* **2012**, *36*, 2042–2047. [CrossRef]

23. Zhang, B.; Yu, L.; Fan, L.; Wang, N.; Hu, L.; Yang, W. Indolo[3,2-b]carbazole and benzofurazan based narrow band-gap polymers for photovoltaic cells. *New J. Chem.* **2014**, *38*, 4587–4593. [CrossRef]
24. Meng, D.; Fu, H.; Xiao, C.; Meng, X.; Winands, T.; Ma, W.; Wei, W.; Fan, B.; Huo, L.; Doltsinis, N.L.; et al. Three-bladed rylene propellers with three- dimensional network assembly for organic electronics. *J. Am. Chem. Soc.* **2016**, *138*, 10184–10190. [CrossRef] [PubMed]
25. Huang, H.; Yang, L.; Facchetti, A.; Marks, T.J. Organic and polymeric semiconductors enhanced by noncovalent conformational locks. *Chem. Rev.* **2017**, *117*, 10291–10318. [CrossRef]
26. Brabec, C.J.; Winder, C.; Sariciftci, N.S.; Hummelen, J.C.; Dhanabalan, A.; van Hal, P.A.; Janssen, R.A.J. A low-bandgap semiconducting polymer for photovoltaic devices and infrared emitting diodes. *Adv. Funct. Mater.* **2002**, *12*, 709–712. [CrossRef]
27. Hou, J.; Inganas, O.; Friend, R.H.; Gao, F. Organic solar cells based on non-fullerene acceptors. *Nat. Mater.* **2018**, *17*, 119–128. [CrossRef]
28. Wang, K.; Azouz, M.; Babics, M.; Cruciani, F.; Marszalek, T.; Saleem, Q.; Pisula, W.; Beaujuge, P.M. Solvent annealing effects in dithieno[3,2-b:2′,3′-d]pyrrole-5,6-Difluorobenzo[c][1,2,5]thiadiazole small molecule donors for bulkheterojunction solar cells. *Chem. Mater.* **2016**, *28*, 5415–5425. [CrossRef]
29. Babics, M.; Liang, R.; Wang, K.; Cruciani, F.; Kan, Z.; Wohlfahrt, M.; Tang, M.; Laquai, F.; Beaujuge, P.M. Solvent vapor annealing-mediated crystallization directs charge generation, recombination and extraction in BHJ solar cells. *Chem. Mater.* **2018**, *30*, 789–798. [CrossRef]
30. Zhang, S.; Zhang, J.; Abdelsamie, M.; Shi, Q.; Zhang, Y.; Parker, T.C.; Jucov, E.V.; Timofeeva, T.V.; Amassian, A.; Bazan, G.C.; et al. Intermediate-sized conjugated donor molecules for organic solar cells: Comparison of benzodithiophene and benzobisthiazole-based cores. *Chem. Mater.* **2017**, *29*, 7880–7887. [CrossRef]
31. Yi, S.; Deng, W.; Sun, S.; Lan, L.; He, Z.; Yang, W.; Zhang, B. Trifluoromethyl-substituted large band-gap polytriphenylamines for polymer solar cells with high open-circuit voltages. *Polymers* **2018**, *10*, 52. [CrossRef]

Sample Availability: Samples of the PIDT-O, PIDTT-O, PIDT-S and PIDTT-S are available from the authors.

© 2020 by the authors. Licensee MDPI, Basel, Switzerland. This article is an open access article distributed under the terms and conditions of the Creative Commons Attribution (CC BY) license (http://creativecommons.org/licenses/by/4.0/).

Article

Innovative Green Chemistry Approach to Synthesis of Sn^{2+}-Metal Complex and Design of Polymer Composites with Small Optical Band Gaps

Shujahadeen B. Aziz [1,2,*], Muaffaq M. Nofal [3], Mohamad A. Brza [4], Niyaz M. Sadiq [1], Elham M. A. Dannoun [5], Khayal K. Ahmed [1], Sameerah I. Al-Saeedi [6], Sarkawt A. Hussen [1] and Ahang M. Hussein [1]

[1] Hameed Majid Advanced Polymeric Materials Research Lab., Physics Department, College of Science, University of Sulaimani, Qlyasan Street, Sulaimani 46001, Kurdistan Regional Government, Iraq; niyaz.sadiq@univsul.edu.iq (N.M.S.); khayal.ahmed@univsul.edu.iq (K.K.A.); sarkawt.hussen@univsul.edu.iq (S.A.H.); ahang.hussein@univsul.edu.iq (A.M.H.)
[2] Department of Civil Engineering, College of Engineering, Komar University of Science and Technology, Sulaimani 46001, Kurdistan Regional Government, Iraq
[3] Department of Mathematics and Science, Prince Sultan University, P.O. Box 66833, Riyadh 11586, Saudi Arabia; muaffaqnofal69@gmail.com
[4] Medical Physics Department, College of Medicals & Applied Science, Charmo University, Chamchamal, Sulaimania 46023, Iraq; mohamad.brza@gmail.com
[5] Department of Mathematics and Science, Woman Campus, Prince Sultan University, P.O. Box 66833, Riyadh 11586, Saudi Arabia; elhamdannoun1977@gmail.com
[6] Department of Chemistry, College of Science, Princess Nourah Bint Abdulrahman University, P.O. Box 84428, Riyadh 11671, Saudi Arabia; sialsaeedi@pnu.edu.sa
* Correspondence: shujahadeenaziz@gmail.com

Abstract: In this work, the green method was used to synthesize Sn^{2+}-metal complex by polyphenols (PPHs) of black tea (BT). The formation of Sn^{2+}-PPHs metal complex was confirmed through UV-Vis and FTIR methods. The FTIR method shows that BT contains NH and OH functional groups, conjugated double bonds, and PPHs which are important to create the Sn^{2+}-metal complexes. The synthesized Sn^{2+}-PPHs metal complex was used successfully to decrease the optical energy band gap of PVA polymer. XRD method showed that the amorphous phase increased with increasing the metal complexes. The FTIR and XRD analysis show the complex formation between Sn^{2+}-PPHs metal complex and PVA polymer. The enhancement in the optical properties of PVA was evidenced via UV-visible spectroscopy method. When Sn^{2+}-PPHs metal complex was loaded to PVA, the refractive index and dielectric constant were improved. In addition, the absorption edge was also decreased to lower photon. The optical energy band gap decreases from 6.4 to 1.8 eV for PVAloaded with 30% (v/v) Sn^{2+}-PPHs metal complex. The variations of dielectric constant versus wavelength of photon are examined to measure localized charge density (N/m^*) and high frequency dielectric constant. By increasing Sn^{2+}-PPHs metal complex, the N/m^* are improved from 3.65×10^{55} to 13.38×10^{55} m^{-3} Kg^{-1}. The oscillator dispersion energy (E_d) and average oscillator energy (E_o) are measured. The electronic transition natures in composite films are determined based on the Tauc's method, whereas close examinations of the dielectric loss parameter are also held to measure the energy band gap.

Keywords: Sn^{2+}-PPHs metal complex; UV-Vis; XRD and FTIR analyses; optical property; bandgap analysis

1. Introduction

According to a recent study, the optical properties of polymer composites (PCs) have piqued the interest of a lot of academics because of their extensive application in a variety of sectors, including solar cells, optoelectronic device and light-emitting diode (LED) [1,2].

Inorganic particles are commonly found in polymers, which are thought to be an outstanding host material. Studies on the optical characteristics of PVA based on metal complexes have been conducted in the literature [3,4].The insertion of inorganic particles into the host polymer might result in a significant alteration in the host's characteristics due to their high surface to bulk ratio [5].Green techniques have been widely reported as potential approaches for the synthesis of inorganic particles, with the results being shown to be safe and environmentally benign [6].

Alkaloids, amino acids, catechins, theavins, isomers of theavins, and other elements make polyphenols (PPHs) in black tea (BT). The most obvious molecular or chemical structures of the ingredients of BT have also been described in other investigations [7,8]. Dryan et al. [8] recently published a study which discovered that PPHs components are abundant in the BT aqueous mixture. PPHs conjugate, PPHs, and polymerized phenolic structure are the key elements of BT. In addition, black, green, and white tea all have a unique blend of conjugated flavonoids [9]. In earlier researches, it was found that extract solutions of black and green tea play a key role in lowering the polar polymers optical band gap including PMMA and PVA [10,11]. The extract tea solution contained PPHs, carboxylic acid groups, and hydroxyl group, according to the FTIR study [11]. As a result of the discoveries of experimental studies, the tea extract solution contains a large number of active ligands and functional groups, which are essential for complex formation with polymers and/or transition metal salts.

As a green technique, BT plant extract solutions can be utilized to synthesize Sn^{2+}-PPHs metal complexes. These solutions are high in PPHs, which have a significant interaction with the Sn^{2+} ion, forming a Sn^{2+}-PPHs metal complex. Zielinski et al. described the primary ingredients and uses of tea leaves, for instance, PPHs and caffeine [12].

Earlier research has shown that functional groups and PPHs in tea extract solutions can capture the cations of heavy metals to crate metal complexes [3,4]. Metal complexes are combined with PVA polymer to create PCs with high-performance optical characteristics in the current work. This process is a new green technique to make PCs with adjustable optical band gaps. Electrical and optical properties of polymers have attracted researchers' interest in recent years due to their widespread application in optical systems and their superior interference, reflection, anti-reflection, and polarization capabilities [13].In recent studies, it has been discovered that PCs with low band gap energy (E_g) and large absorption play a key role in photonics and optoelectronic device applications [14]. Hasan et al. [15] reported that the use of nanotube-PCs in photonics is due to the composites' good optical absorptions, which cover a wide spectrum range from UV to near IR [15]. Organic–inorganic hybrid (PCs) serve as an active or passive layer in optoelectronic devices for instance large refractive index films, protective coatings, thin films, LEDs, solar cells, transistors, and waveguide materials play a vital role in various applications [16].

The goal of this research is to create PCs with a low energy bandgap (E_g). Because of the good optical properties, the green method might be used to make PCs with low E_g. The findings of this research can be regarded as a novel PCs approach. The optical dielectric function was accurately used in this study to experimentally detect the different types of optical transition between the conduction band and the valence band. Sn^{2+}-polyphenol complex has a strong effect on the decrease of optical band gap in comparison with the other fillers for example nanoparticles (NPs). Aziz et al. [17] prepared PCs based on polystyrene. In their research, copper (Cu) powder was loaded into the polystyrene from 0 to 6 wt.%. Upon the incorporation of 6 wt.% Cu, the E_g decreased from 4.05 to 3.65 eV. Aziz et al. [18], in another work, added copper monosulfide (CuS) NPs into methyl cellulose (MC) polymer to prepare polymer nanocomposites based on MC. The E_g of MC decreased from 6.2 to 2.3 eV by the incorporation of 0.08 M of CuS NPs. In the current work, we observed that the E_g decreased from 6.4 to 1.8 eV for PVA loaded with 30% (v/v) of Sn^{2+}-polyphenol complex. Thus, based on the band gap analysis result, the green method is an appropriate for fabricating PCs with low value of E_g.

2. Methodology

2.1. Materials

Sigma-Aldrich provided PVA powder (MW ranging from 85,000 to 124,000) and Tin(II) chloride ($SnCl_2$) (MW = 189.6 g/mol). The BT leaf was bought from a nearby market.

2.2. Sample Preparation

The use of distilled water (D.W.) in the extraction of tea leaves is required. The steps are as follows: In the absence of sunshine, 50 g of BT leaf was placed in 250 mL D.W. at almost 90 °C. The resultant extract solution was filtered by (Whatman paper 41, cat. no. 1441) with a pore radius (20 µm) to thoroughly remove the residues after standing for 10 min. 200 mL HCl was diluted into 400 mL of D.W. and then used it to dissolve 10 g of $SnCl_2$ in a separate flask. The Sn^{2+}-PPHs metal complex was then made by adding $SnCl_2$ solution to the extract tea leaf solution and stirring for 10 min at 80 °C. The complexation between Sn^{2+}-metal ions and PPHs was confirmed by the color change of the extract solution from dark to green at the top of the beaker and formation of sediment as clouds at the bottom of the beaker. The complex solution was allowed to cool to room temperature. These complexes were detached in 100 mL of D.W. after numerous washings of the Sn^{2+}-PPHs metal complexes with D.W. The solution cast approach was used to produce composite samples made up of PVA loaded with Sn^{2+}-PPHs metal complex. To begin, a PVA solution was made by adding 1 g of PVA to 40 mL of D.W., stirring for 1 h at roughly 80 °C, then cooling to room temperature. Different volumes of the complex solution, ranging from 0 to 30% (v/v), were added to the homogenous PVA solution in 15% (v/v) increments. The resulting solutions were stirred for approximately 50 min. PVSN0, PVSN1 and PVSN2 were used to represent 0% (v/v), 15% (v/v) and 30% (v/v) of the loaded complex solution, respectively. To cast the manufactured films, the contents of the mixture were poured into petri dishes and allowed to dry at ambient temperature. The samples were dried more using blue silica gel desiccant prior to characterization. Pure PVA and composite films have thicknesses ranging from 0.012 to 0.015 cm. A pictorial sample preparation of PCs consists of Sn^{2+}-PPHs metal complex and PVA is shown in Scheme 1.

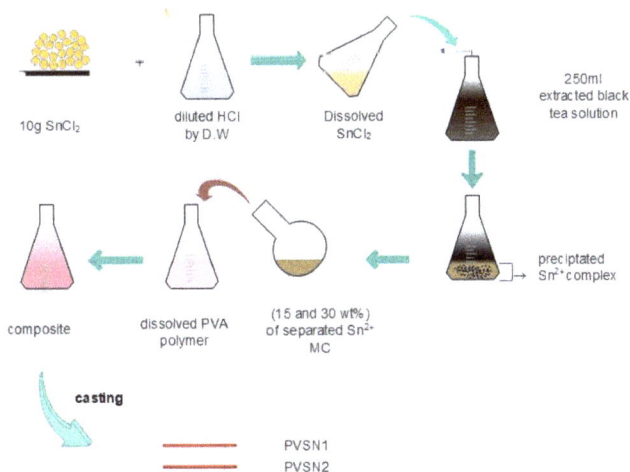

Scheme 1. Schematic diagram of sample preparation.

2.3. Measurement Techniques

X-ray diffraction (XRD) patterns were analyzed at room temperature using a Bruker AXS diffractometer (Billerica, MA, USA) with a 40-kV voltage and 45-mA current. The composite films were examined using a Nicolet iS10 FTIR spectrophotometer (Perkin Elmer,

Yokohama, Japan) with a resolution of 2 cm^{-1} in the range of 450 and 4000 cm^{-1}. A Jasco V-570 UV-vis-NIR spectrophotometer (JASCO, Tokyo, Japan) was used to record the samples UV-vis absorption spectra. For measuring UV-Vis for the liquid samples (Sn^{2+}-PPHs complexes), firstly, two cuvettes filled with distilled water were used for correcting background and then one of the cuvettes was removed while another cuvette was left and used as a reference sample. The absorbance of the liquid samples (Sn^{2+}-PPHs complexes) was measured in comparison to the reference sample. For measuring FTIR for the liquid samples (Sn^{2+}-PPHs complex), Sn^{2+}-PPHs complexes were coated on the standard glass slides and then dried at room temperature until evaporated. The dried Sn^{2+}-PPHs complexes were scratched on the glass slides to create powders form. Then potassium bromide (KBr) (100 mg) was added to the Sn^{2+}-PPHs (1 mg) powders and then the powders were combined in a mortar and finally turned to pellets in a sample holder.

When Sn^{2+}-PPHs complexes were added to the dissolved PVA, PVA composite films were created. For measuring UV-Vis and FTIR for the solid composite films, firstly the UV-Vis spectroscopy and FTIR devices with air and without any samples were corrected for background and then the UV-Vis and FTIR spectra were measure for the composite films.

3. Results and Discussion

3.1. UV-Vis and FTIR Study of Sn^{2+}-PPHs Metal Complex

It's worth noting that coordination chemistry involves complicated coordinated systems, complex molecules, or simply complexes. The lights and empty orbital metallic core that are coordinated by donors of electron pairs are examples of coordination compounds [19]. Coordination chemistry produces metal complexes that have a significant absorption characteristic in visible areas.

Figure 1 shows the absorption spectrum of the complicated colloidal suspension (Sn^{2+}-PPHs metal complex), which is equivalent to that of organometallic-based materials and semiconductors [20]. The absorption spectrum is notable for covering the whole visible range. The Sn^{2+}-complex displays absorption even at high wavelength ranges to near-infrared, as shown in the inset of Figure 1, meaning that increase the optical absorption and light harvesting. Such absorption of a broader spectrum of solar radiation shows the use of such material for various applications. The current UV-Vis data for metal-PPHs complexes generated using green methods are similar to those shown by other studies [21].

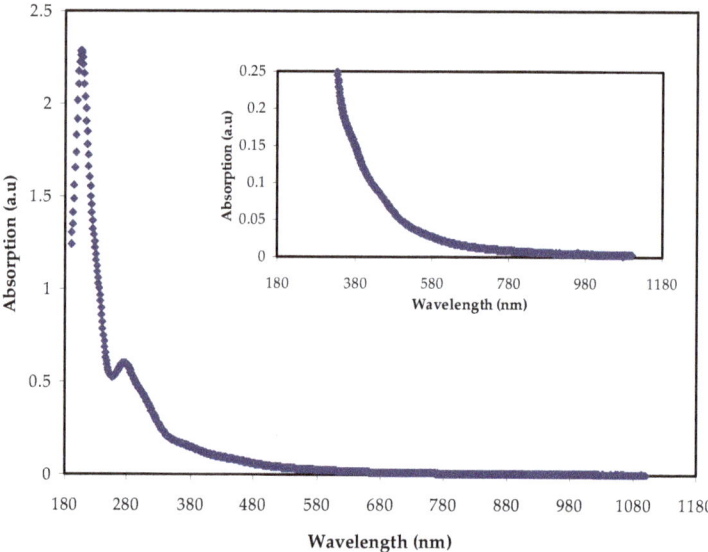

Figure 1. UV-visabsorption spectrum for Sn^{2+}-PPHs metal complex.

The transition of electron of n–π* of methylxanthines, catechins, and caffeine emerges as an absorbance band between 200 and 350 nm. The C=O chromophore in caffeine has a band absorbance of 278 nm [22,23]. Surface plasmon resonance (SPR) absorption band in UV-visible range is required for metallic's with diameters in the nano range [24]. Nevertheless, the lack of this band in the current Sn^{2+}-PPHs complex suggests that the PPHs capping inhibited the complex system's metal properties from forming on particle surfaces. Cu NPs in chitosan-based PEs produce an SPR band in the region of 500 and 800 nm, according to earlier study [25].

3.2. FTIR Study of BT and Sn^{2+}-PPHs Metal Complex

The FTIR spectra of the extracted BT's are shown in Figure 2a. The emergence of many peaks is viewed as the FTIR spectrum's main characteristic. The C-H stretching of carboxylic acid and aliphatic group is responsible for the current peaks in the range of 2913–2847 cm^{-1} [26]. The existence of a band at 1623 cm^{-1} could also be used to identify the aromatic ring's C=C stretch [16,19,26]. It is worth mentioning that the existing FTIR spectrum's overall characteristics match those found in prior investigations [27,28]. The caffeine spectrum has recently been discovered to feature several changes in the range between 1700 and 400 cm^{-1} (Figure 2a). The existence of a range of functional groups with stretching and binding movements, for instance carbonyl, methyl, imidazole, and pyrimidine fragments, can be seen in these changes [29]. The key functional groups in tea, as shown by the FTIR spectra, are PPHs, carboxylic acid, and amino acids. PPHs have been demonstrated to interact with the metal cation to produce colloidal metal-PPHs complex solutions, according to the literature [30].

Figure 2b shows the FTIR spectrum of the Sn^{2+}-PPHs metal compound. A sequence of peaks in the region between 1700 and 400 cm^{-1} can be noticed in both Figure 2a,b; their intensities were nearly modified as common characteristics.

FTIR is used to investigate the colloidal Sn^{2+}-PPHs metal as one of the characteristics of the Sn^{2+}-complex. Wang et al. [21] investigated the use of eucalyptus leaf extract in the production of Fe-PPHs complexes. The interaction between Fe^{2+} and PPHs was stressed as the mechanism for forming the complex.

In the FTIR spectrum of Sn^{2+}-PPHs metal complexes, the distinctive bands of BT are repeated, but the peak intensities have reduced (see Figure 2b).

When the Sn^{2+}-PPHs metal complexes are formed, the bands of 2914 and 2850 cm^{-1} in BT bands have changed, and currently come into view at 2914 and 2845 cm^{-1}, respectively. This is described on the basis of the generation of coordination interactions between PPHs and Sn^{2+} ions, which results in vibrational decrease and arises in reducing mass. More specifically, the development of coordination bonds among PPHs and Sn^{2+} metal ions are caused by an attraction between the Sn^{2+} ion's empty orbitals and the ligand pairs [31]. In the following section, the mechanism of coordination between the Sn^{2+} ion and the interested ligands is schematically described, as shown in Figure 3. Wang et al. [21] used a range of extracts, containing melaleucanesophila, eucalyptus tereticornis, and rosemarinus, to synthesize and characterize iron-PPHs complexes. The authors have demonstrated that iron ions and PPHs interact together, forming iron-PPHs complexes. FTIR was used by Coinceanainn et al. to investigate the complexation between the aflavinand aluminum (III). The polyphenolic chemicals are ligands observed in BT extract [32].

FTIR analysis can be used to inspect the interaction nature between Sn^{2+} metal cations and caffeine, as well as PPHs in tea extracts, as illustrated in Figure 3. The interaction of the Sn^{2+}-metal ion with BT extract includes the production of a number of complexes (see Figure 3). Metal ion interactions with tea components have already been confirmed [26,33]. Figure 3 depicts three different potential complexes. Sn^{2+}-PPHs metal complex, as expected (see Figure 3A), and Sn^{2+}-caffeine are also expected (Figure 3B). Additionally, as demonstrated in Figure 3C, there is a probability of interaction between Sn^{2+} and both caffeine and PPHs in a complex. The EPR technique had previously been used to study the development of complexes by PPHs in BT extract and metal ions [33]. The nature of metal complex production was investigated utilizing FTIR in the current work.

The use of FTIR spectroscopy to measure interactions among ions or atoms in a PE or PC systems is crucial. The interactions that occur can cause a shift in the polymer electrolyte's vibrational modes [34]. Figure 4 shows the FTIR of PVA and PVA loaded samples, respectively. C–H rocking of PVA is attributed to the band at 821 cm^{-1} [34]. For samples containing 15% (v/v) and 30% (v/v) of dopant material, this peak changes to 818 and 838 cm^{-1}, respectively.

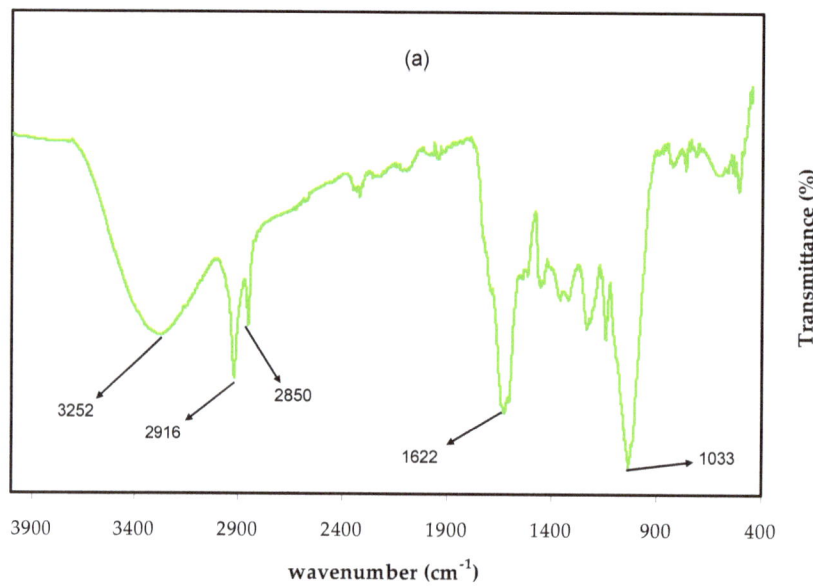

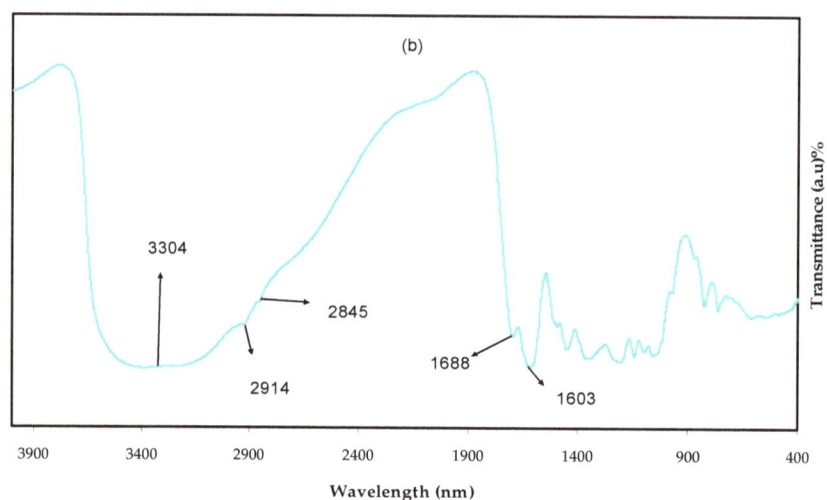

Figure 2. Spectra of the FTIR for the (**a**) extracted BTleaf and (**b**) colloidal Sn^{2+}-PPHs metal complex.

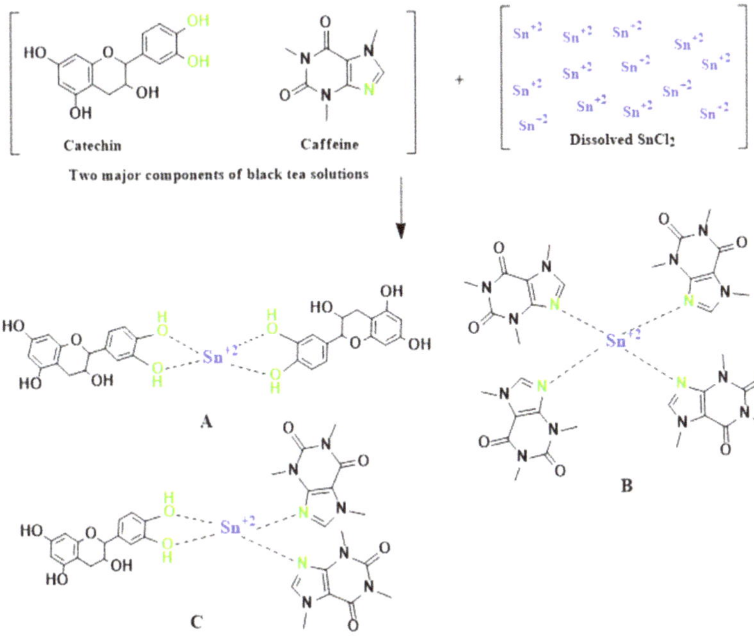

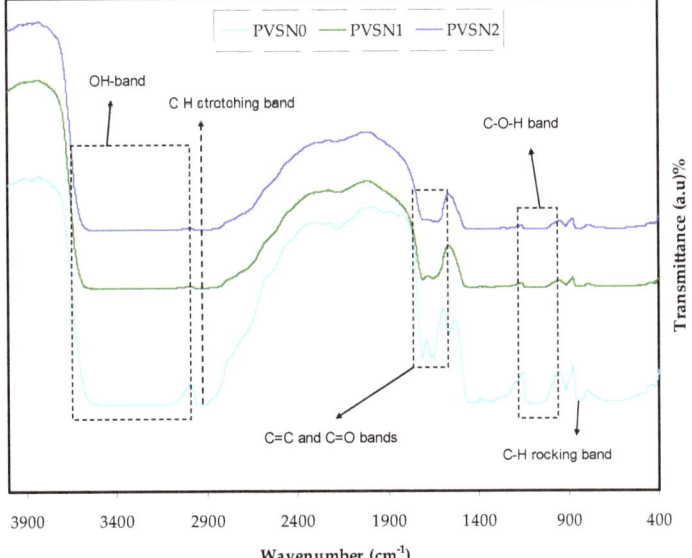

Figure 3. The proposed chemical structure for the Sn^{2+}-PPHs metal complex formation mechanism. (**A**) Sn^{2+}-catechin metal complex, (**B**) Sn^{2+}-caffeine complex, and (**C**) interaction between Sn^{2+} and both caffeine and catechin.

Figure 4. FTIR spectra for pure PVA and loaded films.

Pure PVA absorption maxima at 1313 and 1410 cm^{-1} have been ascribed to C–OH plane bending and CH$_2$ wagging, respectively [35]. The peak at 1316 cm^{-1} vanishes in doped samples, while the peak at 1410 cm^{-1} changes to 1422 cm^{-1} and 1468 cm^{-1}

for samples including 15 and 30% (v/v), respectively. The O-H stretching vibration is responsible for a broad and intense absorption peak centered at 3339 cm^{-1} [36]. It is seen that the absorption peak at the 3339 cm^{-1} is a saturated flat pattern rather than a shoulder peak. This might be related to the thickness of the films as the FTIR is thickness dependent. The strong intra and inter type hydrogen bonding can be associated with the high intensity of this band [34]. This band shifts and its intensity are considerably reduced in doped materials. For the doped samples, the peak at 1644 cm^{-1}, which is attributed to C=O stretching of the acetate groups, that is the remaining component of PVA, is changed to 1607 cm^{-1} [35]. At 2905 cm^{-1}, the band analogous to C-H asymmetric stretching occurs [36]. For the loaded films, there is a noticeable change and substantial drop in this absorption band. The peak at 1076 cm^{-1} in Figure 4, which is a typical stretching vibration of –C–O– in pure PVA [37], is displaced to 1090 cm^{-1} and its strength diminishes.

3.3. XRD Analysis

The XRD patterns of pure PVA and PVA doped with 30% (v/v) and 40% (v/v) of Sn^{2+}-PPHs complex are shown in Figure 5. Pure PVA's XRD pattern revealed a large peak about 20° that corresponded to the semi-crystalline structure of pure PVA [10]. A side from the major peak, two broad peaks may be found at 2θ = 23.4° and 41.18°. Based on the literature, the (101), (200) and (111) crystalline planes of PVA are responsible for the typical diffraction peaks at 2θ = 20°, 23.43° and 41.15°, respectively [37], and their shifts in the doped PVA sample are due to the complex formation between the functional groups of PVA and surface groups of the Sn^{2+}-PPHs metal complex.

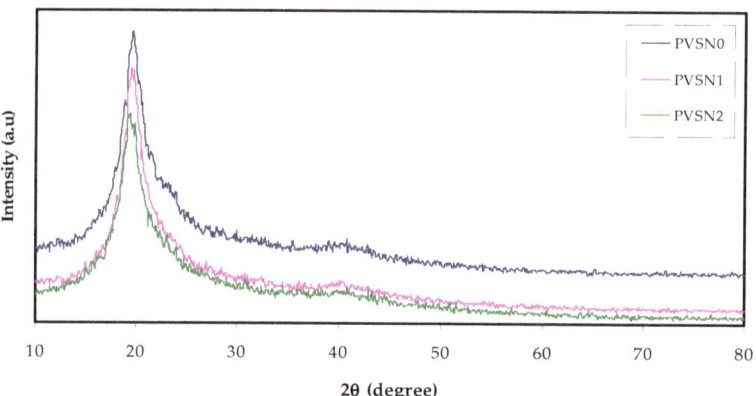

Figure 5. XRD spectra for pure PVA and doped films.

3.4. Absorption Study

The reaction of a substance to electromagnetic radiation, predominantly visible light, is referred to as an "optical property." Sometimes it's more practical to consider electromagnetic radiation (e.m.r) from the perspective of quantum physics, in which the e.m.r is considered as energy packets, i.e., as photons, instead of waves. The following relationship is used to quantify and characterize the energy E of a photon.

$$E = h\nu = hc/\lambda \tag{1}$$

where h stands for Planck constant (6.63 × 10^{-34} J/s), c is denotes to the light speed in free space (3 × 10^8 m/s), and λ is the photon wavelength. Figure 6 illustrates the results. It is clear that the PCs absorption spectra include substantially all of the relevant areas of the UV-visible to NIR ranges. It is well known that the majority of the metal-complex compounds have excellent optical absorption and emission, with wavelengths extending from 600 to 700 nm [38]. This can be explained by the creation of orbital overlaps, which is aided by ligands (functional groups). As a result, electrons can transfer energy via the structure, which is what causes the absorption spectra [39]. The photon is not absorbed and the substance would be transparent to the photon when the incident photon energy is smaller than the energy difference between two levels of electrons. Absorption happens at higher photon energies (usually in 10^{-15} s) when the valence electrons transition between two electronic energy states [3].

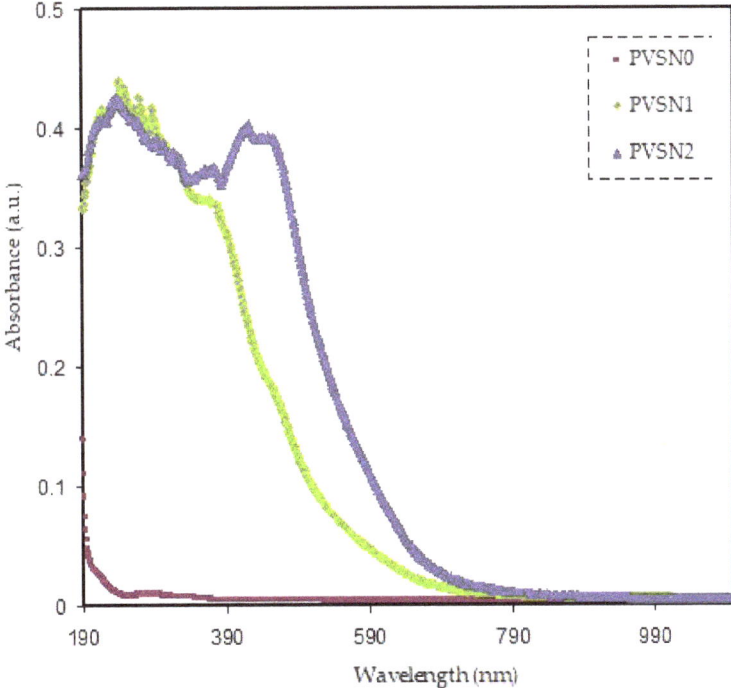

Figure 6. Absorption spectra for pure PVA and loaded films.

The incorporation of metal-complexes into polymers for optoelectronic device and photonic device applications is said to be still under investigation [40]. Organic–inorganic composites have received a lot of attention as a potential material for a novel generation of nonlinear optical, electronic and optical instruments, along with biological labels [41].

3.5. Absorption Edge Study

The optical energy band gap (E_g) of amorphous and crystalline materials can be estimated using the optical absorption spectra. The value and nature of the E_g can be measured using fundamental absorption, which relates to electron excitation from valance to conduction band [42]. It is true that a light wave experiences losses or attenuation when it travels through a substance. The absorption coefficient, often known as the fractional reduction in intensity over distance, is calculated as follows [43]:

$$\alpha = -1/I \times dI/dx = (2.303/d) \times A \qquad (2)$$

where, I denotes to the intensity and A is absorption quantity. The ultraviolet-visible (UV-vis) is valuable method for studying electronic transitions.

When optical transitions begin to occur over a material's fundamental band gap, the absorption edge is formed [44]. When PVA is transformed to tapered band gap polymer hybrid by integrating green produced metal-complex, a new study domain in optical materials is generated. Following that, a sole approach for polymer hybrid production using green technologies is developed. Figure 7 depicts the large absorption edge shift to lower photon energy. For the sample loaded with 30% (v/v) of Sn^{2+}-PPHs metal complex, the value of absorption edge decreased substantially from 6.3 eV to 1.8 eV. (Figure 7). The absorption edge values are shown in Table 1. The absorption coefficient is determined using Equation (2). The intercept of the linear parts of the spectra of absorption coefficient with the axis of photon energy gives the value of the absorption edge. The absorption coefficient values reported in this work are very similar to those obtained for loaded polyacetylene (trans-$(CH)_x$) and polypyrrole [45]. This is connected to the charge transfer complex creation in PC samples. Materials science has found molecule charge transfer materials to be an interesting and a good candidate for assessing molecular CT mechanisms, as well as changes in transport, magnetic, optical, dielectric, and structural properties. CT complexes have fascinating electrical, optical, and photoelectrical properties, and they have a good role in a variety of electro-physical and optical processes [46]. PMMA was doped with Alq3 by Duvenhageetet al. for application in optoelectronics [47]. There was a decrease in device performance and efficiency due to the quick breakdown of organometallic and conjugated polymers [47]. The color of the hybrid samples necessitates a change in the hybrid films' band structure. Because there is enough evidence for a link between color and electrical structure in conductive polymers, polypyrrole's small E_g can be predicted from its blackish color [45]. As a result, the optical property of polymeric materials is deduced from their color.

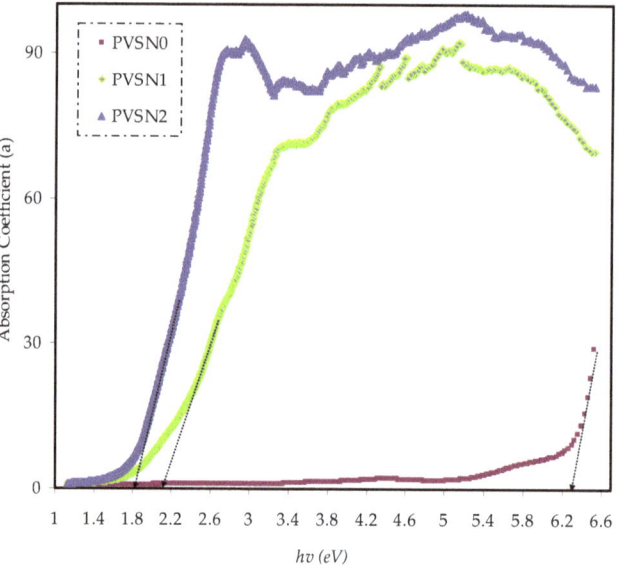

Figure 7. Absorption coefficient vs. hv for PVA and loaded films.

Table 1. Absorption edge for PVA and loaded films.

Sample Code	Absorption Edge (eV)
PVSN0	6.3
PVSN1	2.1
PVSN2	1.8

3.6. Refractive Index Study

The refractive index (*n*) and its dispersion behavior are two of the most essential features of an optical material. In optical communication and spectrum dispersion device design, refractive index dispersion is a vital element [48]. Some models are used to measure the optical E_g using the dispersion area of *n*, as shown in the next section. Figure 8 depicts the value of (*n*) in relation to wavelength. It has been verified that higher *n* values are associated with integrated films that show significant dopant dispersion. It is seen that, as the % (v/v) of the Sn^{2+}-PPHs metal complex rises, the value of *n* rises with it. The *n* is a function of both polarizability and density of a medium at constant temperature and pressure [49]. As a result, a material's refractive index is one of the most important factors in measuring its optical efficiency. The following is an illustration of a samples complex refractive index:

$$n^*(\lambda) = n(\lambda) + k(\lambda) \tag{3}$$

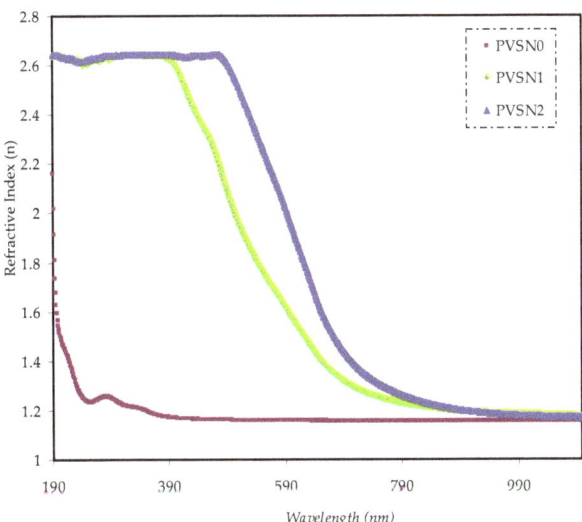

Figure 8. Refractive index spectra versus wavelength for PVA and loaded films.

The *k* and *n* relationship is formulated as follow [35]:

$$n = \left[\frac{(1+R)}{(1-R)}\right] + \sqrt{\frac{4 \times R}{(1-R)^2} - K^2} \tag{4}$$

K is the extinction coefficient and is equal to $\alpha\lambda/4\pi t$ in Equations (3) and (4), where t is the sample thickness.

As photons are decelerated as they pass into a material because of interaction with electrons, the *n* is greater than one. The greater the *n* of a material, the more photons are retarded while passing through it. In general, any method that enhances a material's electron density also enhances its refractive index [50]. Moreover, when compared to pure PVA, the dispersion behavior of refractive index verses wavelength can be seen for all doped films. This is the result of the doped samples' density growing. Two methodologies are considered to improve the *n* value of polymers, depending on the method of synthesis: heavy atoms, for instance, polymers loaded with halogens and/or sulfur atoms [51], and the integration of metal or inorganic NPs into polymers to produce compounds with relatively high *n* values [52]. In all circumstances, there are two primary obstacles when manipulating *n*. To begin, the first way faces two challenges: the technological and financial difficulties of incorporating heavy atoms into polymer matrices [53]. Second, when inorganic NPs (ZrO$_2$, TiO$_2$ or Au NPs) are combined with nanofillers, aggregation occurs [52,54]. As a result, significant surface energy is formed, along with low compatibility with the polymer. In this work, Sn^{2+}-PPHs metal complex was injected into the PVA polymer in order to modify the (n) value.

The single oscillator model proposed by Wemple and DiDomenico [55] is used to study refractive index dispersion (n_o) in the normal dispersion zone. A dispersion energy parameter (E_d) was incorporated into this model to represent the n_o. It is a measure of the strength of the inter band optical transition. This parameter is directly related to chemical bonding and connects the charge distribution and coordination number through each unit cell [56]. Therefore, the energy of an oscillator is proportional to a single oscillator parameter (E_o). This semi-empirical formula can be used to connect the refractive index to the photon energy below the interband absorption edge.

$$n^2 - 1 = \frac{E_d E_o}{\left[E_o^2 - (h\nu)^2\right]} \tag{5}$$

Plotting $1/(n^2 - 1)$ against $(h\nu)^2$ yields the values of (E_d) and (E_o) from the slope and intercept of the linear fitted lines, as shown in Figure 9. Table 2 shows the E_o and E_d values that were calculated. The single oscillator energy (E_o) declines as the % (v/v) of Sn^{2+}-PPHs metal complex increases, whereas the dispersion energy (E_d) rises. The static refractive index at zero energy n_0 is measured from the linear part extrapolation of Figure 9 to intersect the ordinates or is measured by $n_0 = \sqrt{1 + \frac{E_d}{E_o}}$. The oscillator energy E_o is a "average" energy gap that, to a reasonable degree, is experimentally related with the lowest direct band gap [57]. As shown in Table 2, the overall image gained is consistent with the fact that refractive index and energy gap are inversely proportional.

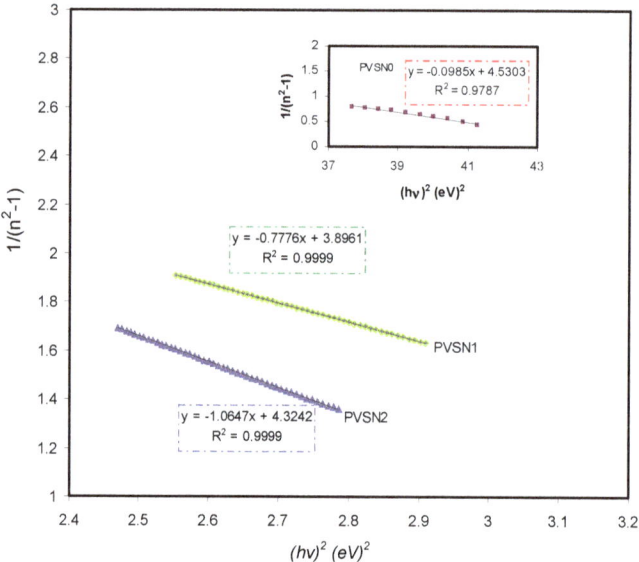

Figure 9. $1/(n^2 - 1)$ against $(h\nu)^2$ for PVA and loaded films.

Table 2. E_o and E_d for the PVA and loaded films.

Sample	E_d	E_o	n_o
PVSN0	1.49	6.74	1.221
PVSN1	0.57	2.22	1.256
PVSN2	0.46	1.98	1.232

3.7. Complex Dielectric Function Study

PC materials are one of the most reliable ways for modifying the dielectric constant (ε_r) value of polymers. Several methods are now being used to improve the ε_r of polymers, which can then be used in photonic or optoelectronic device applications. As demonstrated below, the ε_r is approved in a connection that includes both values of (k) and (n) [58]:

$$\varepsilon_r = n^2 - k^2 \quad (6)$$

Figure 10 shows the ε_r spectra against wavelength for PVA and PC samples. It is clear that, when the % (v/v) of Sn^{2+}-PPHs metal complexes rises, the values of ε_r rise as well. This is associated with the production of density of states inside the polymers, which is forbidden gap [58].

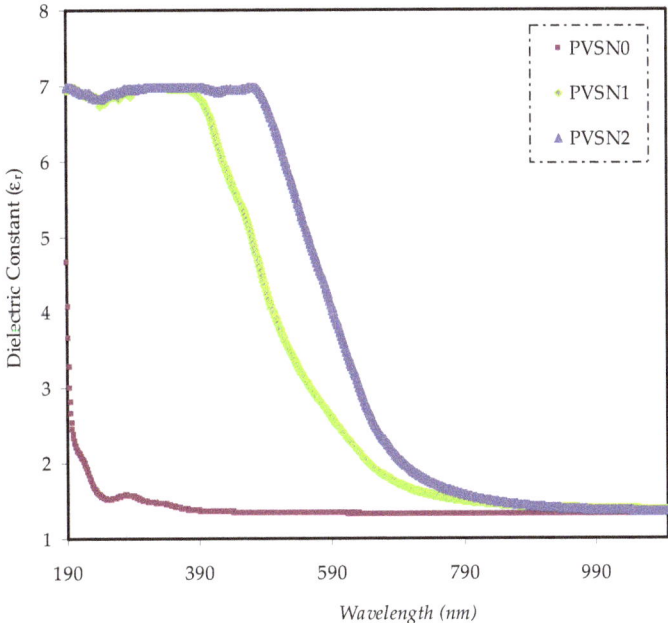

Figure 10. Dielectric constant versus wavelength for PVA and loaded films.

It is seen that the fundamental optical transition in PCs is caused by changes in the ε_r. The response of this feature is reflected in the real (ε_r) and imaginary (ε_i) regions of the spectra. Contrastingly, the actual part determines a material's ability to reduce the speed of an e.m.r wave. The imaginary part, on the other hand, indicates the level of energy absorption efficiency by materials as a result of polarization.

E_r is wavelength dependent. As seen in Figure 10, dielectric constant is high at the low wavelength, while it has a low value at the long wavelength as more photons are absorbed at the low wavelength. Conducting polymers are expensive in comparison with the insulating polymers. In this research, Sn^{2+}-complexes were added to the PVA polymer to increase the dielectric constant, as the Sn^{2+}-complex has more functional groups to interact with the PVA polymer for increasing the dielectric constant and decreasing the E_g. In addition to that Sn^{2+}-complexes create trap energy states within the band gap that cause an increase in the value of the dielectric constant.

The n and wavelength connection, which is on the basis of the Spitzer–Fan model, can be used to specify the dielectric response (ε_∞) of a substance at high frequency (i.e., short wavelength) [59]:

$$\varepsilon_r = n^2 - k^2 = \varepsilon_\infty - \left(\frac{e^2}{4\pi^2 C^2 \varepsilon_o}\right) \times \left(\frac{N}{m^*}\right)\lambda^2 \qquad (7)$$

where ε_0 means the free space dielectric constant, N denotes the number of charge carrier, m^* signifies the effective mass, which is presumed to be 1.16 m_e, and c and e have their normal definitions [60].

In the visible wavelength area, the relationship between the values of ε_r against λ^2 is a straight line, as seen in Figure 11. Using the parameters in Table 3, one may calculate the ε_∞ and N/m^* from the intercept and slope of the line with the vertical axis, correspondingly. Equation (7) can be used to approximate the N/m^*, ε_∞ and N, as shown in Table 4.

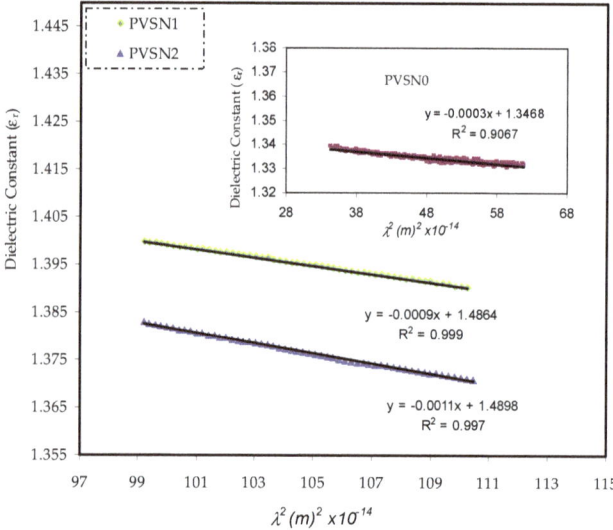

Figure 11. Shows the relationship of ε_r versus λ^2 for pure PVA and doped films.

Table 3. The physical quantities used to determine N/m^* for PVA loaded Sn^{2+} metal complex.

Physical Parameters	Values
m_e	9.109×10^{-31} Kg
E	1.602×10^{-19} coulombs
ε_o	8.85×10^{-12} F/m
π	3.14
C	2.99×10^8 m/s
m^*	10.566×10^{-31} Kg

Table 4. Presents the values of N/m^* and ε_∞ for PVA loaded Sn^{2+}-complex.

Film Code	$N/m^* \times 10^{55}$ (m^{-3}/kg)	ε_∞
PVSN0	3.65	1.346
PVSN1	10.94	1.486
PVSN2	13.38	1.489

Table 4 shows that as the volume of the metal complexes increases, the charge carriers/m* of the parent PVA film increases, from 3.65×10^{55} to 13.38×10^{55} m^{-3} Kg^{-1} and the ε_∞ increases from 1.346 to 1.489. These increases in charge carriers/m* and the ε_∞ is interpreted as indicative of a rise in the number of free charge carrier involved in the polarization mechanism. The calculated N/m^* in this research are in good agreement with those documented in the previous reports by Equation (7) [61].

3.8. Band Gap Study

The clarification of atomic spectra particularly that of the simplest atom, hydrogen, was the first significant achievement of quantum theory. Quantum physics offered a vital concept: atoms could only immerse well-defined energy levels, and these energy states were exceedingly sharp for solitary atoms. Atoms cannot be seen as separate units in a crystalline solid because they are chemically connected to their nearest neighbor since they are in close proximity to one another. The nature of the chemical bond indicates that electrons on close adjacent atoms can exchange with one another, creating the spreading of discrete atomic energy states into energy 'bands' in the solid [62]. When considering a solid, it must take into account the contributions of numerous electronic energy band processes to the optical characteristics. Intraband (IBD) processes, for example, correspond to electronic conduction by free charge carriers and are more relevant in conducting materials such as semimetals, metals, and degenerate semiconductors. The classical Drude theory, or the Boltzmann equation, or the quantum mechanical density matrix method, can explain these IBD phenomena in their most basic terms [63]. Solid-state materials' optical properties are useful for analyzing magnetic excitations, lattice vibrations, energy band structure, localized defects, impurity levels, and excitons. An electron is excited from a full valence band state to an empty conduction band state by a photon. An IBD transition is a quantum mechanical phenomenon [63]. Because of their scientific value and prospective application in energy conversion and harvesting, essential understanding of the charge separation and transfer procedures elaborate in photovoltaic systems is an exciting study topic that is gathering more and more attention [64]. The optical E_g is the most essential property of organic and inorganic materials (E_g)

Tauc's model [65] was used to calculate the energy band gap of the films.

$$(\alpha h v) = B(hv - E_g)^\gamma \quad (8)$$

where B is a transition probability factor that is constant through the visible frequency ranges, and the index is utilized to measure the kind of electronic transition and takes 1/2 or 3/2 for direct transitions, while it is equal to 2 or 3 for indirect transitions, based on whether they are permitted or prohibited [66]. The plot of $(\alpha hv)^{1/\gamma}$ against (hv) for pure PVA and doped films is shown in Figures 12–15.

When the Sn^{2+}-PPHs was added to the PVA polymer, the optical energy bandgap decreased noticeably, as the Sn^{2+}-PPHs are enriched with more functional groups to interact with the functional groups of PVA. Thus, the optical energy bandgap is noticeably decreased. For example, when 15 wt.% Sn^{2+}-PPHs metal complex was added to pure PVA, the BGP reduced. For 30% (v/v) of inserted Sn^{2+}-PPHs metal complex, a considerable modification in the energy band gap may be attained, lowering the BGP of PVA solid films to 1.8 eV. From the interception of the extrapolated linear component of the $(\alpha hv)^{1/\gamma}$ on the photon energy axis, the optical energy band gap for all solid films was obtained (abscissa). Table 5 lists the optical BGP values. In insulator materials, the E_g is too big that no free carriers can thermally excite over it at room temperature. This means there is not any carrier absorption. IBD transitions seem to be essential only at rather high photon energy, as a result (above the visible). Many insulator materials are optically transparent as a result of this. The findings show that PCs with low bandgap energies (1–2 eV) may be made, which has piqued scientists' attention because to their potential applications in visible and infrared detectors, optical parametric oscillators, up converters, and solar cells [67].

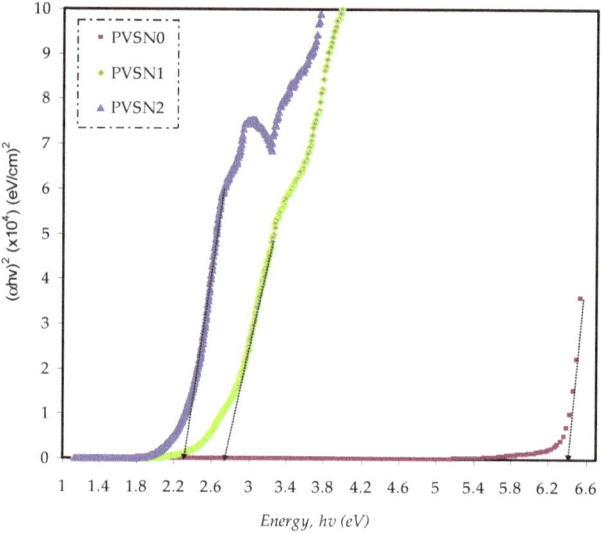

Figure 12. Plot of $(\alpha h\upsilon)^2$ vs. $h\upsilon$ for pure PVA and PC films.

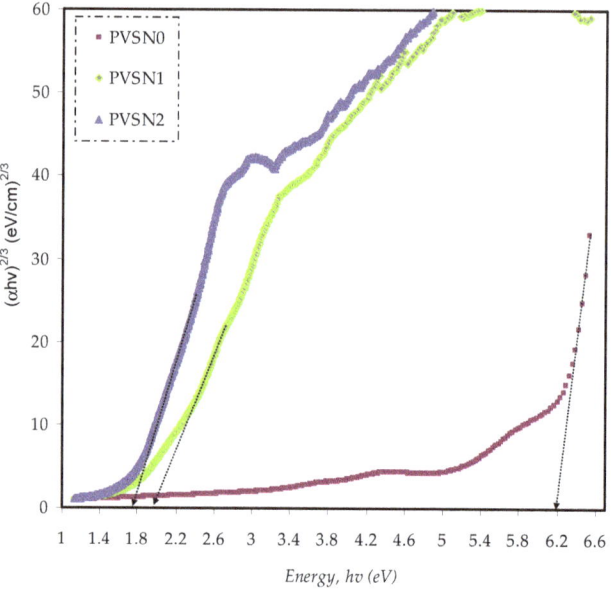

Figure 13. Plot of $(\alpha h\upsilon)^{2/3}$ vs. $h\upsilon$ for pure PVA and PC films.

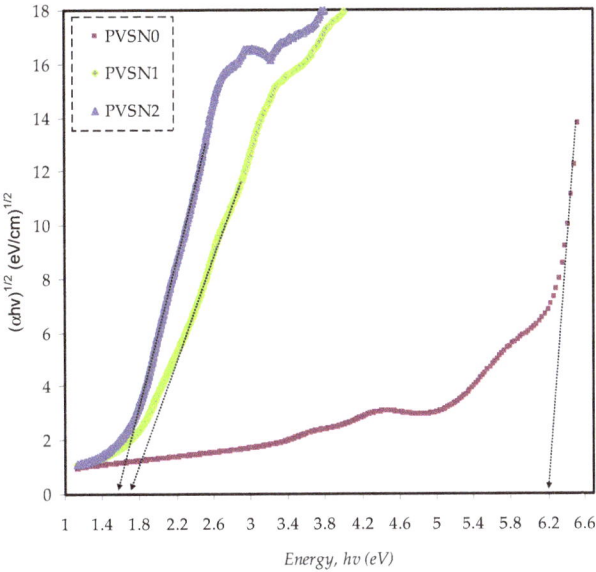

Figure 14. Plot of $(\alpha h\nu)^{1/2}$ vs. $h\nu$ for pure PVA and PC films.

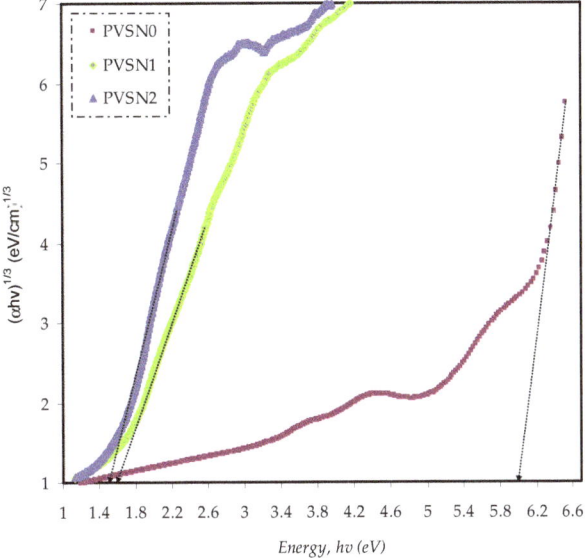

Figure 15. Plot of $(\alpha h\nu)^{1/3}$ vs. $h\nu$ for pure PVA and PC films.

Table 5. Opticalbandgap from Tauc'smodel and dielectric loss plot.

Sample Code	$\gamma = 1/2$	$\gamma = 3/2$	$\gamma = 2$	$\gamma = 3$	E_g From ε_i
PVSN0	6.4	6.19	6.08	6	6.4
PVSN1	2.74	2	1.82	1.6	2.1
PVSN2	2.3	1.78	1.6	1.56	1.8

The complex dielectric function, which is connected with other optical characteristics (i.e., n, reflectivity, and absorption coefficient) by simple equations, is the best way to characterize the optical properties of solids [68]. Electronic transition and charge transport complexes in semiconducting/conducting polymers are not studied well. The transitions are made possible by the incident photon and phonon giving sufficient energy and momentum [69]. Tauc's model and optical dielectric loss have already been shown to be active in determining the E_g and electronic transition types, respectively. This is due to the optical dielectric function is mostly independent of the materials band structure. Simultaneously, study of optical dielectric function utilizing UV–vis spectroscopy have proven to be relatively valued in foreseeing the materials band structure [10,25]. Aside from IBD (free carrier) activities, interband processes occur when electrons in a filled level below the Fermi state absorbs electromagnetic radiation, causing a transition to the unfilled level in a higher band. This IBD process is fundamentally a quantum mechanics method that is explained using quantum mechanical terminology [63].

The ε_i can be determined experimentally from the given n and k data using the following relationships,

$$\varepsilon_2 = 2\,n\,k \tag{9}$$

The refractive index is n, while the extinction coefficient is k. Previous research has shown that the peaks in the ε_i spectra are linked to the interband transitions [34–36]. The real E_g can thus be calculated by taking the intersection of linear sections of ε_i spectrawith the $h\nu$ axis (see Figure 16). This is because the optical dielectric function is intimately linked to the photon–electron interaction and relates the physical process of IBD transition through the structure of electronic materials. The dielectric function's imaginary part(ε_i) primarily describes the electron transition from filled to unfilled levels [70]. Former work documented that studying the ε_i allowed for a detailed understanding of the optical transition mechanism [71]. An electron is excited by a photon from a valence band occupied state to a conduction band unoccupied state. This is referred as IBD transitions. A photon is absorbed in this method, which results in the formation of a hole and an excited electronic level. Quantum mechanics governs this process [63]. The optical dielectric loss is significantly connected to the filled and unfilled electronic levels within a solid from a quantum mechanics (microscopic) standpoint. The peak in the imaginary component of the dielectric function correlates to strong IBD transitions, which is well documented microscopically (quantum mechanically) [70]. The study of the complex dielectric function ($\varepsilon^* = \varepsilon_r - i\varepsilon_i$), which defines the material linear response to e.m.r, will help you better comprehend the optical properties of a solid. The imaginary component ε_i represents the material's optical absorption, which is tightly linked to the valence (filled) and conduction (unfilled) bands, and is given by [18]:

$$\varepsilon_2(\omega) = \frac{2e^2\pi}{\Omega\varepsilon_o}\sum_{K,V,C}\left|\psi_K^C\left|\vec{u}\cdot\vec{r}\right|\psi_K^V\right|^2\delta\left(E_K^C - E_K^V - \hbar\omega\right) \tag{10}$$

where ω, Ω, e, ε_o, $\vec{r}$ and $\vec{u}$ are the incident photon frequency, crystal volume, electron charge, free space permittivity, position vector, and a vector determined by the incident e.m.r. wave polarization, respectively. ψ_k^c and ψ_k^v are the conduction band wave function and valence band wave function, respectively, at k. The optical dielectric constant is characterized by a complex function of frequency based on theoretical models, which necessitates a large-scale computer effort to calculate [71,72]. By comparing the ε'' of Figure 16 to those extracted from Tauc's method, the types of electronic transition is recognized (Figures 12–15). It is feasible to determine that the kind of electronic transition in pure PVA and PC samples is direct allowed ($\gamma = 1/2$) and direct forbidden ($\gamma = 3/2$) transitions, respectively. Table 5 summarizes the band gap calculated using optical dielectric loss and the Tauc technique for more clarity.

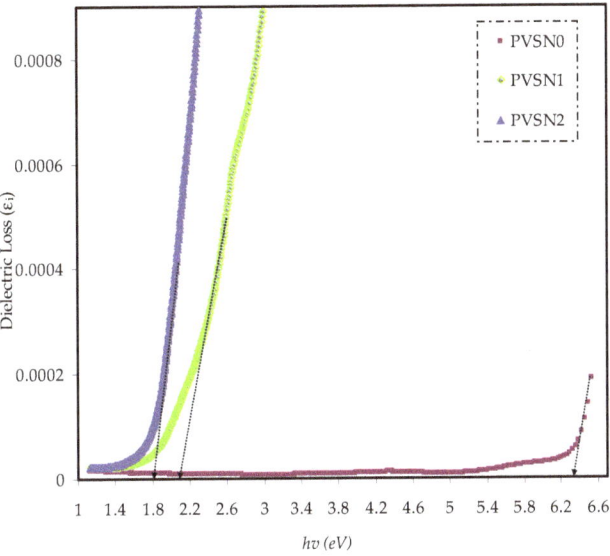

Figure 16. Plot of ε_i vs. $h\nu$ for PVA and PC films.

4. Conclusions

In conclusion, the PVA films doped with Sn^{2+}-PPHs metal complex were synthesized with low optical energy band gaps by solution casting procedure. The FTIR showed that the BT contained sufficient PPHs and functional groups to fabricate Sn^{2+}-PPHs metal complex. The UV-vis and FTIR methods confirmed the formation of Sn^{2+}-PPHs metal complex. The UV-visible method showed the effect of the Sn^{2+}-PPHs metal complex on the optical property of PVA. Furthermore, XRD and FTIR analyses showed the formation of complexation between PVA and Sn^{2+}-PPHs metal complex. The improvement of the amorphous structure is reflected in the broadness and decrease in the XRD intensity. The shifts and decreases in the intensity of the FTIR peaks of the composite films established the interaction between Sn^{2+}-PPHs metal complex and PVA. The absorption edge shifted to lower $h\nu$ by increasing the load of the Sn^{2+}-PPHs metal complex. The refractive index and dielectric constant tuned by loading of Sn^{2+}-PPHs metal complex to PVA. The E_o, E_d and n_0 were calculated for the films. The dielectric constant versus photon wavelength were studied to measure N/m^* and high frequency dielectric constant. Tauc's model was used to measure the type of electronic transition in the films. To estimate the energy gap, the dielectric loss parameter was analyzed. Because of the low optical band gap of the films, these films have good potential for optoelectronic device applications.

Author Contributions: Conceptualization, S.B.A. and A.M.H.; Formal analysis, M.A.B. and S.A.H.; Funding acquisition, M.M.N. and S.I.A.-S.; Investigation, S.A.H.; Methodology, M.A.B. and K.K.A.; Project administration, S.B.A., M.M.N. and E.M.A.D.; Validation, M.M.N., N.M.S., E.M.A.D., K.K.A., S.I.A.-S. and A.M.H.; Writing–original draft, S.B.A.; Writing–review and editing, M.M.N., M.A.B., N.M.S., E.M.A.D., S.I.A.-S., S.A.H. and A.M.H. All authors have read and agreed to the published version of the manuscript.

Funding: We would like to acknowledge all support for this work by the University of Sulaimani, Prince Sultan University and Komar University of Science and Technology. The authors express their gratitude to the support of Princess Nourah bint Abdulrahman University, Researchers Supporting Project number (PNURSP2022R58), Princess Nourah bint Abdulrahman University, Riyadh, Saudi Arabia. The authors would like to acknowledge the support of Prince Sultan University for paying the Article Processing Charges (APC) of this publication and for their financial support.

Institutional Review Board Statement: Not applicable.

Informed Consent Statement: Not applicable.

Data Availability Statement: Not applicable.

Conflicts of Interest: The authors declare no conflict of interest.

References

1. Bockstaller, M.R.; Thomas, E.L. Optical Properties of Polymer-Based Photonic Nanocomposite Materials. *J. Phys. Chem. B* **2003**, *107*, 10017–10024. [CrossRef]
2. Roppolo, I.; Sangermano, M.; Chiolerio, A. Optical Properties of Polymer Nanocomposites. In *Functional and Physical Properties of Polymer Nanocomposites*; John Wiley & Sons, Ltd.: Hoboken, NJ, USA, 2016; pp. 139–157. [CrossRef]
3. Brza, M.A.; Aziz, S.B.; Anuar, H.; Al Hazza, M.H.F. From Green Remediation to Polymer Hybrid Fabrication with Improved Optical Band Gaps. *Int. J. Mol. Sci.* **2019**, *20*, 3910. [CrossRef]
4. Brza, M.A.; Aziz, S.B.; Anuar, H.; Ali, F.; Dannoun, E.M.A.; Mohammed, S.J.; Abdulwahid, R.T.; Al-Zangana, S. Tea from the drinking to the synthesis of metal complexes and fabrication of PVA based polymer composites with controlled optical band gap. *Sci. Rep.* **2020**, *10*, 18108. [CrossRef]
5. Vodnik, V.; Božanić, D.; Džunuzović, E.; Vuković, J.; Nedeljković, J. Thermal and optical properties of silver–poly(methylmethacrylate) nanocomposites prepared by in-situ radical polymerization. *Eur. Polym. J.* **2010**, *46*, 137–144. [CrossRef]
6. Reverberi, A.P.; Vocciante, M.; Lunghi, E.; Pietrelli, L.; Fabiano, B. New trends in the synthesis of nanoparticles by green methods. *Chem. Eng. Trans.* **2017**, *61*, 667–672. [CrossRef]
7. Li, S.; Lo, C.-Y.; Pan, M.-H.; Lai, C.-S.; Ho, C.-T. Black tea: Chemical analysis and stability. *Food Funct.* **2013**, *4*, 10–18. [CrossRef]
8. Drynan, J.W.; Clifford, M.N.; Obuchowicz, J.; Kuhnert, N. The chemistry of low molecular weight black tea polyphenols. *Nat. Prod. Rep.* **2010**, *27*, 417–462. [CrossRef]
9. van der Hooft, J.J.J.; Akermi, M.; Ünlü, F.Y.; Mihaleva, V.; Roldan, V.G.; Bino, R.J.; de Vos, R.C.H.; Vervoort, J. Structural Annotation and Elucidation of Conjugated Phenolic Compounds in Black, Green, and White Tea Extracts. *J. Agric. Food Chem.* **2012**, *60*, 8841–8850. [CrossRef]
10. Aziz, S.B. Modifying Poly(Vinyl Alcohol) (PVA) from Insulator to Small-Bandgap Polymer: A Novel Approach for Organic Solar Cells and Optoelectronic Devices. *J. Electron. Mater.* **2016**, *45*, 736–745. [CrossRef]
11. Aziz, S.B.; Abdullah, O.G.; Hussein, A.M.; Ahmed, H.M. From Insulating PMMA Polymer to Conjugated Double Bond Behavior: Green Chemistry as a Novel Approach to Fabricate Small Band Gap Polymers. *Polymers* **2017**, *9*, 626. [CrossRef]
12. Zielinski, A.; Haminiuk, C.; Alberti, A.; Nogueira, A.; Demiate, I.M.; Granato, D. A comparative study of the phenolic compounds and the in vitro antioxidant activity of different Brazilian teas using multivariate statistical techniques. *Food Res. Int.* **2014**, *60*, 246–254. [CrossRef]
13. El-kader, F.H.A. Structural, optical and thermal characterization of ZnO nanoparticles doped in PEO/PVA blend films. *Nano Sci. Nano Technol. Indian J.* **2013**, *7*, 608–619.
14. Zhang, M.; Howe, R.C.T.; Woodward, R.I.; Kelleher, E.J.R.; Torrisi, F.; Hu, G.; Popov, S.V.; Taylor, J.R.; Hasan, T. Solution processed MoS2-PVA composite for sub-bandgap mode-locking of a wideband tunable ultrafast Er:fiber laser. *Nano Res.* **2015**, *8*, 1522–1534. [CrossRef]
15. Hasan, T.; Sun, Z.; Wang, F.; Bonaccorso, F.; Tan, P.H.; Rozhin, A.G.; Ferrari, A.C. Nanotube—Polymer Composites for Ultrafast Photonics. *Adv. Mater.* **2009**, *21*, 3874–3899. [CrossRef]
16. Yu, Y.-Y.; Chien, W.-C.; Chen, S.-Y. Preparation and optical properties of organic/inorganic nanocomposite materials by UV curing process. *Mater. Des.* **2010**, *31*, 2061–2070. [CrossRef]
17. Aziz, S.B.; Hussein, S.; Hussein, A.M.; Saeed, S.R. Optical Characteristics of Polystyrene Based Solid Polymer Composites: Effect of Metallic Copper Powder. *Int. J. Met.* **2013**, *2013*, 123657. [CrossRef]
18. Aziz, S.B.; Rasheed, M.A.; Ahmed, H.M. Synthesis of Polymer Nanocomposites Based on [Methyl Cellulose]$(1-x)$:(CuS)x (0.02 M $\leq x \leq$ 0.08 M) with Desired Optical Band Gaps. *Polymers* **2017**, *9*, 194. [CrossRef]
19. Smits, J.G.; Boom, G. Resonant diaphragm pressure measurement system with ZnO on Si excitation. *Sens. Actuators* **1983**, *4*, 565–571. [CrossRef]
20. Xu, H.; Chen, R.; Sun, Q.; Lai, W.; Su, Q.; Huang, W.; Liu, X. Recent progress in metal–organic complexes for optoelectronic applications. *Chem. Soc. Rev.* **2014**, *43*, 3259–3302. [CrossRef]
21. Wang, Z.; Fang, C.; Megharaj, M. Characterization of Iron–Polyphenol Nanoparticles Synthesized by Three Plant Extracts and Their Fenton Oxidation of Azo Dye. *ACS Sustain. Chem. Eng.* **2014**, *2*, 1022–1025. [CrossRef]
22. Wang, X.; Huang, J.; Fan, W.; Lu, H. Identification of green tea varieties and fast quantification of total polyphenols by near-infrared spectroscopy and ultraviolet-visible spectroscopy with chemometric algorithms. *Anal. Methods* **2015**, *7*, 787–792. [CrossRef]
23. López-Martínez, L.; López-De-Alba, P.L.; García-Campos, R.; De León-Rodríguez, L.M. Simultaneous determination of methylxanthines in coffees and teas by UV-Vis spectrophotometry and partial least squares. *Anal. Chim. Acta* **2003**, *493*, 83–94. [CrossRef]
24. Jain, P.K.; Xiao, Y.; Walsworth, R.; Cohen, A.E. Surface Plasmon Resonance Enhanced Magneto-Optics (SuPREMO): Faraday Rotation Enhancement in Gold-Coated Iron Oxide Nanocrystals. *Nano Lett.* **2009**, *9*, 1644–1650. [CrossRef]
25. Aziz, S.B. Morphological and Optical Characteristics of Chitosan$(1-x)$:Cuox ($4 \leq x \leq 12$) Based Polymer Nano-Composites: Optical Dielectric Loss as an Alternative Method for Tauc's Model. *Nanomaterials* **2017**, *7*, 444. [CrossRef]

26. Senthilkumar, S.R.; Sivakumar, T. Green tea (Camellia sinensis) mediated synthesis of zinc oxide (ZnO) nanoparticles and studies on their antimicrobial activities. *Int. J. Pharm. Pharm. Sci.* **2014**, *6*, 461–465.
27. Szymczycha-Madeja, A.; Welna, M.; Zyrnicki, W. Multi-Element Analysis, Bioavailability and Fractionation of Herbal Tea Products. *J. Braz. Chem. Soc.* **2013**, *24*, 777–787. [CrossRef]
28. Li, X.; Zhang, Y.; He, Y. Rapid detection of talcum powder in tea using FT-IR spectroscopy coupled with chemometrics. *Sci. Rep.* **2016**, *6*, 30313. [CrossRef]
29. Ucun, F.; Sağlam, A.; Güçlü, V. Molecular structures and vibrational frequencies of xanthine and its methyl derivatives (caffeine and theobromine) by ab initio Hartree–Fock and density functional theory calculations. *Spectrochim. Acta Part A Mol. Biomol. Spectrosc.* **2007**, *67*, 342–349. [CrossRef]
30. Wu, D.; Bird, M.R. The Interaction of Protein and Polyphenol Species in Ready to Drink Black Tea Liquor Production. *J. Food Process Eng.* **2010**, *33*, 481–505. [CrossRef]
31. Kotrba, P.; Mackova, M.; Macek, T. (Eds.) *Microbial Biosorption of Metals*; Springer: Dordrecht, The Netherlands, 2011. [CrossRef]
32. Ó'coinceanainn, M.; Astill, C.; Schumm, S. Potentiometric, FTIR and NMR studies of the complexation of metals with theaflavin. *Dalton Trans.* **2003**, *5*, 801–807. [CrossRef]
33. Goodman, B.A.; Severino, J.F.; Pirker, K.F. Reactions of green and black teas with Cu(ii). *Food Funct.* **2012**, *3*, 399–409. [CrossRef]
34. Hema, M.; Selvasekerapandian, S.; Sakunthala, A.; Arunkumar, D.; Nithya, H. Structural, vibrational and electrical characterization of PVA–NH4Br polymer electrolyte system. *Phys. B Condens. Matter* **2008**, *403*, 2740–2747. [CrossRef]
35. Malathi, J.; Kumaravadivel, M.; Brahmanandhan, G.; Hema, M.; Baskaran, R.; Selvasekarapandian, S. Structural, thermal and electrical properties of PVA–LiCF3SO3 polymer electrolyte. *J. Non-Cryst. Solids* **2010**, *356*, 2277–2281. [CrossRef]
36. Makled, M.; Sheha, E.; Shanap, T.; El-Mansy, M. Electrical conduction and dielectric relaxation in p-type PVA/CuI polymer composite. *J. Adv. Res.* **2013**, *4*, 531–538. [CrossRef]
37. Jiang, L.; Yang, T.; Peng, L.; Dan, Y. Acrylamide modified poly(vinyl alcohol): Crystalline and enhanced water solubility. *RSC Adv.* **2015**, *5*, 86598–86605. [CrossRef]
38. Yakuphanoglu, F.; Kandaz, M.; Yaraşır, M.N.; Şenkal, F. Electrical transport and optical properties of an organic semiconductor based on phthalocyanine. *Phys. B Condens. Matter* **2007**, *393*, 235–238. [CrossRef]
39. Vergara, M.S.; Rebollo, A.O.; Alvarez, J.; Rivera, M. Molecular materials derived from MPc (M = Fe, Pb, Co) and 1,8-dihydroxiantraquinone thin films: Formation, electrical and optical properties. *J. Phys. Chem. Solids* **2008**, *69*, 1–7. [CrossRef]
40. Kovalchuk, A.; Huang, K.; Xiang, C.; Martí, A.A.; Tour, J.M. Luminescent Polymer Composite Films Containing Coal-Derived Graphene Quantum Dots. *ACS Appl. Mater. Interfaces* **2015**, *7*, 26063–26068. [CrossRef]
41. Woelfle, C.; O Claus, R. Transparent and flexible quantum dot–polymer composites using an ionic liquid as compatible polymerization medium. *Nanotechnology* **2007**, *18*, 025402. [CrossRef]
42. Mohamed, S.A.; Al-Ghamdi, A.; Sharma, G.; El Mansy, M. Effect of ethylene carbonate as a plasticizer on CuI/PVA nanocomposite: Structure, optical and electrical properties. *J. Adv. Res.* **2014**, *5*, 79–86. [CrossRef]
43. Jung, H.-E.; Shin, M. Surface-Roughness-Limited Mean Free Path in Si Nanowire FETs. *arXiv* **2013**, arXiv:1304.5597.
44. Greenwood, J.; Wray, T. High accuracy pressure measurement with a silicon resonant sensor. *Sens. Actuators A Phys.* **1993**, *37–38*, 82–85. [CrossRef]
45. Patil, A.O.; Heeger, A.J.; Wudl, F. Optical properties of conducting polymers. *Chem. Rev.* **1988**, *88*, 183–200. [CrossRef]
46. Yakuphanoglu, F.; Barım, G.; Erol, I. The effect of FeCl3 on the optical constants and optical band gap of MBZMA-co-MMA polymer thin films. *Phys. B Condens. Matter* **2007**, *391*, 136–140. [CrossRef]
47. Duvenhage, M.; Ntwaeaborwa, O.; Swart, H. Optical and Chemical Properties of Alq3:PMMA Blended Thin Films. *Mater. Today Proc.* **2015**, *2*, 4019–4027. [CrossRef]
48. Saini, I.; Rozra, J.; Chandak, N.; Aggarwal, S.; Sharma, P.; Sharma, A. Tailoring of electrical, optical and structural properties of PVA by addition of Ag nanoparticles. *Mater. Chem. Phys.* **2013**, *139*, 802–810. [CrossRef]
49. Yakuphanoglu, F.; Arslan, M. Determination of thermo-optic coefficient, refractive index, optical dispersion and group velocity parameters of an organic thin film. *Phys. B Condens. Matter* **2007**, *393*, 304–309. [CrossRef]
50. Babu, K.E.; Veeraiah, A.; Swamy, D.T.; Veeraiah, V. First-principles study of electronic and optical properties of cubic perovskite CsSrF3. *Mater. Sci.* **2012**, *30*, 359–367. [CrossRef]
51. Olshavsky, M.A.; Allcock, H.R. Polyphosphazenes with high refractive indices: Synthesis, characterization, and optical properties. *Macromolecules* **1995**, *28*, 6188–6197. [CrossRef]
52. Xia, Y.; Zhang, C.; Wang, J.-X.; Wang, D.; Zeng, X.-F.; Chen, J.-F. Synthesis of Transparent Aqueous ZrO2 Nanodispersion with a Controllable Crystalline Phase without Modification for a High-Refractive-Index Nanocomposite Film. *Langmuir* **2018**, *34*, 6806–6813. [CrossRef]
53. Kleine, T.S.; Nguyen, N.A.; Anderson, L.E.; Namnabat, S.; LaVilla, E.A.; Showghi, S.A.; Dirlam, P.T.; Arrington, C.B.; Manchester, M.S.; Schwiegerling, J.; et al. High Refractive Index Copolymers with Improved Thermomechanical Properties via the Inverse Vulcanization of Sulfur and 1,3,5-Triisopropylbenzene. *ACS Macro Lett.* **2016**, *5*, 1152–1156. [CrossRef]
54. Wang, C.; Cui, Q.; Wang, X.; Li, L. Preparation of Hybrid Gold/Polymer Nanocomposites and Their Application in a Controlled Antibacterial Assay. *ACS Appl. Mater. Interfaces* **2016**, *8*, 29101–29109. [CrossRef]
55. Wemple, S.H.; DiDomenico, J.M. Behavior of the Electronic Dielectric Constant in Covalent and Ionic Materials. *Phys. Rev. B* **1971**, *3*, 1338–1351. [CrossRef]

56. Ammar, A. Studies on some structural and optical properties of ZnxCd1−xTe thin films. *Appl. Surf. Sci.* **2002**, *201*, 9–19. [CrossRef]
57. Gasanly, N.M.; Nizami, M.G. Coexistence of Indirect and Direct Optical Transitions, Refractive, Index and Oscillator Parameters in TlGaS2, TlGaSe2, and TlInS2 Layered Single Crystals. *J. Korean Phys. Soc.* **2010**, *57*, 164–168. [CrossRef]
58. Aziz, S.B.; Nofal, M.M.; Brza, M.A.; Hussein, S.A.; Mahmoud, K.H.; El-Bahy, Z.M.; Dannoun, E.M.A.; Kareem, W.O.; Hussein, A.M. Characteristics of PEO Incorporated with CaTiO3 Nanoparticles: Structural and Optical Properties. *Polymers* **2021**, *13*, 3484. [CrossRef]
59. Ali, F.; Kershi, R.; Sayed, M.; AbouDeif, Y. Evaluation of structural and optical properties of Ce 3+ ions doped (PVA/PVP) composite films for new organic semiconductors. *Phys. B Condens. Matter* **2018**, *538*, 160–166. [CrossRef]
60. Spitzer, W.G.; Fan, H.Y. Determination of Optical Constants and Carrier Effective Mass of Semiconductors. *Phys. Rev.* **1957**, *106*, 882–890. [CrossRef]
61. Alsaad, A.; Al-Bataineh, Q.M.; Ahmad, A.; Albataineh, Z.; Telfah, A. Optical band gap and refractive index dispersion parameters of boron-doped ZnO thin films: A novel derived mathematical model from the experimental transmission spectra. *Optik* **2020**, *211*, 164641. [CrossRef]
62. Chani, M.T.S.; Karimov, K.; Khalid, F.A.; Moiz, S.A. Polyaniline based impedance humidity sensors. *Solid State Sci.* **2013**, *18*, 78–82. [CrossRef]
63. Dresselhaus, M.S. *Solid State Physics Part II Optical Properties of Solids*; Massachusetts Institute of Technology: Cambridge, MA, USA, 2001; pp. 15–16.
64. Sun, Y.; An, C. Shaped gold and silver nanoparticles. *Front. Mater. Sci.* **2011**, *5*, 1–24. [CrossRef]
65. Ezat, S.G.; Hussen, S.A.; Aziz, S.B. Structure and optical properties of nanocomposites based on polystyrene (PS) and calcium titanate (CaTiO3) perovskite nanoparticles. *Optik* **2021**, *241*, 166963. [CrossRef]
66. Bao, Q.; Zhang, H.; Yang, J.-X.; Wang, S.; Tang, D.Y.; Jose, R.; Ramakrishna, S.; Lim, C.T.; Loh, K. Graphene-Polymer Nanofiber Membrane for Ultrafast Photonics. *Adv. Funct. Mater.* **2010**, *20*, 782–791. [CrossRef]
67. Feng, J.; Xiao, B.; Chen, J.; Zhou, C.; Du, Y.; Zhou, R. Optical properties of new photovoltaic materials: $AgCuO_2$ and $Ag_2Cu_2O^3$. *Solid State Commun.* **2009**, *149*, 1569–1573. [CrossRef]
68. Logothetidis, S. Optical and electronic properties of amorphous carbon materials. *Diam. Relat. Mater.* **2003**, *12*, 141–150. [CrossRef]
69. Yakuphanoglu, F.; Arslan, M. Determination of electrical conduction mechanism and optical band gap of a new charge transfer complex: TCNQ-PANT. *Solid State Commun.* **2004**, *132*, 229–234. [CrossRef]
70. Yu, L.; Li, D.; Zhao, S.; Li, G.; Yang, K. First Principles Study on Electronic Structure and Optical Properties of Ternary GaAs:Bi Alloy. *Materials* **2012**, *5*, 2486–2497. [CrossRef]
71. Guo, M.; Du, J. First-principles study of electronic structures and optical properties of Cu, Ag, and Au-doped anatase TiO_2. *Phys. B Condens. Matter* **2012**, *407*, 1003–1007. [CrossRef]
72. Biskri, Z.E.; Rached, H.; Bouchear, M.; Rached, D.; Aida, M.S. A Comparative Study of Structural Stability and Mechanical and Optical Properties of Fluorapatite ($Ca_5(PO_4)_3F$) and Lithium Disilicate ($Li_2Si_2O_5$) Components Forming Dental Glass–Ceramics: First Principles Study. *J. Electron. Mater.* **2016**, *45*, 5082–5095. [CrossRef]

Article

Temperature-Responsive Photoluminescence and Elastic Properties of 1D Lead Halide Perovskites *R*- and *S*-(Methylbenzylamine)PbBr₃

Rui Feng [1], Jia-Hui Fan [2], Kai Li [2], Zhi-Gang Li [2], Yan Qin [3], Zi-Ying Li [2], Wei Li [2,*] and Xian-He Bu [1,2]

[1] College of Chemistry & State Key Lab of Elemento-Organic Chemistry, Nankai University, Tianjin 300071, China; fengrui1226@hotmail.com (R.F.); buxh@nankai.edu.cn (X.-H.B.)
[2] School of Materials Science and Engineering & Tianjin Key Laboratory of Metal and Molecule-Based Material Chemistry, Nankai University, Tianjin 300350, China; asfjhh@163.com (J.-H.F.); 1120180353@mail.nankai.edu.cn (K.L.); 1120200436@mail.nankai.edu.cn (Z.-G.L.); 1120210484@mail.nankai.edu.cn (Z.-Y.L.)
[3] School of Physics & Wuhan National Laboratory for Optoelectronics, Huazhong University of Science and Technology, Wuhan 430074, China; qinyan@hust.edu.cn
* Correspondence: wl276@nankai.edu.cn

Abstract: Low-dimensional metal halide perovskites (MHPs) have received much attention due to their striking semiconducting properties tunable at a molecular level, which hold great potential in the development of next-generation optoelectronic devices. However, the insufficient understanding of their stimulus-responsiveness and elastic properties hinders future practical applications. Here, the thermally responsive emissions and elastic properties of one-dimensional lead halide perovskites *R*- and *S*-MBAPbBr₃ (MBA⁺ = methylbenzylamine) were systematically investigated via temperature-dependent photoluminescence (PL) experiments and first-principles calculations. The PL peak positions of both perovskites were redshifted by about 20 nm, and the corresponding full width at half maximum was reduced by about 40 nm, from ambient temperature to about 150 K. This kind of temperature-responsive self-trapped exciton emission could be attributed to the synergistic effect of electron–phonon coupling and thermal expansion due to the alteration of hydrogen bonding. Moreover, the elastic properties of *S*-MBAPbBr₃ were calculated using density functional theory, revealing that its Young's and shear moduli are in the range of 6.5–33.2 and 2.8–19.5 GPa, respectively, even smaller than those of two-dimensional MHPs. Our work demonstrates that the temperature-responsive emissions and low elastic moduli of these 1D MHPs could find use in flexible devices.

Keywords: low-dimensional; metal halide perovskite; photoluminescence; stimulus-responsive; elastic property

Citation: Feng, R.; Fan, J.-H.; Li, K.; Li, Z.-G.; Qin, Y.; Li, Z.-Y.; Li, W.; Bu, X.-H. Temperature-Responsive Photoluminescence and Elastic Properties of 1D Lead Halide Perovskites *R*- and *S*-(Methylbenzylamine)PbBr₃. *Molecules* 2022, 27, 728. https://doi.org/10.3390/molecules27030728

Academic Editors: Tersilla Virgili and Mariacecilia Pasini

Received: 18 December 2021
Accepted: 20 January 2022
Published: 23 January 2022

Publisher's Note: MDPI stays neutral with regard to jurisdictional claims in published maps and institutional affiliations.

Copyright: © 2022 by the authors. Licensee MDPI, Basel, Switzerland. This article is an open access article distributed under the terms and conditions of the Creative Commons Attribution (CC BY) license (https://creativecommons.org/licenses/by/4.0/).

1. Introduction

Metal halide perovskites (MHPs) are attracting considerable interest owing to their excellent optoelectronic properties tunable at a molecular level [1–5]. The merits of a high absorption coefficient, good defect resistance, and ease of synthesis [6–8] have led to their wide application in solar cells [9–12], photodetectors [13–15], and light-emitting diodes [16–21]. Currently, the number of reported three-dimensional (3D) MHPs is very limited due to their structural requirement by the Goldschmidt tolerance factor [6–8]. To overcome this restriction, low-dimensional (LD) MHPs, including zero-dimensional (0D), one-dimensional (1D), and two-dimensional (2D) MHPs, are being widely explored. In comparison to their 3D counterparts, LD-MHPs possess higher environmental and thermal stability, as well as larger chemical and structural diversity [22–24]. Accordingly, these LD-MHPs have received intense attention in both synthesis studies and applications [25,26].

In these LD-MHPs, the distortion of PbX₆ octahedra (X = halogen) significantly influences their photoluminescence (PL) behaviors. Hydrogen bonding, as one of the widely available interactions connecting the inorganic and organic parts, plays an important role in determining the magnitude of octahedral distortion [27,28]. By changing strengths of hydrogen bonds upon external stimuli (i.e., temperature and pressure), the emissive processes and properties of LD-MHPs, such as peak position, intensity, and the full width at half maximum (FWHM) of self-trapped excitons (STEs), could be manipulated [29–32]. Although there have been a handful of reports about the influence of hydrogen bonding on the PL properties of LD-MHPs upon external stimulation, more efforts should be devoted to elucidating the underlying mechanism. In addition, the elastic properties of materials are of vital importance since they not only determine the long-term reliability and endurance in service but also regulate the manufacturing and processing [33,34]. However, very little attention has been paid to the understanding of the elastic properties of LD-MHPs [35,36].

In this work, the temperature-responsive PL of a pair of 1D MHPs, R- and S-MBAPbBr$_3$, was systematically investigated by variable-temperature optical spectroscopy. Our results indicate that both perovskites exhibit typical yellow emission under ambient conditions ascribed to the STE emission. Their emission peaks show a remarkable redshift and a significant enhancement of intensity with decreasing temperature. In addition, the elastic properties of S-MBAPbBr$_3$ were comprehensively studied via density functional theory (DFT) calculations.

2. Results and Discussion

2.1. Crystal Structures

Both R- and S-MBAPbBr$_3$ crystallize in the chiral $P2_12_12_1$ space group, which is consistent with reports in the literature [37]. Taking S-MBAPbBr$_3$ as an example, its cell parameters at 100 K are a = 7.8835(3) Å, b = 8.0680(3) Å, and c = 20.1237(8) Å. The asymmetric unit of the structure consists of a methylbenzylamine cation and a [PbBr$_3$]$^-$ unit (Figure 1c). The six-coordinated Pb atoms are coordinated by six Br atoms to form a PbBr$_6^-$ octahedron, and adjacent PbBr$_6$ octahedra are face-shared to form an infinite inorganic chain along the a-axis. Each inorganic chain interacts with surrounding organic amine cations via electrostatic forces and N–H$\cdots$Br hydrogen bonding in a hexagonal manner, forming a 1D organic–inorganic assembly with a chemical formula of S-MBAPbBr$_3$ (Figure 1e). Adjacent 1D organic–inorganic assemblies are connected by intermolecular CH$\cdots$π interactions with distances of 3.383 Å, giving rise to a 3D supramolecular structure. To evaluate the structural change upon temperature, the structure was collected at 293 K and compared with that at 100 K. Specifically, the lengths of Pb–Br bonds of S-MBAPbBr$_3$ are in the range of 2.857–3.062 Å and 2.852–3.070 Å at 100 and 293 K, respectively. The distances between N and Br atoms in N–H$\cdots$Br hydrogen bonds are 3.387–3.499 Å and 3.428–3.558 Å at 100 and 293 K. As mentioned above, hydrogen bonding plays an important role in the octahedral distortion degree. As shown in the distance of N–Br (Table S1), the hydrogen bonding becomes stronger at lower temperature, causing distinct octahedral distortion in the c-direction. With the temperature increase, the increased vibrations of MBA molecules weaken the hydrogen bonding, thus reducing the distortion degree, and S-MBAPbBr$_3$ expands in the c-direction. Combined with the cell parameters at 100 and 293 K of S-MBAPbBr$_3$ (Table S1, Figure S1), the c-axis shows the highest coefficient of thermal expansion, which is consistent with the above analysis, indicating that the distortion of inorganic chains can be adjusted by varied hydrogen bonding upon thermal stimulus. The degree of [PbBr$_6$] distortion can be quantified by the mean octahedral quadratic elongation (λ) and variance of the octahedral angle parameters (σ^2), defined as follows [38]:

$$\lambda = \frac{1}{6}\sum_{i=1}^{6}(d_i/d_0)^2, \tag{1}$$

$$\sigma^2 = \frac{1}{11}\sum_{i=1}^{12}(\alpha_i - 90)^2, \tag{2}$$

where d_i denotes the six individual bond lengths of Pb–Br, d_0 denotes the average distance of the bond length of Pb–Br, and α_i denotes the individual bond angle of Br–Pb–Br. The calculated λ and σ^2 for S-MBAPbBr$_3$ are 1.003 and 221.58, and 1.003 and 197.04, at 100 and 293 K, respectively. The above results suggest that the distortion of octahedra is mainly manifested as the change of bond angles, and the structure at lower temperature is more distorted due to the alteration of hydrogen bonds. This could lead to temperature-responsive emission, as we discuss below.

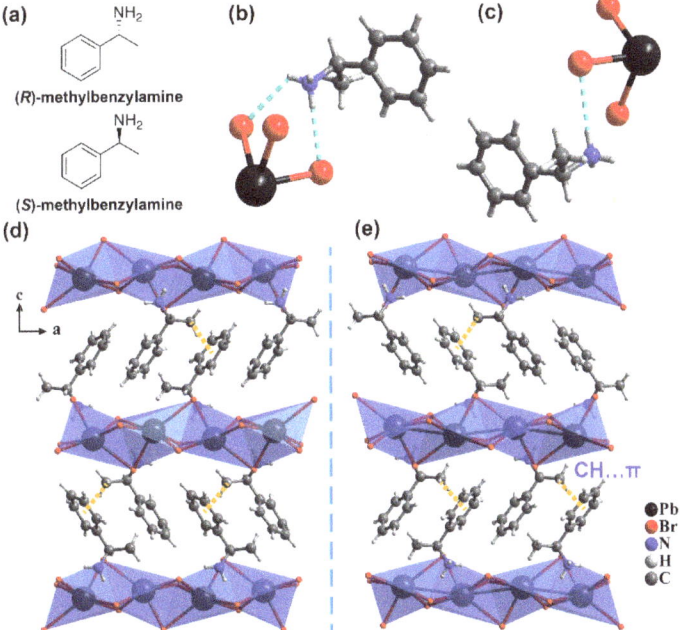

Figure 1. The structure of MBAPbBr$_3$. (**a**) The molecular structure scheme of methylbenzylamine. (**b**,**c**) Hydrogen bonds between the [PbBr$_3$]$^-$ chain and methylbenzylamine in MBAPbBr$_3$. (**d**,**e**) The structures of R-MBAPbBr$_3$ (**d**) and S-MBAPbBr$_3$ (**e**) along the b-axis.

2.2. Electronic Structures

To investigate the electron structural properties, the electronic band structures and density of states of both R- and S-MBAPbBr$_3$ were calculated via DFT (Figure S2); the two structures have almost identical electronic band structures. The valence band maximum (VBM) and conduction band minimum (CBM) of R- and S-MBAPbBr$_3$ are located at (0.236842, 0.5, 0.5) and (0, 0, 0) in k-space, showing indirect bandgaps of 3.571 and 3.573 eV, respectively. The partial density of states was subsequently calculated to identify the orbital contribution during the excitation process. The VBMs of R- and S-MBAPbBr$_3$ are mainly contributed by the 4p orbital of Br atoms, and the two CBMs are mainly derived from the 6p orbital of Pb atoms. The above results indicate that the band edges of the two perovskites are mainly contributed by inorganic PbBr$_6$ octahedra [39].

2.3. PXRD and TGA Measurements

The phase purities of both R- and S-MBAPbBr$_3$ were confirmed by powder X-ray diffraction (PXRD). The cell parameters of the observed crystal were refined with the TOPAS-v6 software using a Le Bail algorithm (Figure S3). The peak positions of both R- and S-MBAPbBr$_3$ are almost the same, and the variant peak intensity can be attributed to the difference of exposed crystal surface after grinding. The TGA curves show a plateau below 225 °C and a weight loss of 35.5% between 225 and 230 °C, identifying their stability (Figure S4). The mass loss near 230 °C can be attributed to the removal of vaporization of methylbenzylamine (21.3%) and HBr (14.2%). The good stability of MBAPbBr$_3$ warrants its further characterization.

2.4. Optical Properties

UV–Vis absorption spectra were determined to characterize the excitation behavior (Figure S7). The absorptions of the two 1D MHPs are almost identical as expected for enantiomeric structures, with exciton absorption peaks at 330 nm. The diffuse reflectance measurements were converted to the Kubelka–Munk method, and the bandgaps were calculated using the Kubelka–Munk function $F(R) = (1 - R)^2/2R$, where R represents the reflection coefficient. The bandgaps for *R*- and *S*-MBAPbBr$_3$ were estimated to be 3.59 eV and 3.67 eV, respectively, which are consistent with the calculated values of about 3.57 eV from DFT.

Under irradiation with UV light, the crystals of *R*- and *S*-MBAPbBr$_3$ show yellow emission at room temperature (Figure 2a). Both perovskites have two broad emission peaks extending across the cyan color to the near-infrared region. The maximum emission wavelengths of *R*- and *S*-MBAPbBr$_3$ are 594 and 616 nm, and 592 and 618 nm, respectively. The FWHMs of *R*- and *S*-MBAPbBr$_3$ are estimated to be 181.3 and 178.1 nm, respectively. The appearance of two emission peaks may be attributed to two kinds of exciton paths. To illustrate the PL color at room temperature clearly, the Commission Internationale de L'Eclairage (CIE) chromaticity diagram and color temperatures of PL are illustrated in Figure 2b. The CIE coordinates of *R*- and *S*-MBAPbBr$_3$ are (0.498, 0.471) and (0.510, 0.463) with the color temperatures of 2647 and 2470 K, respectively.

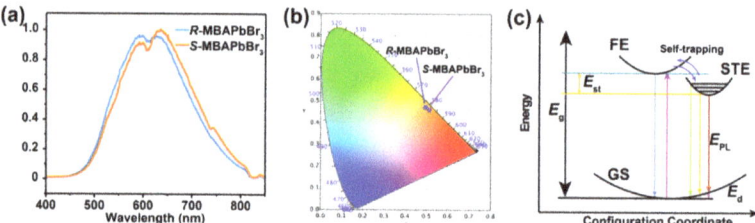

Figure 2. The PL properties of *R*- and *S*-MBAPbBr$_3$. (**a**) PL spectra at 296 K excited by a 325 nm laser. (**b**) The CIE coordinates of PL. (**c**) The configuration coordinate models of PL. FE: free exciton, GS: ground state, E_g: bandgap, E_{st}: self-trapped energy, E_d: lattice distortion energy, E_{PL}: emission energy.

The diagram of the PL process is shown in Figure 2c. Upon UV light irradiation, the electrons in the ground state are excited to form free excitons. Some free excitons radiate photons and return to the ground state directly, which is known as free-exciton emission. Due to lattice distortion caused by strong electron–phonon coupling, some excitons become self-trapped, emitting photons with reduced energy before returning to the ground state [40]. This STE radiative process leads to the broad emission spectra of the two 1D MHPs.

To further explore the properties of the STE emission behavior, PL spectra at various temperatures were collected (Figure 3). As the temperature decreases from 296 to 146 K, the broad emission peaks gradually redshift by approximately 20 nm with decreased FWHM from 181.3 to 142.2 nm for *R*-MBAPbBr$_3$ and 178.1 to 140.6 nm for *S*-MBAPbBr$_3$, respectively. It is interesting that the PL intensity is increased by about two orders of magnitude with the reduction in temperature. The variation in FWHM could arise from the synergistic effect of electron–phonon coupling and thermal expansion, which is influenced by the strength change of hydrogen bonding. The higher intensity and narrower peak width at low temperatures can be attributed to the suppression of nonradiative complexation of excitons [41–43].

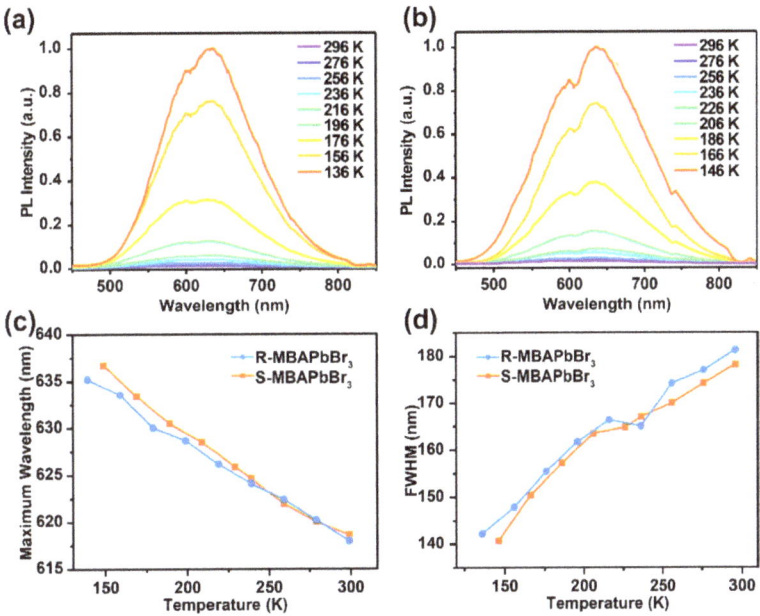

Figure 3. (**a**,**b**) The PL spectra of *R*-MBAPbBr$_3$ (**a**) and *S*-MBAPbBr$_3$ (**b**) at various temperatures. (**c**) The maximum wavelengths of *R*- and *S*-MBAPbBr$_3$ at different temperatures. (**d**) The FWHMs of *R*- and *S*-MBAPbBr$_3$ at different temperatures.

2.5. Elastic Properties

To investigate the elastic properties, the elastic constants (C_{ij}) and bulk modulus (K) of *S*-MBAPbBr$_3$ were calculated by DFT, and the obtained results are listed in Table S7. According to its C_{ij}, the maximal and minimal values of Young's moduli (E) and shear moduli (G) were extracted using the ELATE software [44] as presented in Table S7. The representative 3D and 2D plots of E are shown in Figure 4a,b. The maximum value of E (E_{max}) for this perovskite is 33.2 GPa along the <101> direction due to the large Br–Pb–Br bond angle (154.9°) in this direction. In addition, its E reaches the minimum value (E_{min}) of 6.5 GPa along the <011> direction, which could be attributed to the compliant nature of organic cations packing along this orientation. Accordingly, these two values give an elastic anisotropy ($A_E = E_{max}/E_{min}$) of 5.1, which is relatively larger than that of some 2D MHPs, such as (benzylammonium)$_2$PbBr$_4$ (4.9) [45] and (4-methoxyphenethyammonium)$_2$PbI$_4$ (3.2) [46]. Moreover, the extracted 3D and 2D plots of G for *S*-MBAPbBr$_3$ are shown in Figure 4c,d. It can be observed that the maximal G (G_{max}) is 19.5 GPa along the <010> direction when the (001) plane is sheared, which can be ascribed to the rigid [PbBr$_3$]$^-$ inorganic chains that can significantly resist deformation under the shear force. However, the minimal G (G_{min}) of 2.8 GPa occurs along the <100> inorganic chain direction when the same plane is sheared, which arises from the facile sliding of the 1D inorganic chains under shearing. The obtained elastic anisotropy ($A_G = G_{max}/G_{min}$) of *S*-MBAPbBr$_3$ is 7.0, which is larger than that of 2D (benzylammonium)$_2$PbBr$_4$ (6.5) and (4-methoxyphenethyammonium)$_2$PbI$_4$ (4.0).

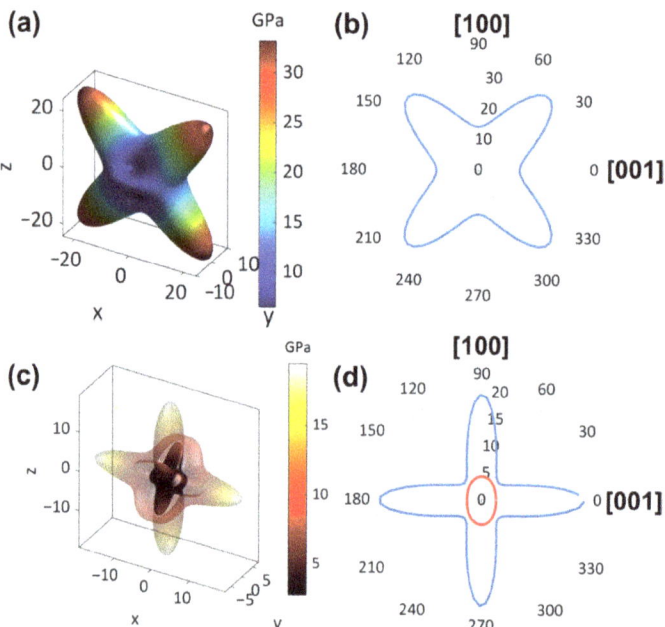

Figure 4. 3D and 2D representations of Young's moduli (**a,b**) and shear moduli (**c,d**) of S-MBAPbBr$_3$. In (**c**), the transparent outer layer and the nontransparent inner layer denote the maximum and minimum values, respectively. The blue outer line and red inner line in (**d**) denote the maximum and minimum values, respectively.

The calculated K of S-MBAPbBr$_3$ is 7.3 GPa, which is significantly smaller than the reported values of 2D MHP (benzylammonium)$_2$PbBr$_4$ (13.6 GPa) and (4-methoxyphenethyammonium)$_2$PbI$_4$ (9.8 GPa), indicating that S-MBAPbBr$_3$ with a 1D structure is more prone to hydrostatic deformation compared with 2D MHPs. According to Pugh's criterion [47], the brittleness of materials can be quantified by the ratio of K/G. The materials with $K/G < 1.75$ are called brittle. The K/G ratio of S-MBAPbBr$_3$ in the range of 0.17–1.99, implying that this MHP would be fairly brittle along certain directions. The low elastic modulus of S-MBAPbBr$_3$ implies that these 1D MHPs could be more desirable for applications in flexible devices, in comparison to 2D and 3D MHPs, although their fragile nature along certain crystallographic directions needs to be taken into account.

3. Materials and Methods

The synthetic method of chiral R-MBAPbBr$_3$ is described in the literature [37,39]. (R)-Methylbenzylamine (C$_8$H$_{11}$N, 0.15 g, 1 mmol, Figure 1a) and lead bromide (PbBr$_2$, 0.239 g, 0.5 mmol) were added to a mixture of acetonitrile (5 mL) and hydrobromic acid (HBr, 5 mL) in a beaker. The mixture was stirred and sonicated to obtain a colorless solution, and the solution was slowly evaporated overnight. The colorless crystal was washed with methanol and dried under vacuum (melting point: 208 °C). The synthetic method of chiral S-MBAPbBr$_3$ is similar to that of R-MBAPbBr$_3$ except (R)-methylbenzylamine was replaced by (S)-methylbenzylamine. Melting point: 209 °C. The mass spectra of R- and S-MBAPbBr$_3$ are shown in Figures S5 and S6.

The single-crystal X-ray diffraction (SC-XRD) tests of S-MBAPbBr$_3$ were performed using a Rigaku XtaLAB PPO MM007 CCD diffractometer with a Cu-Kα target radiation source ($\lambda = 1.54184$ Å) at 293 K and MoKα ($\lambda = 0.71073$ Å) at 100 K, respectively. Using Olex2 [48], the structure was directly solved by ShelXT [49] and refined anisotropically for all nonhydrogen atoms by full-matrix least squares on all F^2 data using ShelXL [50]. All hydrogen atoms were added according to the theoretical model with isotropic displacement parameters and allowed to ride on parent atoms.

Powder X-ray diffraction (PXRD) tests were performed using a Rigaku MiniFlex 600 diffractometer. The samples of *R*- and *S*-MBAPbBr$_3$ were tested in the range of 3–50° with a step size of 0.02° and a speed of 3°·min^{-1}.

Thermogravimetric analysis (TGA) was performed using a Thermo plus EVO2 TG-DTA 9121 thermoanalyzer under N$_2$ atmosphere with a flow rate of 50 mL·min^{-1}. The measurement temperature ranged from 25 °C to 800 °C with a change rate of 10 °C·min^{-1}.

The electronic structure was calculated taking the generalized gradient approximation with a Perdew–Burke–Ernzerh (GGA-PBE) exchange-correlation functional [51] by VASP [52–54]. The plane-wave cutoff energy was set to 450 eV, and a Monkhorst–Pack K-point sampling of 3 × 3 × 1 was used to sample the Brillouin zone. During the geometry optimization step, the cell parameters and atom positions were fully relaxed. The total energy and residual force on each atom converged to 10^{-6} eV and 0.01 eV·Å^{-1}, respectively. The elastic stiffness constants C_{ij} were obtained by the stress–strain method with 0.015 Å of the maximum strain amplitude and seven steps for each strain.

The UV–Vis spectra were measured using a Solidspec 3700 UV–Vis–NIR spectrophotometer with a standard reference of BaSO$_4$ at room temperature. The wavelength range was set to 200–800 nm. Variable temperature photoluminescence experiments were performed using a Horiba LabRAM HR 800 Raman spectrometer excited by a 325 nm He–Cd laser. The photoluminescence (PL) spectra were dispersed by a 600 groove per millimeter diffraction grating and accumulated two times with 2 s of exposure.

4. Conclusions

In summary, the temperature-responsive PL properties and elastic properties of 1D MHPs, *R*- and *S*-MBAPbBr$_3$, were systematically investigated via combined experimental and theoretical approaches. Both *R*- and *S*-MBAPbBr$_3$ exhibit yellow emissions covering a wide wavelength range. With decreasing temperature, the STE emission peaks of both perovskites exhibit narrowed widths and redshifted positions. In addition, the temperature reduction leads to an intensity enhancement of about two orders of magnitude, which can be ascribed to the synergistic effect of electron–phonon coupling and thermal expansion influenced by the alteration of hydrogen bonding. In addition, our DFT calculations reveal that *S*-MBAPbBr$_3$ exhibits a relatively large elastic anisotropy and small bulk modulus, compared with 2D and 3D MHPs. This work demonstrates the temperature-responsive emissions and low elastic properties of LD-MHPs could be useful for making smart optoelectronic devices.

Supplementary Materials: The following supporting information can be downloaded online: Table S1. The cell parameters of *S*-MBAPbBr$_3$ at different temperatures; Figure S1. The change of cell parameters of *S*-MBAPbBr$_3$ at different temperatures and the diagram of thermal expansion; Table S2. The crystal data and structure refinement for *S*-MBAPbBr$_3$ at 100 K and 293 K; Table S3. Bond lengths for *S*-MBAPbBr$_3$-100 K; Table S4. Bond angles for *S*-MBAPbBr$_3$-100 K; Table S5. Bond lengths for *S*-MBAPbBr$_3$-293 K; Table S6. Bond angles for *S*-MBAPbBr$_3$-293 K; Table S6. The electronic structures of *R*- and *S*-MBAPbBr$_3$; Figure S3. The PXRD fitting of MBAPbBr$_3$; Figure S4. The TGA curves of *R*- and *S*-MBAPbBr$_3$; Figure S5. The mass spectrum of *R*-MBAPbBr$_3$; Figure S6. The mass spectrum of *S*-MBAPbBr$_3$; Figure S7. The UV–Vis absorption spectra of *R*- and *S*-MBAPbBr$_3$; Table S7. Summary of the elastic properties of *S*-MBAPbBr$_3$.

Author Contributions: R.F., J.-H.F. and K.L. wrote the manuscript with the cooperation of Y.Q. and Z.-Y.L.; conceptualization, W.L. and X.-H.B.; methodology, R.F. and J.-H.F.; software, R.F. and Z.-G.L.; validation, K.L., Z.-Y.L., W.L. and X.-H.B.; formal analysis, R.F., K.L. and W.L.; investigation, R.F. and J.-H.F.; writing—original draft preparation, R.F. and J.-H.F.; writing—review and editing, K.L., Z.-G.L., Y.Q., Z.-Y.L. and W.L.; visualization, R.F. and J.-H.F.; supervision, W.L. and X.-H.B.; project administration, W.L. and X.-H.B.; funding acquisition, W.L. and X.-H.B. All authors have read and agreed to the published version of the manuscript.

Funding: This research was funded by the National Natural Science Foundation of China (Nos. 21975132 and 21991143).

Institutional Review Board Statement: Not applicable.

Informed Consent Statement: Not applicable.

Data Availability Statement: The crystal data of S-MBAPbBr$_3$ are deposited in The Cambridge Crystallographic Data Centre (Nos. 2132892–2132893).

Conflicts of Interest: The authors declare no conflict of interest.

Sample Availability: Samples of the compounds R- and S- MBAPbBr$_3$ are available from the authors.

References

1. Kieslich, G.; Sun, S.; Cheetham, A.K. Solid-state principles applied to organic-inorganic perovskites: New tricks for an old dog. *Chem. Sci.* **2014**, *5*, 4712–4715. [CrossRef]
2. Gao, F.F.; Li, X.; Qin, Y.; Li, Z.G.; Guo, T.M.; Zhang, Z.Z.; Su, G.D.; Jiang, C.; Azeem, M.; Li, W.; et al. Dual-stimuli-responsive photoluminescence of enantiomeric two-dimensional lead halide perovskites. *Adv. Opt. Mater.* **2021**, *9*, 2100003. [CrossRef]
3. Chen, S.; Chen, C.; Bao, C.; Mujahid, M.; Li, Y.; Chen, P.; Duan, Y. White light-emitting devices based on inorganic perovskite and organic materials. *Molecules* **2019**, *24*, 800. [CrossRef] [PubMed]
4. Ernsting, D.; Billington, D.; Millichamp, T.E.; Edwards, R.A.; Sparkes, H.A.; Zhigadlo, N.D.; Giblin, S.R.; Taylor, J.W.; Duffy, J.A.; Dugdale, S.B. Vacancies, disorder-induced smearing of the electronic structure, and its implications for the superconductivity of anti-perovskite MgC$_{0.93}$Ni$_{2.85}$. *Sci. Rep.* **2017**, *7*, 10148. [CrossRef]
5. Burger, S.; Grover, S.; Butler, K.T.; Boström, H.L.B.; Grau-Crespo, R.; Kieslich, G. Tilt and shift polymorphism in molecular perovskites. *Mater. Horiz.* **2021**, *8*, 2444–2450. [CrossRef]
6. Zhou, X.; Qiao, J.; Xia, Z. Learning from mineral structures toward new luminescence materials for light-emitting diode applications. *Chem. Mater.* **2021**, *33*, 1083–1098. [CrossRef]
7. Vishnoi, P.; Zuo, J.L.; Strom, T.A.; Wu, G.; Wilson, S.D.; Seshadri, R.; Cheetham, A.K. Structural diversity and magnetic properties of hybrid ruthenium halide perovskites and related compounds. *Angew. Chem. Int. Ed.* **2020**, *59*, 8974–8981. [CrossRef]
8. Boström, H.L.B.; Kieslich, G. Influence of metal defects on the mechanical properties of ABX$_3$ perovskite-type metal-formate frameworks. *J. Phys. Chem. C* **2021**, *125*, 1467–1471. [CrossRef]
9. Akman, E.; Akin, S. Poly(N,N′-bis-4-butylphenyl-N,N′-bisphenyl)benzidine-based interfacial passivation strategy promoting efficiency and operational stability of perovskite solar cells in regular architecture. *Adv. Mater.* **2021**, *33*, 2006087. [CrossRef]
10. Guo, H.; Zhang, H.; Shen, C.; Zhang, D.; Liu, S.; Wu, Y.; Zhu, W.-H. A coplanar π-extended quinoxaline based hole-transporting material enabling over 21% efficiency for dopant-free perovskite solar cells. *Angew. Chem. Int. Ed.* **2021**, *60*, 2674–2679. [CrossRef] [PubMed]
11. Hu, L.; Zhao, Q.; Huang, S.; Zheng, J.; Guan, X.; Patterson, R.; Kim, J.; Shi, L.; Lin, C.-H.; Lei, Q.; et al. Flexible and efficient perovskite quantum dot solar cells via hybrid interfacial architecture. *Nat. Commun.* **2021**, *12*, 466. [CrossRef] [PubMed]
12. Jiang, Q.; Zeng, X.; Wang, N.; Xiao, Z.; Guo, Z.; Lu, J. Electrochemical lithium doping induced property changes in halide perovskite CsPbBr$_3$ crystal. *ACS Energy Lett.* **2018**, *3*, 264–269. [CrossRef]
13. Vishnoi, P.; Zuo, J.L.; Cooley, J.A.; Kautzsch, L.; Gómez-Torres, A.; Murillo, J.; Fortier, S.; Wilson, S.D.; Seshadri, R.; Cheetham, A.K. Chemical control of spin-orbit coupling and charge transfer in vacancy-ordered ruthenium(iv) halide perovskites. *Angew. Chem. Int. Ed.* **2021**, *60*, 5184–5188. [CrossRef] [PubMed]
14. Wang, H.-P.; Li, S.; Liu, X.; Shi, Z.; Fang, X.; He, J.-H. Low-dimensional metal halide perovskite photodetectors. *Adv. Mater.* **2021**, *33*, 2003309. [CrossRef] [PubMed]
15. Zhang, X.; Zhu, T.; Ji, C.; Yao, Y.; Luo, J. In-situ epitaxy growth of centimeter-sized lead-free (BA)$_2$CsAgBiBr$_7$/Cs$_2$AgBiBr$_6$ heterocrystals for self-driven X-ray detection. *J. Am. Chem. Soc.* **2021**, *143*, 20802–20810. [CrossRef] [PubMed]
16. Zhao, M.; Xia, Z.; Huang, X.; Ning, L.; Gautier, R.; Molokeev, M.S.; Zhou, Y.; Chuang, Y.-C.; Zhang, Q.; Liu, Q.; et al. Li substituent tuning of LED phosphors with enhanced efficiency, tunable photoluminescence, and improved thermal stability. *Sci. Adv.* **2019**, *5*, eaav0363. [CrossRef]
17. Zhou, Y.; Su, B.; Huang, J.; Zhang, Q.; Xia, Z. Broad-band emission in metal halide perovskites: Mechanism, materials, and applications. *Mater. Sci. Eng. R-Rep.* **2020**, *141*, 100548. [CrossRef]
18. Fu, Y.; Zhu, H.; Chen, J.; Hautzinger, M.P.; Zhu, X.Y.; Jin, S. Metal halide perovskite nanostructures for optoelectronic applications and the study of physical properties. *Nat. Rev. Mater.* **2019**, *4*, 169–188. [CrossRef]
19. Li, K.; Dong, L.-Y.; Xu, H.-X.; Qin, Y.; Li, Z.-G.; Azeem, M.; Li, W.; Bu, X.-H. Electronic structures and elastic properties of a family of metal-free perovskites. *Mater. Chem. Front.* **2019**, *3*, 1678–1685. [CrossRef]
20. Wei, W.J.; Jiang, X.X.; Dong, L.Y.; Liu, W.W.; Han, X.B.; Qin, Y.; Li, K.; Li, W.; Lin, Z.S.; Bu, X.H.; et al. Regulating second-harmonic generation by van der waals interactions in two-dimensional lead halide perovskite nanosheets. *J. Am. Chem. Soc.* **2019**, *141*, 9134–9139. [CrossRef] [PubMed]

21. Nakada, K.; Matsumoto, Y.; Shimoi, Y.; Yamada, K.; Furukawa, Y. Temperature-dependent evolution of raman spectra of methylammonium lead halide perovskites, $CH_3NH_3PbX_3$ (X = I, Br). *Molecules* 2019, 24, 626. [CrossRef] [PubMed]
22. Yue, C.-Y.; Sun, H.-X.; Liu, Q.-X.; Wang, X.-M.; Yuan, Z.-S.; Wang, J.; Wu, J.-H.; Hu, B.; Lei, X.-W. Organic cation directed hybrid lead halides of zero-dimensional to two-dimensional structures with tunable photoluminescence properties. *Inorg. Chem. Front.* 2019, 6, 2709–2717. [CrossRef]
23. Li, J.; Wang, H.; Li, D. Self-trapped excitons in two-dimensional perovskites. *Front. Optoelectron.* 2020, 13, 225–234. [CrossRef]
24. Jiang, Q.; Chen, M.; Li, J.; Wang, M.; Zeng, X.; Besara, T.; Lu, J.; Xin, Y.; Shan, X.; Pan, B.; et al. Electrochemical doping of halide perovskites with ion intercalation. *ACS Nano* 2017, 11, 1073–1079. [CrossRef]
25. Pedesseau, L.; Sapori, D.; Traore, B.; Robles, R.; Fang, H.-H.; Loi, M.A.; Tsai, H.; Nie, W.; Blancon, J.-C.; Neukirch, A.; et al. Advances and promises of layered halide hybrid perovskite semiconductors. *ACS Nano* 2016, 10, 9776–9786. [CrossRef]
26. Mao, L.; Wu, Y.; Stoumpos, C.C.; Wasielewski, M.R.; Kanatzidis, M.G. White-light emission and structural distortion in new corrugated two-dimensional lead bromide perovskites. *J. Am. Chem. Soc.* 2017, 139, 5210–5215. [CrossRef] [PubMed]
27. Li, K.; Li, Z.-G.; Xu, J.; Qin, Y.; Li, W.; Stroppa, A.; Butler, K.T.; Howard, C.J.; Dove, M.T.; Cheetham, A.K.; et al. Origin of Ferroelectricity in Two Prototypical Hybrid Organic–Inorganic Perovskites. *J. Am. Chem. Soc.* 2022, 144, 816–823. [CrossRef] [PubMed]
28. Li, J.; Wang, J.; Ma, J.; Shen, H.; Li, L.; Duan, X.; Li, D. Self-trapped state enabled filterless narrowband photodetections in 2D layered perovskite single crystals. *Nat. Commun.* 2019, 10, 806. [CrossRef] [PubMed]
29. Peng, Y.; Yao, Y.; Li, L.; Wu, Z.; Wang, S.; Luo, J. White-light emission in a chiral one-dimensional organic–inorganic hybrid perovskite. *J. Mater. Chem. C* 2018, 6, 6033–6037. [CrossRef]
30. Gautier, R.; Paris, M.; Massuyeau, F. Exciton self-trapping in hybrid lead halides: Role of halogen. *J. Am. Chem. Soc.* 2019, 141, 12619–12623. [CrossRef]
31. Qi, Z.; Chen, Y.; Guo, Y.; Yang, X.; Gao, H.; Zhou, G.; Li, S.L.; Zhang, X.M. Highly efficient self-trapped exciton emission in a one-dimensional face-shared hybrid lead bromide. *Chem. Commun.* 2021, 57, 2495–2498. [CrossRef]
32. Jung, M.H. Broadband white light emission from one-dimensional zigzag edge-sharing perovskite. *New J. Chem.* 2020, 44, 171–180. [CrossRef]
33. Ji, L.-J.; Sun, S.-J.; Qin, Y.; Li, K.; Li, W. Mechanical properties of hybrid organic-inorganic perovskites. *Coord. Chem. Rev.* 2019, 391, 15–29. [CrossRef]
34. Sen, S.; Guo, G.-Y. Electronic structure, lattice dynamics, and magnetic properties of ThXAsN (X = Fe, Co, Ni) superconductors: A first-principles study. *Phys. Rev. B* 2020, 102, 224505. [CrossRef]
35. Sanchez-Palencia, P.; Garcia, G.; Wahnon, P.; Palacios, P. The effects of the chemical composition on the structural, thermodynamic, and mechanical properties of all-inorganic halide perovskites. *Inorg. Chem. Front.* 2021, 8, 3803–3814. [CrossRef]
36. Ma, L.; Li, W.; Yang, K.; Bi, J.; Feng, J.; Zhang, J.; Yan, Z.; Zhou, X.; Liu, C.; Ji, Y.; et al. A- or X-site mixture on mechanical properties of APbX(3) perovskite single crystals. *APL Mater.* 2021, 9, 041112. [CrossRef]
37. Billing, D.G.; Lemmerer, A. Synthesis and crystal structures of inorganic-organic hybrids incorporating an aromatic amine with a chiral functional group. *CrystEngComm* 2006, 8, 686–695. [CrossRef]
38. Robinson, K.; Gibbs, G.V.; Ribbe, P.H. Quadratic elongation: A quantitative measure of distortion in coordination polyhedra. *Science* 1971, 172, 567–570. [CrossRef]
39. Dang, Y.; Liu, X.; Sun, Y.; Song, J.; Hu, W.; Tao, X. Bulk Chiral Halide Perovskite Single Crystals for Active Circular Dichroism and Circularly Polarized Luminescence. *J. Phys. Chem. Lett.* 2020, 11, 1689–1696. [CrossRef] [PubMed]
40. Cai, X.M.; Lin, Y.; Li, Y.; Chen, X.; Wang, Z.; Zhao, X.; Huang, S.; Zhao, Z.; Tang, B.Z. BioAIEgens derived from rosin: How does molecular motion affect their photophysical processes in solid state? *Nat. Commun.* 2021, 12, 1773. [CrossRef]
41. Booker, E.P.; Thomas, T.H.; Quarti, C.; Stanton, M.R.; Dashwood, C.D.; Gillett, A.J.; Richter, J.M.; Pearson, A.J.; Davis, N.; Sirringhaus, H.; et al. Formation of Long-Lived Color Centers for Broadband Visible Light Emission in Low-Dimensional Layered Perovskites. *J. Am. Chem. Soc.* 2017, 139, 18632–18639. [CrossRef] [PubMed]
42. Fang, Y.; Zhang, L.; Yu, Y.; Yang, W.; Wang, K.; Zou, B. Manipulating Emission Enhancement and Piezochromism in Two-Dimensional Organic-Inorganic Halide Perovskite $[(HO)(CH_2)_2NH_3)]_2PbI_4$ by High Pressure. *CCS Chem.* 2021, 3, 2203–2210. [CrossRef]
43. Qin, Y.; Lv, Z.; Chen, S.; Li, W.; Wu, X.; Ye, L.; Li, N.; Lu, P. Tuning Pressure-Induced Phase Transitions, Amorphization, and Excitonic Emissions of 2D Hybrid Perovskites via Varying Organic Amine Cations. *J. Phys. Chem. C* 2019, 123, 22491–22498. [CrossRef]
44. Gaillac, R.; Pullumbi, P.; Coudert, F.-X. ELATE: An open-source online application for analysis and visualization of elastic tensors. *J. Phys. Condens. Matter* 2016, 28, 275201. [CrossRef]
45. Feng, G.; Qin, Y.; Ran, C.; Ji, L.; Dong, L.; Li, W. Structural evolution and photoluminescence properties of a 2D hybrid perovskite under pressure. *APL Mater.* 2018, 6, 114201. [CrossRef]
46. Fan, J.-H.; Qin, Y.; Azeem, M.; Zhang, Z.-Z.; Li, Z.-G.; Sun, N.; Yao, Z.-Q.; Li, W. Temperature-responsive emission and elastic properties of a new 2D lead halide perovskite. *Dalton Trans.* 2021, 50, 2648–2653. [CrossRef]
47. Pugh, S.F. XCII. Relations between the elastic moduli and the plastic properties of polycrystalline pure metals. *Lond. Edinb. Dubl. Phil. Mag.* 2009, 45, 823–843. [CrossRef]

48. Dolomanov, O.V.; Bourhis, L.J.; Gildea, R.J.; Howard, J.A.K.; Puschmann, H. OLEX2: A complete structure solution, refinement and analysis program. *J. Appl. Crystallogr.* **2009**, *42*, 339–341. [CrossRef]
49. Sheldrick, G.M. SHELXT—Integrated space-group and crystal-structure determination. *Acta Crystallogr. Sect. A* **2015**, *71*, 3–8. [CrossRef]
50. Sheldrick, G.M. Crystal structure refinement with SHELXL. *Acta Crystallogr. Sect. C-Struct. Chem.* **2015**, *71*, 3–8. [CrossRef]
51. Perdew, J.P.; Burke, K.; Ernzerhof, M. Generalized gradient approximation made simple. *Phys. Rev. Lett.* **1996**, *77*, 3865–3868. [CrossRef] [PubMed]
52. Kresse, G.; Furthmüller, J. Efficiency of ab-initio total energy calculations for metals and semiconductors using a plane-wave basis set. *Comput. Mater. Sci.* **1996**, *6*, 15–50. [CrossRef]
53. Kresse, G.; Furthmüller, J. Efficient iterative schemes for ab initio total-energy calculations using a plane-wave basis set. *Phys. Rev. B* **1996**, *54*, 11169–11186. [CrossRef] [PubMed]
54. Kresse, G.; Hafner, J. Ab initio molecular dynamics for liquid metals. *Phys. Rev. B* **1993**, *47*, 558–561. [CrossRef] [PubMed]

Article

Reductive Amination Reaction for the Functionalization of Cellulose Nanocrystals

Omar Hassan Omar [1], Rosa Giannelli [2], Erica Colaprico [2], Laura Capodieci [3], Francesco Babudri [2] and Alessandra Operamolla [4,*]

[1] Istituto di Chimica dei Composti Organo Metallici (ICCOM), Section of BARI, Consiglio Nazionale delle Ricerche (CNR), Via Edoardo Orabona 4, I-70126 Bari, Italy; hassan@ba.iccom.cnr.it
[2] Dipartimento di Chimica, Università degli Studi di Bari Aldo Moro, Via Edoardo Orabona 4, I-70126 Bari, Italy; rossella-giannelli@libero.it (R.G.); ericacolaprico@gmail.com (E.C.); francesco.babudri@uniba.it (F.B.)
[3] Laboratory for Functional Materials and Technologies for Sustainable Applications (SSPT-PROMAS-MATAS), ENEA-Italian National Agency for New Technologies, Energy and Sustainable Economic Development, S.S. 7 "Appia" km 706, I-72100 Brindisi, Italy; laura.capodieci@enea.it
[4] Dipartimento di Chimica e Chimica Industriale, Università di Pisa, Via Giuseppe Moruzzi 13, I-56124 Pisa, Italy
* Correspondence: alessandra.operamolla@unipi.it; Tel.: +39-050-2219342

Citation: Hassan Omar, O.; Giannelli, R.; Colaprico, E.; Capodieci, L.; Babudri, F.; Operamolla, A. Reductive Amination Reaction for the Functionalization of Cellulose Nanocrystals. *Molecules* 2021, *26*, 5032. https://doi.org/10.3390/molecules26165032

Academic Editors: Tersilla Virgili and Mariacecilia Pasini

Received: 29 June 2021
Accepted: 17 August 2021
Published: 19 August 2021

Publisher's Note: MDPI stays neutral with regard to jurisdictional claims in published maps and institutional affiliations.

Copyright: © 2021 by the authors. Licensee MDPI, Basel, Switzerland. This article is an open access article distributed under the terms and conditions of the Creative Commons Attribution (CC BY) license (https://creativecommons.org/licenses/by/4.0/).

Abstract: Cellulose nanocrystals (CNCs) represent intriguing biopolymeric nanocrystalline materials, that are biocompatible, sustainable and renewable, can be chemically functionalized and are endowed with exceptional mechanical properties. Recently, studies have been performed to prepare CNCs with extraordinary photophysical properties, also by means of their functionalization with organic light-emitting fluorophores. In this paper, we used the reductive amination reaction to chemically bind 4-(1-pyrenyl)butanamine selectively to the reducing termini of sulfated or neutral CNCs (S_CNC and N_CNC) obtained from sulfuric acid or hydrochloric acid hydrolysis. The functionalization reaction is simple and straightforward, and it induces the appearance of the typical pyrene emission profile in the functionalized materials. After a characterization of the new materials performed by ATR-FTIR and fluorescence spectroscopies, we demonstrate luminescence quenching of the decorated N_CNC by copper (II) sulfate, hypothesizing for these new functionalized materials an application in water purification technologies.

Keywords: cellulose nanocrystals; nanocellulose; pyrene; organic light-emitting material; reductive amination; nanocellulose functionalization; nanocellulose fluorescence

1. Introduction

The nanostructures isolated from native cellulose, such as cellulose nanofibers (CNFs), cellulose nanocrystals (CNCs) and bacterial cellulose (BC) [1–3], have recently attracted tremendous interest, thanks to their outstanding chemical and mechanical properties [4–8]. These nanomaterials of biologic origin are very appealing for numerous applications and have raised the industrial interest, since cellulose is abundant and can be isolated in crystalline nanostructures with low environmental impact [9]. Besides, nanocelluloses are colorless or even transparent and harmless for human health [10,11]. Their potential in industrial application is mainly related to their sustainable nature, which makes them good candidates as substitutes of oil derivatives for the manufacture of goods, in order to reduce pollution and better commit to the principles of green and blue economies [2]. The current research stream is exploring their use in high-volume industrial applications, such as in the fields of water treatment, in the paint and coating industry, in the building industry, in the hygiene sector and in the paper industry [12].

Siding the industrially oriented research, new emerging fields of application of nanocellulose include hydrogels and aerogels [13,14], emulsion stabilizers [15], biocatalyst immobilizers [16], biosensors [17], drug delivery systems [18], adsorbents for contaminants [19,20], nanocomposites for environmental remediation [21], photonic films and transparent substrates for optoelectronic devices, as well as new nanostructured electroactive materials [22–26].

Among the various typologies of nanocellulose, CNCs are defined, according to the standards of the Technical Association of Pulp and Paper Industry (TAPPI), as crystalline rods featuring a high aspect ratio, a diameter between 5 and 50 nm and a length ranging from 100 to 500 nm. CNCs are the smallest nanostructures which may be isolated from native cellulose and display a high content of crystalline domains. CNCs were discovered in the 1950s by Rånby and Battista [27,28], who independently isolated them by acid hydrolysis. However, only in the last 15 years was the high potentiality of cellulose nanocrystals re-evaluated in light of the new intriguing applications listed above.

CNCs display a high specific surface area (SSA) and expose pending hydroxyl groups on their surface: these chemical functionalities may be exploited to carry out surface functionalization of the nanocrystals. For this reason, the concept of topochemical functionalization of nanocellulose, i.e., a functionalization involving only the surface of the nanocrystals, has become crucial with respect to bulk functionalization, i.e., the complete dissolution of the cellulose crystalline structure due to complete derivatization of the cellulose polymer [29]. In this respect, only mild reaction protocols, with controlled reaction times, can prevent crystalline phase transitions or, even worse, dissolution of the cellulose polymer [30]. This approach sometimes brings some complications, such as the necessity to perform a sequence of chemical reactions on nanocellulose to convert the hydroxyl into an activated group, with the difficulty of obtaining a fine control over the surface functionalities, their number and homogeneity.

An alternative and straightforward strategy for nanocellulose functionalization relies on a reaction used in polysaccharide technology, the reductive amination reaction [31], which modifies only the reducing ends of the cellulose nanocrystals, leaving the remaining nanocrystal surface unfunctionalized or available for further chemical manipulation. According to a model proposed so far [32], CNCs correspond to elementary fibrils and are rigid rods composed of 36 parallel cellulose polymer chains, kept together by intermolecular hydrogen bonds. The number 36 comes from the nature of the enzymatic complex which synthesizes cellulose in plants, the cellulose synthase (CESA) complex, composed by 6 rosettes each made of 6 enzymes which polymerize D-glucopyranose from UDP-glucose [33]. The trans-membrane enzymatic complex extrudes the 36 cellulose chains in parallel organization, and the chains associate into a cellulose I crystal structure [34] as soon as they leave the cell wall. Once cellulose nanocrystals have been recovered from a cellulose source by acid hydrolysis, they present the same crystalline association as in their native structure with the reducing termini of the polymer chains on the same side of the cellulose nanorod (as depicted in Figure 1).

The extensive functionalization of CNCs' surface may drastically change their surface properties, for instance modifying their suspension-forming ability in water or the way they interact with oleophilic or oleophobic molecules. Conversely, an incomplete functionalization on the CNCs' surface, for instance with a degree of substitution (DS) ≤ 0.37 [35], would not allow to understand the exact position of the new functional groups, since most of the reactions which may be performed on the cellulose polymer are only selective, but not specific towards the primary alcohol group [31]. This, for some applications, could be tricky. For this reason, an issue arises in which a mild reaction strategy allows specific functionalization of nanocellulose. In the present work, we use the reducing end functionalization to introduce 1-pyrenyl units as substituents on cellulose nanocrystals: the organic fluorophores are introduced to endow the nanocrystals with an additional property, i.e., pyrene-luminescence, while their hydrophilicity is preserved, being the pyrene units introduced only at the reducing end of the nanocrystal, as depicted in Figure 1. We demonstrate that pyrene-emission (and its excimer) is detectable in the functionalized nanocrystals in spite of the low amount of dye linked to the nanocrystals. Second, we demonstrate pyrene luminescence quenching in the presence of Cu(II) salts. These experiments demonstrate that, using the reductive amination reaction, it is possible to introduce an additional property on CNCs while leaving the remaining pending hydroxyl groups available for interaction with the solution or for further functionalization. In our view, this strategy is important in order to have the surface of the nanocrystal available for anchoring on filtration devices.

Figure 1. A cellulose nanorod, represented according to the model proposed in [32], with 1-pyrenyl units attached to the reducing ends (highlighted in yellow) by reductive amination.

2. Results

2.1. Synthesis of the Pyrene Dye

For performing the reductive amination reaction, we needed a pyrene dye functionalized with a pending primary amine functionality. A good precursor of the 1-pyrenyl butanamine **4** was the commercially available 4-(1-pyrenyl)buthanol **1**. The reaction sequence leading to compound **4** starting from **1** and the relative yields are reported in the Scheme 1. We first prepared the tosylate derivative **2** via a room temperature tosylation reaction of the alcohol group with a yield of 55%; following this, we converted the tosyl group into an azide by a nucleophilic substitution performed using sodium azide in DMSO at room temperature. The azide is a good precursor of primary amines, and was isolated in 87% yield. Finally, **4** was afforded in satisfactory yield (50%) by submitting the pyrene azide **3** to a hydrogenation reaction in the presence of palladium, supported on charcoal as a catalyst and EtOH as a solvent. The catalytic hydrogenation was much more efficient in furnishing **4** than the Staudinger reaction, which yielded a mixture of the primary amine with phosphine oxide. Our synthetic sequence slightly differs from other synthetic sequences reported by others [36,37], but allowed the isolation of **4** in good overall yield and good purity. The identity of **4** was assessed by ^{1}H- and ^{13}C-NMR spectroscopies and by LCMS-IT-TOF spectrometry (compare Figures S1–S6, Supplementary Materials).

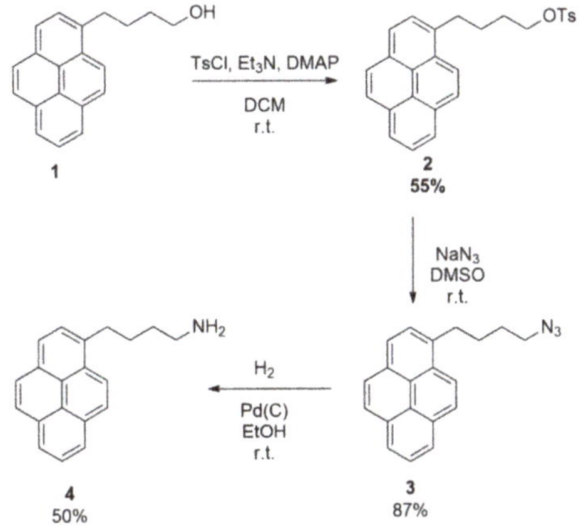

Scheme 1. Synthetic sequence leading to 4-(1-pyrenyl)butanamine **4**.

2.2. Synthesis of Sulfated and Neutral CNCs

Cellulose nanocrystals were prepared by acid hydrolysis with either sulfuric acid or hydrochloric acid at controlled reaction conditions. The starting material was Avicel PH-10.1, a cotton linter-based commercial stationary phase for column chromatography. As extensively reported in the literature, sulfuric acid hydrolysis yields surface negatively charged S_CNC, due to a parallel process of surface sulfatation involving the primary hydroxyl groups on the surface glucopiranose units of the nanocrystals [38]. As explained in the next sections, the surface sulfatation determined aggregation of the cellulose nanocrystals in the presence of the salts used in the following experiments. For this reason, we also prepared neutral cellulose nanocrystals (neutral = no bonded surface charge) by hydrochloric acid hydrolysis from the same starting material. The uncharged nanocrystals were named N_CNC. N_CNC displayed a higher aggregation tendency in distilled water, as an effect of the absence of any surface charge. However, they could be characterized by FE-SEM microscopy by deposition from DMSO suspensions, in which they showed better stability. Figure 2 shows the FE-SEM micrographs recorded on (a) S_CNC deposited from water and (b) N_CNC deposited from DMSO on silicon substrates. Thanks to these microscopic observations, we could measure an average length of 149 ± 32 nm and 151 ± 17 nm for S_CNC and N_CNC respectively, while in both cases. the thickness was detected by AFM to be 10 ± 2 nm.

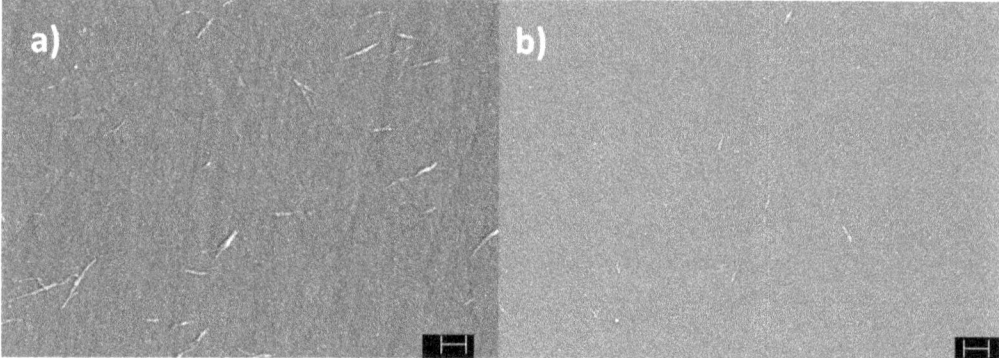

Figure 2. FE-SEM micrographs of (**a**) S_CNC deposited from 10 mg L^{-1} water suspension, and (**b**) N_CNC deposited from 1 mg L^{-1} DMSO suspension. Scale bar: 200 nm.

2.3. Synthesis of Pyrene Derivatives of CNCs

The functionalization of cellulose nanocrystals with the dye **4** was conveniently performed in water solvent, dispersing the cellulose nanocrystals in a pH 6.0 and 50 mM phosphate buffer. A 1% molar ratio between the dye and the nanocellulose (mmol of **4** vs. mmol of glucopyranose) was used to perform the reaction. The functionalization was carried out in the presence of a molar excess of sodium cyanoborohydride. The synthesis of the pyrene-functionalized cellulose nanocrystals is depicted in Scheme 2. At the end of the reaction time, the nanocrystals were purified by washing with methanol, followed by dialysis against mQ water to eliminate traces of unreacted dye and excess salts. Finally, the samples were freeze-dried and used in the next experiments. An AFM check of the preserved nanocrystals' morphology was performed on their thin films. Figure S7, reported in the Supplementary Materials, demonstrates the preserved nanorod aspect of the S_CNC sample after the reductive amination.

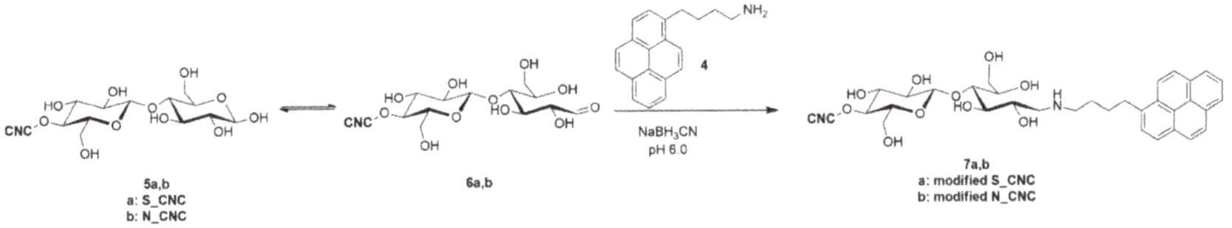

Scheme 2. Reductive amination between **4** and sulfated or neutral cellulose nanocrystals.

ATR-FTIR spectroscopy performed on the freeze-dried samples was very useful to have an indication of the success of the functionalization. Figure 3 shows the ATR-FTIR spectra of the 4-(1-pyrenyl)butanamine **4** (blue trace), of S_CNC (black trace) and of the modified S_CNC sample **7a** (red trace). The spectrum of the cellulose nanocrystals displayed the common features of cellulose I, with a strong absorption band at 3340 cm^{-1} attributed to -OH groups' stretching, forming intramolecular hydrogen bonds. Absorption bands attributed to C-H stretching modes below 3000 cm^{-1} were also clearly visible and centered in the nanocelluloses at 2900 cm^{-1}. The signal centered at 1644 cm^{-1} in CNCs could be attributed to residual water contained in the cellulose nanocrystals. The intense band between 920 and 1220 cm^{-1} in the nanocrystals was attributed to stretching modes of single C-O-C bonds of the acetal units.

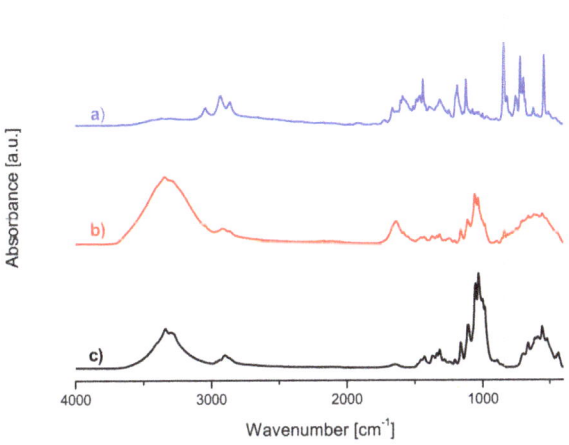

Figure 3. ATR-FTIR spectra of (**a**) 4-(1-pyrenyl)butanamine **4**, (**b**) pyrene-modified S_CNC **7a** and (**c**) unmodified S_CNC.

Butanamine **4** presented distinctive signals, such as a broad absorption at ~3350 cm^{-1} (stretching of -NH$_2$ group), characteristics of absorption of aromatic C-H stretching at 3044 cm^{-1}, -NH$_2$ group bending modes at ~1574 and in the region below 900 cm^{-1}, besides aromatic stretching modes in the region between 1600 and 1400 cm^{-1} and intense aromatic C-H bending modes below 900 cm^{-1}. In particular, the intense absorption band centered at 841 cm^{-1} could be attributed to aromatic C-H bending and produced a small overtone at ~1670 cm^{-1}.

The modified S_CNC **7a** showed typical features of the cellulose nanocrystals in their spectrum, such as the broad absorption band attributed to the -OH stretching, but with the appearance of features representative of the butanamine **4**. First, the increase in intensity of the -OH stretching band with respect to the acetal stretching band pointed at an increase in the sample of hydrogen bond-donating groups (hence the presence of secondary amine functionalities and additional hydroxyl groups in the reducing termini introduced by the reaction), with the contemporary loss in intensity of acetal absorption due to the reducing ends turned from hemiacetal into secondary amines. The broad band of -OH stretching also presented a small shoulder corresponding to the resonance frequency of aromatic C-H stretching at 3044 cm^{-1}. Presence of the aromatic units was also evident from the increase in intensity of the band attributed to crystallized water, overlapped to the overtone of the bending absorption of aromatic C-H. Finally, the strong absorption at 841 cm^{-1}, attributed to aromatic C-H bending, appeared as a distinctive absorption in the spectrum of modified S_CNC **7a**.

2.4. Photophysical Characterization

Emission spectra of the pristine S_CNC and N_CNC and of the pyrene-modified nanocrystals **7a** and **7b** are shown in Figure 4, while the positions of the emission maxima are listed in Table 1. Emission spectra were collected in DMSO to evidence the presence or not of the excimer of pyrene, whose emission is centered at ~470 nm. The pyrene butanamine **4** showed a distinctive emission profile, with characteristic emission maxima at 378, 399, 420 and 475 nm (Figure 4e). The last emission peak was absent in water solution and detectable in DMSO (compare Supplementary Figure S8 for the relevant emission spectrum of **4** in water). As expected, modified S_CNC and N_CNC **7a** and **7b** showed the distinctive emission peaks of butanamine **4** at the following wavelengths: 378, 398, 419 and 470 nm. Surprisingly, both pristine S_CNC and N_CNC showed an emission in the UV region, upon excitation at 345 nm, with considerably lower intensity than the pyrene-modified nanocrystals **7a** and **7b**. This emission was not observed in commercial sulfated cellulose nanocrystals (panel f, Figure 4), which were investigated as a control sample. As it will be explained in the Discussion Section, this phenomenon was attributed to the cellulose source used for the preparation of cellulose nanocrystals by acid hydrolysis.

2.5. Luminescence Quenching Experiments

Luminescence quenching experiments were investigated on both pyrene-modified S_CNC and N_CNC **7a** and **7b** in the presence of increasing concentration of diverse salts in water suspensions. The scope of the experiments was to have a preliminary information on whether the pyrene-modified cellulose nanocrystals interacted preferentially with some salts and were therefore suitable to develop sensing systems in water or to being immobilized in membranes for water purification. Several salts were tested: NaCl, FeCl$_3$, CuSO$_4$·5H$_2$O, Hg(OAc)$_2$. The Figure 5 shows the result of CuSO$_4$ slow addition to a water suspension of N_CNC (left panel) and pyrene-modified N_CNC **7b** (right panel). The same experiments were performed on S_CNC, but the pyrene-modified S_CNC **7a** yielded inconsistent response to the addition of each salt (data not shown in this paper), due to aggregation and precipitation of the nanocrystals. Pyrene-modified N_CNC **7b**, instead, yielded a rational response only in the presence of copper sulfate, while were apparently unperturbed by the other salts. These results will be better discussed in the next paragraph.

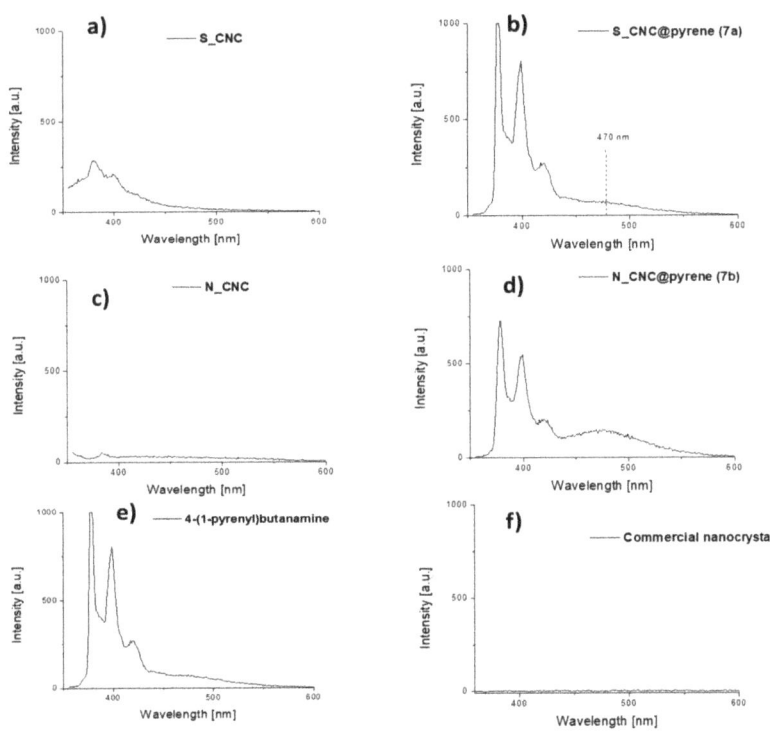

Figure 4. Emission spectra of cellulose nanocrystals recorded in 0.2 mg mL^{-1} DMSO suspensions: (**a**) S_CNC, (**b**) pyrene-modified S_CNC **7a**, (**c**) N_CNC, (**d**) pyrene-modified N_CNC **7b**, (**e**) spectrum of 4-(1-pyrenyl)butanamine **4** in the same solvent and (**f**) commercial sulfated CNCs. Excitation wavelength = 345 nm. Emission spectra of S_CNC and N_CNC were recorded using a 1 cm cuvette. Emission spectra of pyrene-modified S_CNC and N_CNC **7a** and **7b** were recorded using a 0.1 cm cuvette.

Table 1. Emission maxima of all samples measured in DMSO.

Sample	$\lambda_{em,DMSO}$ (nm)
4	378, 399, 420, 475 [1]
S_CNC	378, 400
N_CNC	383
Commercial sulfated nanocrystals	-
Pyrene-modified S_CNC **7a**	378, 398, 419, 470
Pyrene-modified N_CNC **7b**	378, 398, 419, 470

[1] Emission peaks of 4-(1-pyrenyl)butanamine **4** in water: 378, 399, 420 nm (Supplementary, Figure S8). Relevant absorption spectra of **4** in DMSO, CHCl$_3$ and water are reported in the Supplementary, Figures S9 and S10. UV-Vis Absorption spectra of S_CNC and pyrene-modified S_CNC **7a** are reported in the Supplementary, Figures S11 and S12.

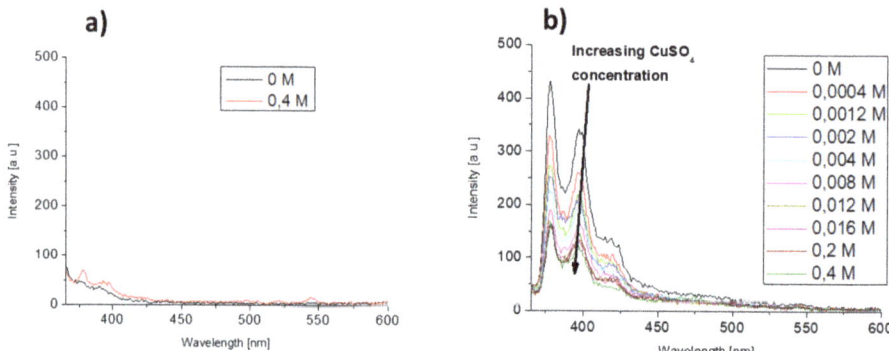

Figure 5. Emission spectra of N_CNC recorded in 0.2 mg mL^{-1} water solutions in the presence of Cu^{2+} salts: (**a**) pristine N_CNC and (**b**) pyrene-modified N_CNC **7b**. The insets show the molar concentration of salts used.

3. Discussion

The reductive amination reaction allowed us to carry out a covalent functionalization of the cellulose nanocrystals with 4-(1-pyrenylbutanamine) **4**, availing of a very mild reaction protocol performed in water. Purification of the sample was straightforward, as it was performed by dialysis in water. To assess the effectiveness of the functionalization, we acquired the ATR-FTIR spectra of S_CNC and their pyrene-modified derivative **7a**, identifying the signals of the pyrene aromatic system in the spectrum of the hybrid system. In order to confirm the effectiveness of the reaction, we also coupled an aliphatic amine with D-glucopyranose in the same conditions, successfully isolating the corresponding amination product (data not shown in this paper). This allowed us to understand that the proposed synthetic protocol is reliable for the effective conversion of the imine intermediate (the Schiff's base generated as a result of the nucleophile attack of the primary amine to the aldose unit) into a secondary amine by sodium cyanoborohydride. The amount of dye bonded to the nanocrystals was very low, as the reductive amination reaction was performed with a molar ratio of **4**:glucopiranose units of 1:100, and the linked pyrene dyes were not detectable by UV-Vis spectroscopy, as they were covered by the scattering produced by the nanocrystals (see Supplementary Figures S10 and S11 acquired for the S_CNC samples).

Then, we started the luminescence characterization. Unexpectedly, recording the reference spectra for S_CNC and N_CNC, we found an emission from both samples, upon excitation of the relevant suspensions at the wavelength of 345 nm, that peaked at 378 and 400 nm for S_CNC and 383 nm for N_CNC (compare Table 1). The relevant spectra for DMSO suspensions are the ones reported in Figure 4a,c. This emission was found on spectra recorded in DMSO and in water (for the relevant emission spectrum recorded from N_CNC water suspensions, please refer to Figure 5a, black trace). Suspecting a possible contamination of the products during their preparation in our laboratories, we repeated the preparation of both typologies of cellulose nanocrystals, by hydrolysis with sulfuric acid or hydrochloric acid, several times in the same and also in different laboratories, each time recording the same result. To finally understand the possible origin of this emission from our cellulose nanocrystals, in the same conditions, we analyzed commercial CNCs purchased from CelluForce. The relevant emission spectrum is reported in Figure 4f. The commercial cellulose nanocrystals did not show any emission feature. Commercial CNCs from cellulose are produced, as declared by the company, by sulfuric acid hydrolysis from wood pulp. Hence, we started to hypothesize whether the emission from the CNCs produced in our laboratories was dependent on the nature of the starting cellulose material, the microcrystalline cellulose Avicel PH-10.1. This microcrystalline cellulose derives from cotton by hydrolysis with diluted mineral acids [39], has a high cellulose I content with very high crystallinity index and it is often used as a reference for XRD or Raman assessment of the crystallinity index of cellulose [40,41].

The luminescent properties of cellulose are the subject of controversial literature reports. Actually, our CNCs displayed a similar emission to the one observed from Avicel by Castellan et al. [42], attributed to heteroaromatics of the furan and pyron type, possibly formed in traces by degradation of the polysaccharide chain. A similar hypothesis was formed by Kalita et al. [43], who attributed the autofluorescence at ~400 nm found in their synthetic CNCs to the presence of fluorescent subunits identified in the FTIR spectra as lignin fragments, such as p-coumeric and ferulic acid or phenolic structures. Ding et al. [44] attributed the autofluorescence of cellulose to a partial double-bond character displayed by the glycosidic bond, whose presence was also hypothesized in disaccharides such as cellobiose and maltose. However, it was demonstrated by Arndt and Stevens [45], using UV-circular dichroism, that mono- and di-saccharides absorb only between 150 and 190 nm. A final interpretation was that by Gan et al. [46], who hypothesized that the emission recorded from cellulose nanocrystals was the consequence of a Stokes scattering [47,48] enhanced by CNC-oriented assembly [49]. In our case, it is very difficult to understand which can be the origin of the luminescence observed in CNCs, as the FTIR spectra of our pristine CNCs also did not reveal the presence of aromatic structures. However, the native emission intensity was very low if compared to the luminescence of pyrene-modified nanocrystals, and it was not responsive to the titration experiments that we performed with salts (compare Figure 5a).

Pyrene-modified S_CNC and N_CNC **7a–b** displayed the distinctive emission features of 4-(1-pyrenyl)butanamine **4**, with emission peaks at 378, 398, 419 and 470 nm. This result was remarkable considering the very low amount of dye used during the functionalization reaction (1 molar % with respect to glucopyranose units). The emission wavelength of 470 nm observed in DMSO was attributed to pyrene excimer formation. At this point, we decided to investigate how interaction with cations influenced the emission intensity from the cellulose nanocrystals, with the aim to understand whether a luminescence quenching induced by any of the salts used could suggest a possible application of the systems that we had prepared in water purification technology. During this investigation, we had to discard experiments conducted on pyrene-modified S_CNC **7a**, because interaction with cations of these surface-negatively charged nanorods induced their aggregation, preventing any reasonable conclusion on their luminescence behavior. We took this behavior as a demonstration that the terminal functionalization had left the surface of the S_CNC unperturbed, exposed to the external solution and therefore free to interact with the other solutes. Conversely, Zhang et al. [50] reported an extensive surface functionalization of S_CNC with pyrene units, and no aggregation was observed in the presence of a wide number of cations. Pyrene-modified N_CNC **7b** yielded a luminescence quenching response only in the presence of increasing concentrations of copper sulfate, while it was apparently unperturbed by NaCl and Hg(OAc)$_2$. FeCl$_3$ was discarded as a quencher, since it displayed an absorption in the relevant region for the experiments (overlapping absorption of the pyrene dye). Thus, among the salts investigated, the pyrene-modified N_CNC **7b** selectively interacted only with copper sulfate.

The presence of two vicinal -OH and secondary amine groups, formed on the C1 and C2 positions of the terminal polysaccharide unit of the nanocrystal in the pyrene-modified N_CNC **7b**, could potentially act as a bidentate ligand for metal cations, with potential assistance from other neighboring -OH groups, while the free -OH pending groups on the CNC surface remain available for establishing H-bonds, which could be, for instance, used to anchor the nanocrystals on a surface or on paper, to obtain a sustainable filtration device. A similar principle was exploited by Lu et al. [51] to prepare a hydrophilic fluorescent paper sensor for nitroaromatic compounds (NACs) exploiting pyrene sensing ability, towards whom our pyrene-cellulose derivatives could be exploited as well.

4. Materials and Methods

4.1. Synthesis of 4-(1-pyrenyl)Butanamine 4

General: All reagents were purchased at the highest commercial quality and used without further purification. Reactions were carried out under a nitrogen atmosphere in oven-dried glassware, using dry solvents unless otherwise stated. Dichloromethane was distilled immediately prior to use on phosphorus pentoxide. Anhydrous grade dimethylsulfoxide was used and dispensed under nitrogen. Absolute ethanol was used without further purification. Preparative column chromatography was performed using silica gel 60, particle size 40 ÷ 63 μm from Merck. Merck silica gel 60 F254 aluminium sheets were used for TLC analyses. All new compounds were characterized by ^{1}H-NMR, ^{13}C-NMR, FTIR spectroscopy and HMRS spectrometry. GC-MS analyses were performed on a gas chromatograph equipped with an SE-30 (methyl silicone, 30 m × 0.25 mm id) capillary column and an ion trap selective mass detector. ^{1}H-NMR and ^{13}C-NMR spectra were recorded at 500 MHz and 125 MHz, respectively, on a Bruker Avance AM 500, using the residual proton peak of CDCl$_3$ at 7.26 ppm as reference for ^{1}H spectra and the signals of CDCl$_3$ at 77 ppm for ^{13}C spectra. Coupling constants values are given in hertz. High-resolution mass spectra were acquired on an Agilent high performance liquid chromatography-QTOF spectrometer via direct infusion of the samples using methanol or water as the elution solvent. Melting points (uncorrected) were determined on a Stuart Scientific Melting point apparatus SMP3.

4-(1-pyrenyl)butyl tosylate 2: A 2 neck round bottom flask was charged under nitrogen with 0.5 g of 4-(1-pyrenyl)buthanol (1.8 mmol), 0.22 g of dimethylaminopyridine (1.8 mmol), 5 mL of dry dichloromethane and 0.5 mL of triethylamine (3.6 mmol). The system was cooled to 0 °C and a solution of tosyl chloride (0.7 g, 3.6 mmol in 5 mL) was added dropwise. The system was kept at 0 °C for 2 h, then it was allowed to warm to room temperature and stirred until TLC analysis revealed complete disappearance of the starting material. The reaction was quenched with 30 mL of HCl 1.5 N. The mixture was extracted with dichloromethane (3 × 30 mL). The organic extracts were collected and dried over anhydrous Na$_2$SO$_4$, filtered and the solvent was distilled under reduced pressure. The crude material was purified by column chromatography on silica gel using hexane and ethyl acetate in volumetric ratio 8:2 as the eluent. 0.423 g of a yellow wax were isolated (yield 55%). ^{1}H-NMR (CDCl$_3$, 500 MHz): δ8.23-8.15 (m, 3H), 8.10 (d, 2H, J = 8.1 Hz), 8.03 (s, 2H), 8.01 (t, 1H, J = 8.1 Hz), 7.80 (d, 2H, J = 7.8 Hz), 7.76 (d, 2H, J = 7.8 Hz), 7.25 (d, 2H, J = 7.3 Hz), 4.09 (t, 2H, J = 6.2 Hz), 3.31 (t, 2H, J = 7.5 Hz), 2.36 (s, 3H), 1.98–1.72 (m, 2H) ppm. ^{13}C-NMR (CDCl$_3$, 126 MHz): δ144.6, 135.8, 133.10, 131.4, 130.8, 129.9, 129.8, 128.6, 127.8, 127.5, 127.3, 127.15, 126.7, 125.9, 125.10, 125.0, 124.9, 124.8, 123.14, 70.3, 32.7, 28.7, 27.5, 21.5, 14.19 ppm.

1-(4-azidobutyl)pyrene 3: A 100 mL round bottom flask was charged with 0.4 g of 4-(1-pyrenyl)butyl tosylate (0.93 mmol), 0.091 g of sodium azide and 5 mL of dry DMSO under nitrogen atmosphere. The mixture was stirred at room temperature for one night. After this time TLC analysis revealed the complete disappearance of the starting material. The mixture was diluted with 30 mL of brine and extracted with ethyl acetate (3 × 30 mL). The organic extracts were collected and washed with brine (2 × 30 mL), dried over anhydrous Na$_2$SO$_4$, filtered and the solvent was distilled under reduced pressure. 0.241 g of a white solid were isolated (yield 87%). Mp = 78–79 °C. ^{1}H-NMR (CDCl$_3$, 500 MHz): δ8.26 (d, 1H, J = 9.5 Hz), 8.17 (dd, 2H, J_1 = 7.5, J_2 = 4.0 Hz), 8.12 (d, 2H, J = 8.5 Hz), 8.04 (d, 1H, J = 9.5 Hz), 8.02 (d, 1H, J = 9.5 Hz), 7.99 (d, 1H, J = 7.5 Hz), 7.86 (d, 1H, J = 8.0 Hz), 3.38 (t, 2H, J = 7.5), 3.34 (t, 2H, J = 7.5 Hz), 1.96 (quintuplet, 2H, J = 7.5 Hz), 1.78 (quintuplet, 2H, J = 7.5 Hz). ppm. ^{13}C-NMR (CDCl$_3$, 126 MHz): δ136.10, 131.4, 130.9, 129.9, 128.6, 127.5, 127.3, 127.18, 126.7, 125.8, 125.10, 125.0, 124.9, 124.8, 124.7, 123.18, 51.4, 33.0, 28.9, 28.8 ppm.

4-(1-pyrenyl)butanamine 4: A 250 mL round bottom flask was charged with 0.22 g of 1-(4-azidobutyl)pyrene (0.74 mmol), 0.079 g of Pd(C) (10%) (0.074 mmol) and 15 mL of absolute ethanol. The mixture was degassed by nitrogen bubbling for 10 min. Afterwards hydrogen was blown into the mixture using a syringe and a balloon. The mixture was stirred at the room temperature for 48 h. After this time, TLC analysis revealed complete disappearance of the starting material. The solvent was removed under reduced pressure and the crude material was purified by column chromatography on silica gel using dichloromethane, methanol and triethylamine in volumetric ratio 90:10:1 as the eluent. 100 mg of a yellow wax was isolated (yield 50%). ^{1}H-NMR (CDCl$_3$, 500 MHz): δ8.26 (d, 1H, J = 8.3 Hz), 8.16 (dd, 2H, J_1 = 8.16 Hz, J_2 = 4.0 Hz), 8.10 (dd, 2H, J_1 = 8.10 Hz, J_2 = 3.0 Hz), 8.06–7.93 (m, 3H), 7.86 (d, 1H, J = 7.86 Hz), 3.36 (t, 2H, J = 8.0 Hz), 2.76 (t, 2H, J = 7.0 Hz), 1.89 (quintuplet, 2H, J = 8.0 Hz), 1.58 (m, 2H) ppm. ^{13}C-NMR (CDCl$_3$, 126 MHz): δ136.8, 131.4, 130.9, 129.8, 128.6, 127.5, 127.2, 127.17, 126.5, 125.8, 125.10, 125.0, 124.8 (two signals), 124.6, 123.4, 42.10, 33.7, 33.3, 29.13 ppm. LCMS-IT-TOF calculated for $C_{20}H_{19}N$: 273.1517, found (M + H)$^+$: m/z 274.1588.

4.2. Hydrolysis of Cellulose, Nanocrystals Isolation and Functionalization

General Avicel PH-101 was used as the starting material for nanocrystalline cellulose isolation. Sonication of nanocellulose suspensions was carried out with a Branson Sonifier 250 (Danbury, CT, USA). Dialysis was carried out at room temperature against mQ water in nitrocellulose tubes with a cut off of 12,600 Da. Commercial nanocrystals used for acquisition of the emission spectrum in the Figure 4f were a donation from the Pharmacy Department of the University of Pisa, and had been purchased from CelluForce.

Synthesis of sulfated CNCs 5a: This procedure was adapted from Operamolla et al. [52] 47 mL of deionized water were introduced in a 250 mL three necked round-bottom flask equipped with a water condenser and a mechanical stirrer. Then, the flask was cooled in an ice bath and 47 mL of concentrated H_2SO_4 were added. After that, 5 g of Avicel PH-101 were added and the suspension was warmed to 50 °C for 80 min. The system was cooled to room temperature and the mixture was transferred to polypropylene centrifugation tubes. Centrifugation at 4000 rpm was repeated replacing the supernatant solution with fresh deionized water until the pH was approximately 1. Then, the precipitate was suspended in deionized water with the aid of a Branson sonifier 250 (Danbury, CT, USA) equipped with an ultrasonic horn with 3.5 mm diameter (micro tip) operated in pulsed mode, with a power of 40 W, 0.6 s pulses for 10 min and dialyzed against distilled water until neutrality using a cellulose nitrate membrane with a molecular weight cut-off of 12,400 Da. The resulting suspension was transferred to polypropylene centrifugation tubes and centrifuged at 4000 rpm for 20 min. The supernatant solution was kept and water was removed under reduced pressure, yielding 925 mg of cellulose nanocrystals with an average length of 280 ± 70 nm.

Synthesis of neutral CNCs 5b: 50 mL of deionized water were introduced in a 250 mL three necked round-bottom flask equipped with a water condenser and a mechanical stirrer. Then, the flask was cooled in an ice bath and 50 mL of concentrated HCl were added. After that, 5 g of Avicel PH-101 were added and the suspension was warmed to 105 °C for 6 h. The system was cooled to room temperature, diluted with 50 mL of distilled water and the mixture was transferred to polypropylene centrifugation tubes. Centrifugation at 1000 rpm for 10 min was repeated 4 times replacing the supernatant solution with fresh deionized water until the pH was approximately 3. Then the precipitate was dialyzed against distilled water until neutrality using a cellulose nitrate membrane with a molecular weight cut-off of 12,400 Da. The resulting suspension was recovered and water was removed by distillation under reduced pressure, yielding 4.713 g of cellulose nanocrystals.

General Procedure for the reductive amination of cellulose nanocrystals (7a,b): In a 100 mL round bottom flask 500 mg of cellulose nanocrystals (3.1 mmol) were suspended in a 50mM phosphate buffer (pH = 6, 50 mL) using Branson sonifier 250 (Danbury, CT, USA) equipped with an ultrasonic horn with 3.5 mm diameter (micro tip) operated in pulsed mode, with a power of 40 W, 0.6 s pulses for 10 min in the case of sulfated CNCs and operated in constant mode in the case of neutral CNCs. A solution of 4-(1-pyrenyl)butanamine (8 mg, 0.03 mmol) in phosphate buffer (5 mL) was added, followed by 100 mg of $NaBH_3CN$ (14.6 mmol). The mixture was stirred at room temperature for 72 h. After this time, the mixture was diluted with methanol (50 mL) and centrifuged at 4000 rpm for 10 min. The liquid phase was removed and replaced with fresh methanol, the systems were mixed and centrifuged again at 4000 rpm for 10 min. The washing operation was repeated two more times. Then, the precipitate was resuspended in water (30 mL) and dialyzed against distilled water using a cellulose nitrate membrane with a molecular weight cut-off of 12,400 Da. Water was distilled under reduced pressure and a white solid was isolated (0.286 g, 57% yield in the case of sulfated CNCs; quantitative yield in the case of neutral CNCs).

4.3. Characterization

ATR-FTIR Analyses FTIR spectra were recorded on a Spectrum, Perkin-Elmer, Waltham, MA, USA. Two spectrophotometer equipped with an UATR Accessory. The nanocellulose powders were placed in direct contact with the diamond/ZnSe crystal without the aid of any solvent. For the same analysis on 4-(1-pyrenyl)butanamine, a drop of chloroform solution with the concentration of 1 mg/mL was deposited on the crystal and the solvent was allowed to evaporate at room temperature before the measurement was acquired.

FE-SEM images The morphology of the surface of nanocrystalline cellulose was analyzed by FE-SEM, on a Field Emission Scanning Electron Microscope, ZEISS Merlin, Jena, Germany, equipped with a GEMINI IIs column and Beam-Booster, with acceleration voltages between 0.05 and 30 kV and 0.8 nm as the best resolution, four optional detectors for SE and BSE, charge compensation and an in situ sample cleaning system. Silicon slabs were used as substrates for FE-SEM measurements. For S_CNC investigation, a solution 0.01% by weight of nanocrystals in water was drop cast on a silicon slab and allowed to dry overnight. For N_CNC investigation, a solution 0.001% by weight of nanocrystals in DMSO was drop cast on a silicon slab and allowed to dry overnight.

AFM Microscopy Atomic force microscopy topographies (Figure S6) were taken using a XE-100 SPM, Park, Suwon, Korea, system microscope. Images were acquired in the tapping mode using tips (Type PPP-NCHR) on a cantilever of 125 μm length, about 330 kHz resonance frequency, 42 N m^{-1} nominal force constant and 10 nm guaranteed tip curvature radius. Surface areas were sampled with a scan rate of 1 Hz. The topographies were analysed using the software XEI (Park System Corporation, version 1.8.0).

UV-VIS Spectroscopy Analyses were recorded on a Spectrophotometer, Shimadzu, Kyoto, Japan, using the software Spectrum, with 1 cm cuvettes in distilled water or spectrophotometric grade DMSO. Cellulose nanocrystals were analyzed at concentration of 0.2 mg/mL.

Emission Spectroscopy Fluorescence spectra were recorded on Cary Eclipse Instrument, Agilent, Santa Clara, CA, USA, with the software Scan. Cellulose nanocrystals were analysed at a concentration of 0.2 mg/mL, in 0.1 cm cuvettes. Titration experiments with heavy metal salts were performed adding increasing volumes of a mother solution of the salt at a concentration of 100 mg/mL in distilled water.

5. Conclusions

In this paper, we have shown how the reductive amination reaction can be used as an effective synthetic protocol to covalently bind luminescent dyes to cellulose nanocrystals. Pyrene-modified S_CNC and N_CNC **7a,b** displayed the emission profile of 4-(1-pyrenyl) butanamine **4**, with emission peaks at 378, 398, 419 and 470 nm, though a very low amount of dye was used during the functionalization reaction (1 molar % with respect to glucopyranose units). The luminescence behavior of the modified nanocrystals, under excitation at the wavelength of 345 nm, was compared with the behavior of the pristine cellulose nanocrystals, evidencing a low emission from nanocellulose, centered at around 400 nm, whose origin is controversial and under literature debate. However, the emission properties of the pyrene-modified samples were clearly identified, and we could test the effect of interactions with different salts on the luminescence properties. In our experiments, N_CNC were revealed to be best-suited to investigate interaction with salts, as they lack any surface-negatively charged group which could induce nanocrystals' aggregation in the presence of positive cations. Pyrene-modified N_CNC **7b** showed luminescence quenching in the presence of copper sulfate, and this suggests a possible application of these systems in water purification technologies, especially after their immobilization on sustainable devices composed of biodegradable cellulose paper.

Supplementary Materials: The following are available online, Figures S1–S12: ^{1}H-NMR and ^{13}C-NMR spectra of compounds **2**, **3** and **4**, AFM topography of a pyrene-modified S_CNC **7a** film, emission of **4** in water, UV-Vis absorption profiles of 4-(1-pyrenylbutanamine) **4** in chloroform, DMSO and water solution, UV-Vis absorption profile of S_CNC and pyrene-modified S_CNC **7a** in DMSO suspension.

Author Contributions: Conceptualization, O.H.O., F.B. and A.O.; methodology, R.G., E.C. and L.C.; validation, O.H.O. and L.C.; formal analysis, A.O.; investigation, O.H.O., R.G., E.C., L.C. and A.O.; resources, A.O.; data curation, R.G. and E.C.; writing—original draft preparation, O.H.O. and A.O.; writing—review and editing, O.H.O., F.B. and A.O.; supervision, A.O.; project administration, F.B. and A.O.; funding acquisition, F.B. and A.O. All authors have read and agreed to the published version of the manuscript.

Funding: This research received financial support from the University of Pisa through the projects PRA_2020_21 "SUNRISE: Concentratori solari luminescenti NIR riflettenti" and "BIHO 2021—Bando Incentivi di Ateneo Horizon e Oltre" (D.d. 408, Prot. n. 0030596/2021).

Institutional Review Board Statement: Not applicable.

Informed Consent Statement: Not applicable.

Data Availability Statement: Data are contained in the article and Supplementary Materials.

Acknowledgments: The authors are grateful to Lorenzo Veronico for performing preliminary experiments and to Angela Punzi for acquiring high-resolution mass spectra on compound **4**. Andrea Mezzetta is acknowledged for furnishing to the authors a sample of commercial sulfated nanocrystals as a kind contribution.

Conflicts of Interest: The authors declare no conflict of interest.

Sample Availability: Samples of the compounds shown in this paper are available upon request from the corresponding author.

References

1. Dufresne, A. *Nanocellulose*; Walter de Gruyter GmbH: Berlin/Boston, Germany, 2018.
2. Thomas, B.; Raj, M.C.; Athira, K.B.; Rubiyah, M.H.; Joy, J.; Moores, A.; Drisko, G.L.; Sanchez, C. Nanocellulose, a Versatile Green Platform: From Biosources to Materials and Their Applications. *Chem. Rev.* **2018**, *118*, 11575–11625. [CrossRef]
3. Kargarzadeh, H.; Ahmad, I.; Thomas, S.; Dufresne, A. *Handbook of Nanocellulose and Cellulose Nanocomposites*; Wiley-VCH Verlag GmbH & Co.: Weinheim, Germany, 2017.
4. Iwamoto, S.; Kai, W.; Isogai, A.; Iwata, T. Elastic Modulus of Single Cellulose Microfibrils from Tunicate Measured by Atomic Force Microscopy. *Biomacromolecules* **2009**, *10*, 2571–2576. [CrossRef]
5. Nishino, T.; Takano, K.; Nakamae, K. Elastic modulus of the crystalline regions of cellulose polymorphs. *J. Polym. Sci. Part B Polym. Phys.* **1995**, *33*, 1647–1651. [CrossRef]
6. Henriksson, M.; Berglund, L.A.; Isaksson, P.; Lindström, T.; Nishino, T. Cellulose Nanopaper Structures of High Toughness. *Biomacromolecules* **2008**, *9*, 1579–1585. [CrossRef]
7. Zhu, H.; Zhu, S.; Jia, Z.; Parvinian, S.; Li, Y.; Vaaland, O.; Hu, L.; Li, T. Anomalous scaling law of strength and toughness of cellulose nanopaper. *Proc. Natl. Acad. Sci. USA* **2015**, *112*, 8971. [CrossRef] [PubMed]
8. Zhu, Z.; Fu, S.; Lavoine, N.; Lucia, L.A. Structural reconstruction strategies for the design of cellulose nanomaterials and aligned wood cellulose-based functional materials—A review. *Carbohydr. Polym.* **2020**, *247*, 116722. [CrossRef] [PubMed]
9. Sharma, P.R.; Joshi, R.; Sharma, S.K.; Hsiao, B.S. A Simple Approach to Prepare Carboxycellulose Nanofibers from Untreated-Biomass. *Biomacromolecules* **2017**, *18*, 2333–2342. [CrossRef] [PubMed]
10. Roman, M. Toxicity of Cellulose Nanocrystals: A Review. *Ind. Biotechnol.* **2015**, *11*, 25–33. [CrossRef]
11. Colombo, L.; Zoia, L.; Violatto, M.B.; Previdi, S.; Talamini, L.; Sitia, L.; Nicotra, F.; Orlandi, M.; Salmona, M.; Recordati, C.; et al. Organ Distribution and Bone Tropism of Cellulose Nanocrystals in Living Mice. *Biomacromolecules* **2015**, *16*, 2862–2871. [CrossRef]
12. Shatkin, J.A.; Wegner, T.H.; Bilek, E.M.; Cowie, J. Market projections of cellulose nanomaterial-enabled products—Part 1: Applications. *TAPPI J.* **2014**, *13*, 9–16. [CrossRef]
13. Wu, Z.-Y.; Liang, H.-W.; Chen, L.-F.; Hu, B.-C.; Yu, S.-H. Bacterial Cellulose: A Robust Platform for Design of Three Dimensional Carbon-Based Functional Nanomaterials. *Acc. Chem. Res.* **2016**, *49*, 96–105. [CrossRef] [PubMed]

14. De France, K.J.; Hoare, T.; Cranston, E.D. Review of Hydrogels and Aerogels Containing Nanocellulose. *Chem. Mater.* **2017**, *29*, 4609–4631. [CrossRef]
15. Tang, C.; Spinney, S.; Shi, Z.; Tang, J.; Peng, B.; Luo, J.; Tam, K.C. Amphiphilic Cellulose Nanocrystals for Enhanced Pickering Emulsion Stabilization. *Langmuir* **2018**, *34*, 12897–12905. [CrossRef] [PubMed]
16. Grishkewich, N.; Mohammed, N.; Tang, J.; Tam, K.C. Recent advances in the application of cellulose nanocrystals. *Curr. Opin. Colloid Interface Sci.* **2017**, *29*, 32–45. [CrossRef]
17. Golmohammadi, H.; Morales-Narváez, E.; Naghdi, T.; Merkoçi, A. Nanocellulose in Sensing and Biosensing. *Chem. Mater.* **2017**, *29*, 5426–5446. [CrossRef]
18. Sunasee, R.; Hemraz, U.D.; Ckless, K. Cellulose nanocrystals: A versatile nanoplatform for emerging biomedical applications. *Expert Opin. Drug Deliv.* **2016**, *13*, 1243–1256. [CrossRef]
19. Zhan, C.; Sharma, P.R.; He, H.; Sharma, S.K.; McCauley-Pearl, A.; Wang, R.; Hsiao, B.S. Rice husk based nanocellulose scaffolds for highly efficient removal of heavy metal ions from contaminated water. *Environ. Sci. Water Res. Technol.* **2020**, *6*, 3080–3090. [CrossRef]
20. Chen, H.; Sharma, S.K.; Sharma, P.R.; Yeh, H.; Johnson, K.; Hsiao, B.S. Arsenic (III) Removal by Nanostructured DialdehydeCellulose–Cysteine Microscale and Nanoscale Fibers. *ACS Omega* **2019**, *4*, 22008–22020. [CrossRef]
21. Zhan, C.; Li, Y.; Sharma, P.R.; He, H.; Sharma, S.K.; Wang, R.; Hsiao, B.S. A study of TiO (2) nanocrystal growth and environmental remediation capability of TiO (2)/CNC nanocomposites. *RSC Adv.* **2019**, *9*, 40565–40576. [CrossRef]
22. Brunetti, F.; Operamolla, A.; Castro-Hermosa, S.; Lucarelli, G.; Manca, V.; Farinola, G.M.; Brown, T.M. Printed Solar Cells and Energy Storage Devices on Paper Substrates. *Adv. Funct. Mater.* **2019**, *29*, 1806798. [CrossRef]
23. Operamolla, A. Recent Advances on Renewable and Biodegradable Cellulose Nanopaper Substrates for Transparent Light-Harvesting Devices: Interaction with Humid Environment. *Int. J. Photoenergy* **2019**, *2019*, 16. [CrossRef]
24. Giannelli, R.; Babudri, F.; Operamolla, A. Chapter 3—Nanocellulose-Based Functional Paper. In *Nanocellulose Based Composites for Electronics*; Thomas, S., Pottathara, Y.B., Eds.; Elsevier: Amsterdam, The Netherlands, 2021; pp. 31–72.
25. Sawalha, S.; Milano, F.; Guascito, M.R.; Bettini, S.; Giotta, L.; Operamolla, A.; Da Ros, T.; Prato, M.; Valli, L. Improving 2D organization of fullerene Langmuir-Schäfer thin films by interaction with cellulose nanocrystals. *Carbon* **2020**, *167*, 906–917. [CrossRef]
26. Milano, F.; Guascito, M.R.; Semeraro, P.; Sawalha, S.; Da Ros, T.; Operamolla, A.; Giotta, L.; Prato, M.; Valli, L. Nanocellulose/Fullerene Hybrid Films Assembled at the Air/Water Interface as Promising Functional Materials for Photo-electrocatalysis. *Polymers* **2021**, *13*, 243. [CrossRef] [PubMed]
27. Rånby, B.G. Fibrous macromolecular systems. Cellulose and muscle. The colloidal properties of cellulose micelles. *Discuss. Faraday Soc.* **1951**, *11*, 158–164. [CrossRef]
28. Battista, O.A. Hydrolysis and Crystallization of Cellulose. *Ind. Eng. Chem.* **1950**, *42*, 502–507. [CrossRef]
29. Dufresne, A. Nanocellulose: A new ageless bionanomaterial. *Mater. Today* **2013**, *16*, 220–227. [CrossRef]
30. Zafeiropoulous, N.E. Engineering the fibre—Matrix interface in natural-fibre composites. In *Properties and Performance of Natural Fibre Composites*; Pickering, K.L., Ed.; CRC Press: Boca Raton, FL, USA, 2008; pp. 127–162.
31. Yalpani, M. *Polysaccharides. Syntheses, Modifications and Structure/Property Relations*, 1st ed.; Studies in Organic Synthesis Series; Elsevier: Amsterdam, The Netherlands, 1988; Volume 36.
32. Ioelovich, M. Recent Findings and the Energetic Potential of Plant Biomass as a Renewable Source of Biofuels—A Review. *Bioresources* **2015**, *10*, 1879.
33. Delmer, D.P.; Amor, Y. Cellulose biosynthesis. *Plant Cell* **1995**, *7*, 987–1000.
34. Gardner, K.H.; Blackwell, J. The structure of native cellulose. *Biopolymers* **1974**, *13*, 1975–2001. [CrossRef]
35. Kim, D.-Y.; Nishiyama, Y.; Kuga, S. Surface acetylation of bacterial cellulose. *Cellulose* **2002**, *9*, 361–367. [CrossRef]
36. Afonso, C.A.M.; Farinha, J.P.S. Synthesis of 4-aryl-butylamine fluorescent probes. *J. Chem. Res.* **2002**, *11*, 584–586. [CrossRef]
37. Battistini, G.; Cozzi, P.G.; Jalkanen, J.-P.; Montalti, M.; Prodi, L.; Zaccheroni, N.; Zerbetto, F.; Battistini, G.; Cozzi, P.G.; Jalkanen, J.-P.; et al. The Erratic Emission of Pyrene on Gold Nanoparticles. *ACS Nano* **2008**, *2*, 77–84. [CrossRef]
38. Lu, P.; Lo Hsieh, Y. Preparation and properties of cellulose nanocrystals: Rods, spheres, and network. *Carbohydr. Polym.* **2010**, *82*, 329–336. [CrossRef]
39. Haafiza, M.K.M.; Eichhorn, S.J.; Hassan, A.; Jawaid, M. Isolation and characterization of microcrystalline cellulose from oilpalm biomass residue. *Carbohydr. Polym.* **2013**, *93*, 628–634. [CrossRef]
40. Park, S.; Baker, J.O.; Himmel, M.E.; Parilla, P.A.; Johnson, D.K. Cellulose crystallinity index: Measurement techniques and their impact on interpreting cellulase performance. *Biotechnol. Biofuels* **2010**, *3*, 10. [CrossRef]
41. Agarwal, U.P. Raman Spectroscopy in the Analysis of Cellulose Nanomaterials. In *Nanocelluloses: Their Preparation, Properties, and Applications*; ACS Symposium Series; American Chemical Society: Washington, DC, USA, 2017; pp. 75–90.
42. Castellan, A.; Ruggiero, R.; Frollini, E.; Ramos, L.A.; Chirat, C. Studies on fluorescence of cellulosics. *Holzforschung* **2007**, *61*, 504–508. [CrossRef]
43. Kalita, E.; Nath, B.K.; Agan, F.; More, V.; Deb, P. Isolation and characterization of crystalline, autofluorescent, cellulose nanocrystals from saw dust wastes. *Ind. Crop. Prod.* **2015**, *65*, 550–555. [CrossRef]
44. Ding, Q.; Han, W.; Li, X.; Jiang, Y.; Zhao, C. New insights into the autofluorescence properties of cellulose/nanocellulose. *Sci. Rep.* **2020**, *10*, 21387. [CrossRef] [PubMed]

45. Arndt, E.R.; Stevens, E.S. Vacuum ultraviolet circular dichroism of simple saccharides. *J. Am. Chem. Soc.* **1993**, *115*, 7849–7853. [CrossRef]
46. Gan, L.; Feng, N.; Liu, S.; Zheng, S.; Li, Z.; Huang, J. Assembly-Induced Emission of Cellulose Nanocrystals for Hiding Information. *Part. Part. Syst. Charact.* **2019**, *36*, 1800412. [CrossRef]
47. Yan, D.; Popp, J.; Pletz, M.W.; Frosch, T. Highly sensitive broadband Raman sensing of antibiotics in step-index hollow-core photonic crystal fibers. *ACS Photonics* **2017**, *4*, 138. [CrossRef]
48. Belli, F.; Abdolvand, A.; Travers, J.C.; Russell, P.S.J. Control of ultrafast pulses in a hydrogen-filled hollow-core photonic-crystal fiber by Raman coherence. *Phys. Rev. A* **2018**, *97*, 013814. [CrossRef]
49. Marchessault, R.H.; Morehead, F.F.; Walter, N.M. Liquid crystal systems from fibrillar polysaccharides. *Nature* **1959**, *184*, 632–633. [CrossRef]
50. Zhang, L.; Li, Q.; Zhou, J.; Zhang, L. Synthesis and Photophysical Behavior of Pyrene-Bearing Cellulose Nanocrystals for Fe^{3+} Sensing. *Macromol. Chem. Phys.* **2012**, *213*, 1612–1617. [CrossRef]
51. Lu, W.; Zhang, J.; Huang, Y.; Theato, P.; Huang, Q.; Chen, T. Self-diffusion driven ultrafast detection of ppm-level nitroaromatic pollutants in aqueous media using a hydrophilic fluorescent paper sensor. *ACS Appl. Mater. Interfaces* **2017**, *9*, 23884. [CrossRef] [PubMed]
52. Operamolla, A.; Casalini, S.; Console, D.; Capodieci, L.; Di Benedetto, F.; Bianco, G.V.; Babudri, F. Tailoring water stability of cellulose nanopaper by surface functionalization. *Soft Matter* **2018**, *14*, 7390–7400. [CrossRef]

Article

Optical Gain in Semiconducting Polymer Nano and Mesoparticles

Mark Geoghegan [1], Marta M. Mróz [2], Chiara Botta [3], Laurie Parrenin [4], Cyril Brochon [4], Eric Cloutet [4], Eleni Pavlopoulou [4], Georges Hadziioannou [4] and Tersilla Virgili [2,*]

[1] Department of Physics and Astronomy, University of Sheffield, Sheffield S3 7RH, UK; mark.geoghegan@newcastle.ac.uk
[2] IFN-CNR, Dipartimento di Fisica, Politecnico di Milano, 20132 Milan, Italy; marta.mroz@polimi.it
[3] SCITEC-CNR, Istituto di Scienze e Tecnologie Chimiche "Giulio Natta", 20133 Milan, Italy; chiara.botta@scitec.cnr.it
[4] Laboratoire de Chimie des Polymères Organiques (LCPO) UMR 5629, CNRS-Université de Bordeaux-Bordeaux INP, CEDEX, 33607 Pessac, France; laurie.parrenin@gmail.com (L.P.); Cyril.Brochon@enscbp.fr (C.B.); Eric.cloutet@enscbp.fr (E.C.); epavlopoulou@iesl.forth.gr (E.P.); hadzii@enscbp.fr (G.H.)
* Correspondence: tersilla.virgili@polimi.it

Abstract: The presence of excited-states and charge-separated species was identified through UV and visible laser pump and visible/near-infrared probe femtosecond transient absorption spectroscopy in spin coated films of poly[N-9"-heptadecanyl-2,7-carbazole-alt-5,5-(4,7-di-2-thienyl-2',1',3'-benzothiadiazole)] (PCDTBT) nanoparticles and mesoparticles. Optical gain in the mesoparticle films is observed after excitation at both 400 and 610 nm. In the mesoparticle film, charge generation after UV excitation appears after around 50 ps, but little is observed after visible pump excitation. In the nanoparticle film, as for a uniform film of the pure polymer, charge formation was efficiently induced by UV excitation pump, while excitation of the low energetic absorption states (at 610 nm) induces in the nanoparticle film a large optical gain region reducing the charge formation efficiency. It is proposed that the different intermolecular interactions and molecular order within the nanoparticles and mesoparticles are responsible for their markedly different photophysical behavior. These results therefore demonstrate the possibility of a hitherto unexplored route to stimulated emission in a conjugated polymer that has relatively undemanding film preparation requirements.

Keywords: transient absorption spectroscopy; conjugated polymers; stimulated emission; nanoparticles; mesoparticles

1. Introduction

Semiconducting polymer films may be a viable alternative to inorganic materials when high performance is less important than cost. Polymers have a significant advantage in that they may be solution processed, which permits film preparation routes such as spin coating, doctor blading, and drop casting. However, polymers need not be dissolved in solution but rather can be dispersed in a liquid medium as latex particles before being cast as films. This is appropriate for certain synthetic routes, such as emulsion polymerization or nanoprecipitation, which result in a particle suspension [1]. These are all compatible with various cross-coupling routes that can be used to synthesize conjugated polymers [2,3]. This wide choice of synthesis and film preparation routes allows fine tuning of the morphological and structural properties of the film. Semiconducting polymer particles can be optimized for light emission [4,5], energy generation [6,7], or phototherapies [8,9], among other purposes [10–12].

The optoelectronic properties of polymer films depend, at least indirectly, on their morphology as well as their electronic structure. Because optical and electronic properties are inextricably linked, electronic and excitonic transport need to be optimized for the required optical properties. For main-chain conjugated polymers, charge transport benefits from both interchain and intrachain phenomena and also from a good degree of crystallization. Optical properties, however, are less affected by crystallization than transport properties but perturbing the conjugation length of even rigid polymers is known to affect them [13]. Indeed, where comparisons between amorphous and crystalline polymers have been made, it is noted that changes in optical properties are due to the concomitant change in conjugation length as opposed to chain ordering [14,15].

Semiconducting polymer films that display stimulated emission often have excellent electroluminescent properties, and so effort spent optimizing optical gain is often performed with wider applications in mind [16]. Although optical gain in semiconducting polymer films has been known for some time [17] and is not rare, it is not commonplace either, and some thought is needed to achieve it. Interchain interactions are often a means of quenching excitations, so control of the photophysics of conjugated polymers can be obtained by enhancing single chain properties through dilution in a matrix of an amorphous polymer [18], encapsulation in nanochannels [19], or through the use of an appropriately structured film [20]. The separation of semiconducting polymers from each other (dilution) does not allow for good electronic transport, but other work [21] showed that, by mixing different polyfluorenes, it was possible to have films of semiconducting polymers with both optical gain and excellent charge transport properties. Further, alternative, approaches include mimicking the effects of dilution by adding a small amount of dimethyl phenylene into polyfluorene chains [22] or by the supramolecular separation of chains by threading cyclodextrin rings around the conjugated polymer to create polyrotaxanes [23].

Films are usually produced by spin coating or doctor blading from a polymer solution, which give rise to good uniformity and control of thickness. Rough films may result in weaker optical properties due to the scattering of light by the film surface. Such scattering may be exacerbated in particle films, particularly for particles of optical dimensions. Nevertheless, it could be reasonably argued that nanoparticle films convey some advantages because confinement inherent in the particles allows some element of dilution by enhancing intrachain properties due to the extended conjugation length with respect to interchain interactions. However, transient absorption spectroscopy revealed no evidence of stimulated emission in an aqueous nanoparticle suspension of poly[N-9''-heptadecanyl-2,7-carbazole-alt-5,5-(4,7-di-2-thienyl-2',1',3'-benzothiadiazole)] (PCDTBT) [24]. In contrast, a mesoparticle suspension of PCDTBT in propanol did show stimulated emission in the infrared region of the spectrum. Here, transient absorption is used to show that the photophysics of PCDTBT mesoparticle films (of diameter ~450 nm) is very different to those of nanoparticle films (~50 nm), and in particular, the mesoparticles exhibit good optical gain. Nanoparticles on the other hand exhibit photoinduced absorption due to the formation of charged states, rather than stimulated emission. Both polymers used for these films exhibit the same optical properties when spin cast from chloroform solution to form a continuous PCDTBT film. PCDTBT films are largely amorphous when spin coated from chlorobenzene solution [25], although there is some evidence of limited order when cast from chloroform [26]. These results show that it is possible to create good optical gain by the preparation route used to synthesize the films, rather than by using elaborate means to dilute semiconducting polymers.

2. Results

Figure 1 shows the absorption and emission spectra of the spin-coated nanoparticle, mesoparticle, and continuous polymer films taken as reference. The absorption spectrum of the continuous polymer film (red line) exhibits two broad transitions with peaks at 390 (3.19 eV) and at 550 nm (2.26 eV). In agreement with earlier work [24,27,28] the two bands are identified as the π–π*-transition of the first and second excited singlet states (S_1 and S_2). The emission spectrum shows a broad and unstructured emission between 620 (2 eV) and 920 nm (1.35 eV) (red line), peaking at 703 nm (1.77 eV). The spectra of the nanoparticles (blue lines) are very similar to that of the continuous polymer film, with a red shift of around 65 meV for both peaks in both absorption and emission. This is likely to indicate a more planar chain conformation and an increased conjugation length [29,30].

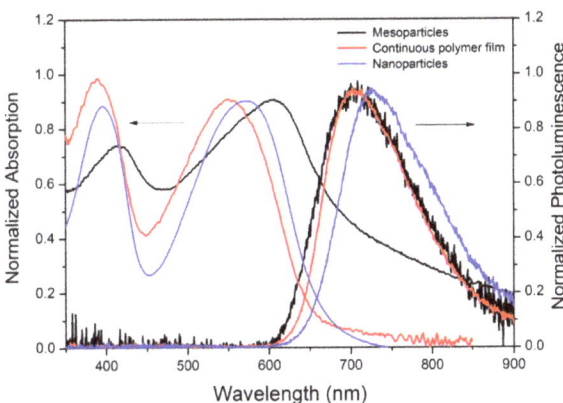

Figure 1. Absorption and emission (rightmost peaks) spectra of the continuous polymer (red line), mesoparticle (black line), and nanoparticle (blue line) spin coated films.

The absorption spectrum of the mesoparticle film (black line in Figure 1), prepared from a lower molar mass polymer, is quite different from those previously reported for PCDTBT [27,28]. In contrast to what would be expected [31], both transition peaks are red-shifted by around 150 meV, and also their relative intensity is changed, with the transition at 607 nm being more intense than that at 415 nm. The long wavelength absorption tail is likely to be due to the Rayleigh scattering of light from the 450 nm particles [32]. Despite the red-shifted absorption, the photoluminescence (PL) emission (black line in Figure 1) is similar to the one of the polymeric film with the main peak at 703 nm, resulting in an evident reduction of the Stokes shift.

The left panel of Figure 2 presents the pump-probe spectra at different probe delays for the continuous polymer film after excitation at 400 nm (Figure 2a) and at 610 nm (Figure 2c). After excitation at 400 nm the pump-probe spectrum shows the presence of an initial bleaching band centred around 550 nm (low energy peak in the absorption spectrum) and a photoinduced absorption band between 650 nm and 780 nm attributed to charged states [24]. In the first ~10 ps a red shift of the bleaching is present (Figure 2a, red line) indicating an energy migration towards the lower energetic sites of the absorption spectrum. In this spectral region an overlap between the positive bleaching signal and the negative one due to charge formation is also present. To highlight the charge formation, the dynamics at 750 nm are considered (Figure 2b), which shows a formation time of around 1 ps.

The low energetic sites at 610 nm were also excited. The pump-probe spectrum shows an instantaneous bleaching of the high and low energetic excited states (see red and black line in Figure 2d) inducing an initial enlargement of the bleaching band of around 60 nm (0.19 eV). Nevertheless, even with this pump excitation the photoinduced absorption (PIA) band due to charged states is present with a low intensity. It is therefore possible to create charges even for excitation of the tail of the polymeric film of the absorption spectrum. In this way fewer charges are created, but a gain region in the continuous polymer film is nevertheless not observed.

The pump-probe spectra are presented at different probe delays for the nanoparticle film after excitation at 400 nm (Figure 3a) and at 610 nm (Figure 3b) and for the mesoparticle film (Figure 3c,d). The difference between the films is evident.

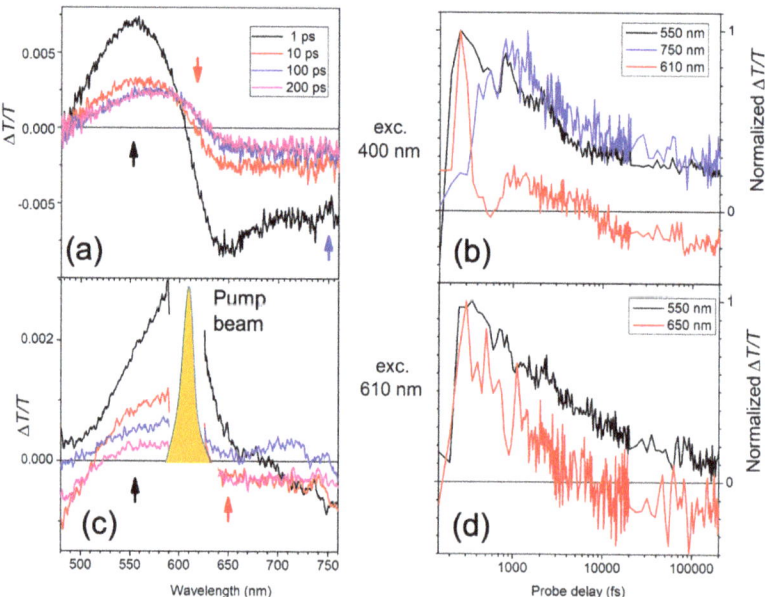

Figure 2. Left panel: Transient absorption spectra at different probe delays for the continuous polymer film at the different wavelengths of pump excitation: 400 nm (**a**) and 610 nm (**b**). The arrows correspond to different wavelengths (black, 550; red, 650; and blue 750 nm) at which the temporal decays are shown in (**b**,**d**). Right panel: Time decays at the two different pump excitation wavelengths: (**b**) 400 and (**d**) 610 nm. The shape of the 610 nm pump beam is also shown in (**c**).

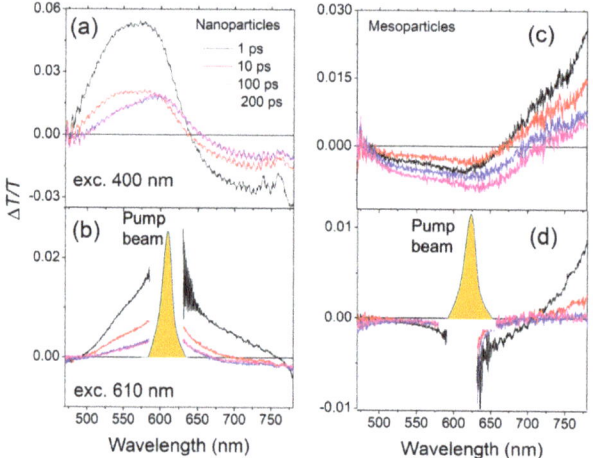

Figure 3. Transient absorption spectra at different probe delays for nanoparticle (**a**,**b**), and mesoparticle (**c**,**d**), films at two different wavelengths of pump excitation: 400 nm (**a**,**c**) and 610 nm (**b**,**d**). The ordinates for all axes are $\Delta T/T$ and the abscissae are wavelength.

The nanoparticle film is first considered. After excitation at 400 nm, the pump-probe spectrum shows the presence of an initial bleaching band centred at around 570 nm (low energy peak in the absorption spectrum) and a photoinduced absorption band at 700–800 nm attributed elsewhere to charged states [24]. The spectrum is similar to the one of the polymer film even if the dynamics show an instantaneous charge formation (blue line in Figure 4a) indicating a more efficient process. Even in this case there is a small red shift of the bleaching band toward low energetic states.

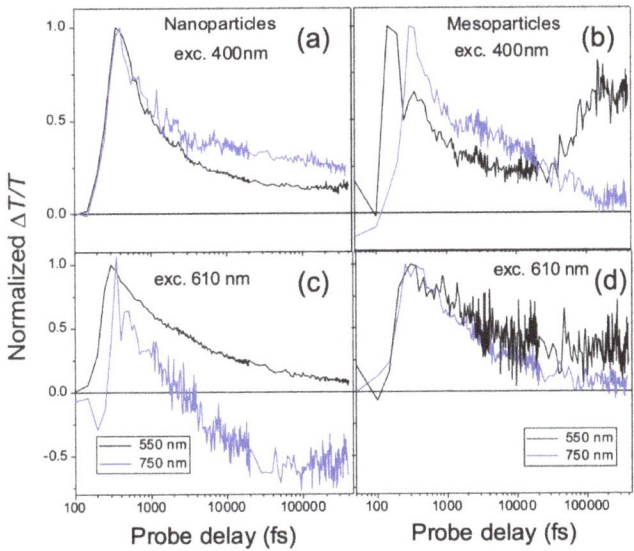

Figure 4. Time decays at different wavelengths for nanoparticle (**a**,**c**), and for mesoparticle (**b**,**d**), spin coated films at two different pump excitation wavelengths of 400 nm (**a**,**b**), and 610 nm (**c**,**d**) are displayed. The ordinates and abscissae for all axes are normalized $\Delta T/T$ and probe delay respectively.

After excitation at 610 nm, a very different behavior was observed compared to that after 400 nm excitation. The positive band extended up to 750 nm indicating the presence of a long-lived gain region that is not present when exciting at 400 nm. It is therefore possible to have optical gain after excitation on the tail of the absorption spectrum. The direct excitation at 610 nm creates few charges (formation time ~10 ps, see Figure 4c), allowing the nanoparticles to have gain (see Figure 5a for the schematic energy diagram after each excitations).

Excitation of the mesoparticle films at 400 nm produces a large PIA band in the visible region and a positive band in the near-infrared region. As already seen in solution [24], the positive band is attributed to stimulated emission, while the instantaneous negative band present in the visible region was not present in the transient transmitted spectra of the solution. Looking at the decay traces (Figure 4b), the signal at 550 nm is instantaneously created. It has a rapid initial decay and then a recovery of the signal after ~200 fs. The positive signal at 750 nm is not instantaneously created and has a similar formation time of ~200 fs. The initial PIA band (PIA_1) at 550 nm is attributed to the S_n–S_m singlet transition, it is followed by a fast internal conversion that increases the population of the state S_1, which in turn produces the stimulated emission (gain) at 750 nm and a signal (PIA_2) from S_1 to S_n (see the schematic energy diagram in Figure 5b) in the visible region. Moreover, after ~50 ps, a negative band centred at 650 nm (PIA_3), attributed to the formation of charged states, overlaps with the initial photoinduced signal [24]. In the mesoparticle film, after excitation at 400 nm, this represents a different photophysics compared to the continuous and nanoparticle films: an internal transition from S_n to S_1, a strong gain signal in the near-infrared region, and slow charge formation are evident.

After excitation of the mesoparticle film at 610 nm, the pump-probe spectra are similar to the 400 nm excitation. A large negative band in the visible region and a positive one (gain) in the near infrared are present. The decay traces show that the signals at 750 and 550 nm are both instantaneously created and behave similarly (Figure 4d). This is then a direct excitation of the S_1 state which produces a stimulated emission in the near infrared and PIA_2 in the visible region (S_1–S_n transition). In this case, charges are not created. In the mesoparticle films, it is possible to have optical gain after excitation at both excitation wavelengths. After excitation at 400 nm, a 200 fs internal conversion brings the excitation to the lower energetic site from which stimulated emission is observed. The direct excitation at 610 nm allows an instantaneous emission with no charge formation.

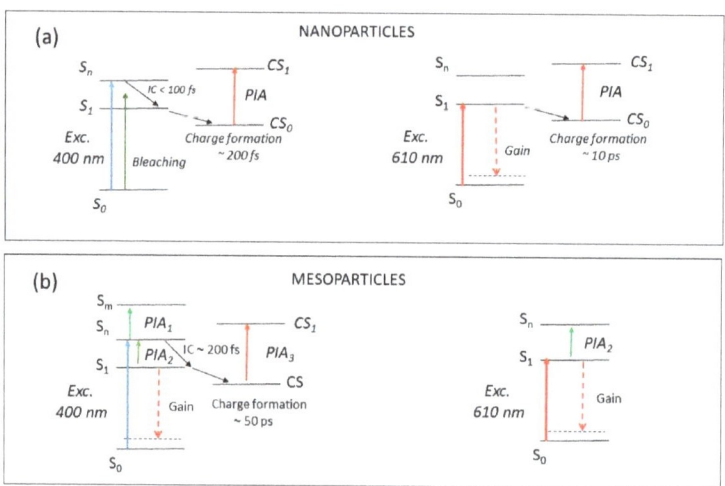

Figure 5. Schematic energy diagrams for the nanoparticle (**a**) and mesoparticle; (**b**) polymer films, where CS indicates the charged state and IC is an internal transition.

3. Discussion

Before comparing the different films, it is important to recall that when these two particles are dissolved in a common solvent, and spin cast to form a continuous film, they exhibit the same photophysical properties [24]. This observation provides strong evidence that differences between the optical properties of the mesoparticle and nanoparticle films are due to size effects, rather than their differing molar masses.

The red shift in the absorption spectrum of the nanoparticle films, with respect to the continuous polymer films, can be associated to an increased conjugation length. The Stokes shift of ~0.49 eV (~150 nm), however, is similar for the two films, with the absorption and emission spectra red-shifted by the same amount. The mesoparticle films, on the other hand, display a much smaller (0.27 eV) Stokes shift, leaving the emission spectrum in the same position as that for the uniform film. The differences in the steady-state absorption-emission spectra of polymeric assemblies are commonly analyzed within an H/J-aggregate model [33,34]. As for the mesoparticle suspension [24] but also for the mesoparticle film absorption spectrum, the π–π*-transition of the first singlet state S_1 is red-shifted with respect to the nanoparticle one while the PL peak is blue-shifted as expected for J-aggregate-like behavior, which is generally associated with stronger intrachain than interchain coupling [20,34]. The weak interchain interactions of the mesoparticle films are responsible for the limited charge formation observed here [27].

Stronger interchain interactions evident in the continuous and nanoparticle films explains the pump-probe spectra. The fast red shift in the bleaching, due to excitonic energy transfer [24], reveals the presence of molecular disorder within the continuous polymer and nanoparticle films. This red shift is not present in the mesoparticle films due to reduced exciton migration and/or greater molecular order in the mesoparticles.

Furthermore, charge formation in the mesoparticle films is suppressed at the longer (610 nm) excitation wavelength. Charge formation is very rapid for both the nanoparticle and continuous films. However, at the higher energy excitation, charge formation is observed in all films, although it is slowest in the mesoparticle film, where formation takes place on the order of 50 ps, compared with 1 ps (continuous film) and 200 fs (nanoparticle film). The reduced efficiency of charge formation in the mesoparticle films indicates of a more ordered system in these films, whereas the rapid charge formation in the other films (nanoparticle and continuous) supports the conclusion of a disordered molecular structure.

Optical gain is absent from the continuous polymer films at both excitation wavelengths, and also from the nanoparticle films at 400 nm. However, long-lived stimulated emission is observed following excitation at 610 nm for the nanoparticle films. This longer wavelength was not tested in the suspension. The mesoparticle films exhibit optical gain in the infrared at both wavelengths.

We cannot exclude optical gain existing in all films, but the nanoparticle and continuous polymer films all exhibit strong photoinduced absorption, which may overlap with the stimulated emission signal. Certainly, it has been possible to detect stimulated emission in the infrared in PCDTBT films spin-cast from chloroform, after excitation at 400 nm, and this signal strongly overlapped with that due to photobleaching [28].

It is unlikely that the particle size is the direct reason for the observed behavior, but rather the effect of the particle size on the chain conformation within the particles. This may also be a consequence of the synthetic route used to prepare the films. In the mesoparticle films, the chains are able to order in a way that gives rise to fewer interparticle interactions competing with intraparticle interactions and consequently less charge formation, which allows for an increased optical gain.

The phenomenological explanation of the optical properties of these films also applies to those of the nanoparticle and mesoparticle suspensions, where it was concluded that the mesoparticles have weaker interchain coupling and the optical properties, including stimulated emission, are likely to be due to a more J-aggregate-like behavior that is not observed for the nanoparticle suspension. Although the nanoparticle films revealed some stimulated emission when fewer charges were created after excitation at 610 nm, the longer wavelength excitation was not performed in the experiments on the suspensions, precluding a direct comparison.

4. Materials and Methods

Two routes to create PCDTBT particles were used. High molar mass nanoparticles were created as described previously [35]. PCDTBT (number average molar mass, M_n = 20.2 kDa with dispersity 2.2) was first synthesized by a Suzuki cross-coupling route [36] before being dissolved in chloroform. This was added to an aqueous solution of sodium dodecyl sulfate and sonicated to form a mini emulsion. The chloroform evaporated after heating at 70 °C, leaving an aqueous nanoparticle dispersion. The mean particle diameter was determined by transmission electron microscopy to be 50 nm.

PCDTBT mesoparticles were created as previously described [37]. PCDTBT (M_n = 4.5 kDa with dispersity 2.1) was synthesized by Suzuki cross-coupling polymerization in propanol solution with poly(vinyl pyrrolidone) added as a surfactant. By adjusting the quantity of surfactant, it was possible to adjust the particle size in the range 0.33–1.3 μm, with a virtually uniform size distribution. The films created in this work were made of 0.45 ± 0.05 μm particles.

For the continuous film, chloroform solutions (40 mg/mL) were prepared by dissolving the mesoparticles and then a film of the polymer was prepared by spin coating.

Film thickness was not a determining factor in these experiments but, in order to get a good optical signal during the transient absorption measurements, the mesoparticle films needed to be less than 1 μm thick. Nanoparticle and continuous films were substantially thinner, and they can be assumed to be of the order of ~100 nm that is typical for transient absorption measurements.

Absorption and emission spectra were acquired using a Shimadzu UV-3600 spectrophotometer (Shimadzu, Marne la-Vallée, France) and a Horiba Scientific Fluoromax-4 spectrofluorometer (Horiba, Palaissau, France), respectively. The excitation wavelength for the emission spectra was 390 nm.

Time-resolved measurements were performed using a homebuilt femtosecond pump–probe setup [24,38]. A Ti:sapphire regenerative amplifier (Libra, Coherent, CA, USA) was used as a laser source, delivering 100 fs pulses at a central wavelength of 800 nm with 4 mJ pulse energy at a repetition rate of 1 kHz. The second harmonic of the fundamental wavelength was used for the excitation at $\lambda = 400$ nm and for the excitation pulse at 610 nm, a single stage non-linear optical parametric amplifier, pumped at 400 nm was used. In order to minimize bimolecular effects, the excitation density was kept very low. White light generated with a 2 mm-thick sapphire plate was used as a probe in the visible-near infrared range from 450 to 780 nm. For spectrally resolved detection of the probe light, spectrographs and CCD arrays were used. The chirp in the white light pulse was taken into account during the analysis and evaluation of the two-dimensional (wavelength and time) $\Delta T(\lambda,\tau)/T$ maps before extraction of the spectral and temporal data using homemade software. Overall, a temporal resolution of at least 150 fs was achieved for all excitation wavelengths.

5. Conclusions

Stimulated emission is a phenomenon which occurs in many conjugated polymers where photoluminescence is observed. However, more often than not, it competes with other processes such as charge creation or photobleaching and its detection and consequent utility is impeded. To this extent, researchers have gone to considerable lengths to create films which have limited interchain optical behavior. In this work, UV-visible absorption and photoluminescence spectroscopy, coupled with time-resolved transient absorption spectroscopy was used to demonstrate unambiguous optical gain in hard latex (450 nm) films of the conjugated polymer PCDTBT. Optical gain was weaker in nanoparticle films. It is concluded that conjugated polymer mini emulsions may be a viable route to creating systems with increased conjugation length. However, a full understanding of why such a preparation route increases the polymer conjugation length, or even whether it does or not, is lacking.

Author Contributions: Conceptualization, M.G. and T.V.; methodology, T.V. and E.C.; formal analysis, T.V.; investigation, T.V., M.M.M., C.B. (Chiara Botta), L.P., E.P., C.B. (Cyril Brochon) and E.C.; writing—original draft preparation, M.G. and T.V.; writing—review and editing, all authors; visualization, M.G. and T.V.; supervision, T.V., M.G., E.C. and E.P.; project administration, M.G.; funding acquisition, T.V. and G.H. All authors have read and agreed to the published version of the manuscript.

Funding: T.V. and M.M.M. acknowledge the project TIMES from Regione Lombardia. In Bordeaux, L.P., C.B., E.C., E.P. and G.H. acknowledge the financial support from the ADEME Project ISOCEL number 1182C0212 and from Region Nouvelle Aquitaine. This work was performed within the framework of the LCPO/Arkema/ANR Industrial Chair HOMERIC ANR-13-CHIN-0002-01 with grant number AC-2013-365.

Data Availability Statement: Data are contained within the article.

Acknowledgments: M.G. is grateful to the Politecnico di Milano and CNR-SCITEC for hosting him in Milan.

Conflicts of Interest: The authors declare no conflict of interest.

Sample Availability: Samples of the compounds are not available from the authors.

References

1. Pecher, J.; Mecking, S. Nanoparticles of Conjugated Polymers. *Chem. Rev.* **2010**, *110*, 6260–6279. [CrossRef]
2. Geoghegan, M.; Hadziioannou, G. *Polymer Electronics*; Oxford University Press: Oxford, UK, 2013.
3. Murad, A.R.; Iraqi, A.; Aziz, S.B.; Abdullah, S.N.; Brza, M.A. Conducting Polymers for Optoelectronic Devices and Organic Solar Cells: A Review. *Polymers* **2020**, *12*, 2627. [CrossRef]

4. Gao, H.; Poulsen, D.A.; Ma, B.; Unruh, D.A.; Zhao, X.; Millstone, J.E.; Fréchet, J.M.J. Site Isolation of Emitters within Cross-Linked Polymer Nanoparticles for White Electroluminescence. *Nano Lett.* **2010**, *10*, 1440–1444. [CrossRef]
5. Piok, T.; Gamerith, S.; Gadermaier, C.; Plank, H.; Wenzl, F.; Patil, S.F.; Montenegro, R.; Kietzke, T.; Neher, D.; Scherf, U.; et al. Organic Light-Emitting Devices Fabricated from Semiconducting Nanospheres. *Adv. Mater.* **2003**, *15*, 800–804. [CrossRef]
6. Kietzke, T.; Neher, D.; Landfester, K.; Montenegro, R.; Güntner, R.; Scherf, U. Novel approaches to polymer blends based on polymer nanoparticles. *Nat. Mater.* **2003**, *2*, 408–412. [CrossRef] [PubMed]
7. Ganzer, L.; Zappia, S.; Russo, M.; Ferretti, A.M.; Vohra, V.; Diterlizzi, M.; Antognazza, M.R.; Destri, S.; Virgili, T. Ultrafast spectroscopy on water-processable PCBM: Rod–coil block copolymer nanoparticles. *Phys. Chem. Chem. Phys.* **2020**, *22*, 26583–26591. [CrossRef] [PubMed]
8. Yang, J.; Choi, J.; Bang, D.; Kim, E.; Lim, E.-K.; Park, H.; Suh, J.-S.; Lee, K.; Yoo, K.-H.; Huh, Y.-M.; et al. Convertible Organic Nanoparticles for Near-Infrared Photothermal Ablation of Cancer Cells. *Angew. Chem. Int. Ed.* **2010**, *50*, 441–444. [CrossRef]
9. Zangoli, M.; Di Maria, F. Synthesis, characterization, and biological applications of semiconducting polythiophene-based nanoparticles. *View* **2021**, *2*, 20200086. [CrossRef]
10. Jana, B.; Ghosh, A.; Patra, A. Photon Harvesting in Conjugated Polymer-Based Functional Nanoparticles. *J. Phys. Chem. Lett.* **2017**, *8*, 4608–4620. [CrossRef] [PubMed]
11. Xu, X.; Liu, R.; Li, L. Nanoparticles made of π-conjugated compounds targeted for chemical and biological applications. *Chem. Commun.* **2015**, *51*, 16733–16749. [CrossRef] [PubMed]
12. Macfarlane, L.R.; Shaikh, H.; Garcia-Hernandez, J.D.; Vespa, M.; Fukui, T.; Manners, I. Functional nanoparticles through π-conjugated polymer self-assembly. *Nat. Rev. Mater.* **2021**, *6*, 7–26. [CrossRef]
13. Kim, J.-S.; Lu, L.; Sreearunothai, P.; Seeley, A.; Yim, K.-H.; Petrozza, A.; Murphy, C.E.; Beljonne, D.; Cornil, J.; Friend, R.H. Optoelectronic and Charge Transport Properties at Organic–Organic Semiconductor Interfaces: Comparison between Polyfluorene-Based Polymer Blend and Copolymer. *J. Am. Chem. Soc.* **2008**, *130*, 13120–13131. [CrossRef]
14. Cadby, A.J.; Lane, P.A.; Mellor, H.; Martin, S.J.; Grell, M.; Giebeler, C.; Bradley, D.D.C.; Wohlgenannt, M.; An, C.; Vardeny, Z.V. Film morphology and photophysics of polyfluorene. *Phys. Rev. B* **2000**, *62*, 15604–15609. [CrossRef]
15. Khan, A.L.T.; Sreearunothai, P.; Herz, L.M.; Banach, M.J.; Köhler, A. Morphology-dependent energy transfer within polyfluorene thin films. *Phys. Rev. B* **2004**, *69*, 085201. [CrossRef]
16. Jiang, Y.; Liu, Y.-Y.; Liu, X.; Lin, H.; Gao, K.; Lai, W.-Y.; Huang, W. Organic solid-state lasers: A materials view and future development. *Chem. Soc. Rev.* **2020**, *49*, 5885–5944. [CrossRef] [PubMed]
17. Moses, D. High quantum efficiency luminescence from a conducting polymer in solution: A novel polymer laser dye. *Appl. Phys. Lett.* **1992**, *60*, 3215–3216. [CrossRef]
18. Virgili, T.; Marinotto, D.; Lanzani, G.; Bradley, D.D. Ultrafast resonant optical switching in isolated polyfluorenes chains. *Appl. Phys. Lett.* **2005**, *86*, 091113. [CrossRef]
19. Martini, I.B.; Craig, I.M.; Molenkamp, W.C.; Miyata, H.; Tolbert, S.H.; Schwartz, B.J. Controlling optical gain in semiconducting polymers with nanoscale chain positioning and alignment. *Nat. Nanotechnol.* **2007**, *2*, 647–652. [CrossRef] [PubMed]
20. Portone, A.; Ganzer, L.; Branchi, F.; Ramos, R.; Caldas, M.J.; Pisignano, D.; Molinari, E.; Cerullo, G.; Persano, L.; Prezzi, D.; et al. Tailoring optical properties and stimulated emission in nanostructured polythiophene. *Sci. Rep.* **2019**, *9*, 7370. [CrossRef] [PubMed]
21. Yap, B.K.; Xia, R.; Campoy-Quiles, M.; Stavrinou, P.N.; Bradley, D.D.C. Simultaneous optimization of charge-carrier mobility and optical gain in semiconducting polymer films. *Nat. Mater.* **2008**, *7*, 376–380. [CrossRef] [PubMed]
22. Zhang, Q.; Wu, Y.; Lian, S.; Gao, J.; Zhang, S.; Hai, G.; Sun, C.; Li, X.; Xia, R.; Cabanillas-Gonzalez, J.; et al. Simultaneously Enhancing Photoluminescence Quantum Efficiency and Optical Gain of Polyfluorene via Backbone Intercalation of 2,5-Dimethyl-1,4-Phenylene. *Adv. Opt. Mater.* **2020**, *8*, 2000187. [CrossRef]
23. Mroz, M.M.; Sforazzini, G.; Zhong, Y.; Wong, K.S.; Anderson, H.L.; Lanzani, G.; Cabanillas-Gonzalez, J. Amplified Spontaneous Emission in Conjugated Polyrotaxanes Under Quasi-cw Pumping. *Adv. Mater.* **2013**, *25*, 4347–4351. [CrossRef] [PubMed]
24. Virgili, T.; Botta, C.; Mróz, M.M.; Parrenin, L.; Brochon, C.; Cloutet, E.; Pavlopoulou, E.; Hadziioannou, G.; Geoghegan, M. Size-Dependent Photophysical Behavior of Low Bandgap Semiconducting Polymer Particles. *Front. Chem.* **2019**, *7*, 409. [CrossRef] [PubMed]
25. Cho, S.; Seo, J.H.; Park, S.H.; Beaupré, S.; Leclerc, M.; Heeger, A.J. A Thermally Stable Semiconducting Polymer. *Adv. Mater.* **2010**, *22*, 1253–1257. [CrossRef] [PubMed]
26. Aïch, R.B.; Blouin, N.; Bouchard, A.; Leclerc, M. Electrical and Thermoelectric Properties of Poly(2,7-Carbazole) Derivatives. *Chem. Mater.* **2009**, *21*, 751–757. [CrossRef]
27. Banerji, N.R.; Cowan, S.; Leclerc, M.; Vauthey, E.; Heeger, A.J. Exciton Formation, Relaxation, and Decay in PCDTBT. *J. Am. Chem. Soc.* **2010**, *132*, 17459–17470. [CrossRef] [PubMed]
28. Etzold, F.; Howard, I.A.; Mauer, R.; Meister, M.; Kim, T.-D.; Lee, K.-S.; Baek, N.S.; Laquai, F. Ultrafast Exciton Dissociation Followed by Nongeminate Charge Recombination in PCDTBT:PCBM Photovoltaic Blends. *J. Am. Chem. Soc.* **2011**, *133*, 9469–9479. [CrossRef]
29. Brown, P.J.; Thomas, D.S.; Köhler, A.; Wilson, J.S.; Kim, J.-S.; Ramsdale, C.M.; Sirringhaus, H.; Friend, R.H. Effect of interchain interactions on the absorption and emission of poly(3-hexylthiophene). *Phys. Rev. B* **2003**, *67*, 064203. [CrossRef]

30. Eggimann, H.J.; Le Roux, F.; Herz, L.M. How β-phase content moderates chain cenergy transfer in polyfluorene films. *J. Phys. Chem. Lett.* **2019**, *10*, 1729–1736. [CrossRef] [PubMed]
31. Shao, B.; Bout, D.A.V. Probing the molecular weight dependent intramolecular interactions in single molecules of PCDTBT. *J. Mater. Chem. C* **2017**, *5*, 9786–9791. [CrossRef]
32. McQuarrie, D.A. *Statistical Mechanics*; University Science Books: Sausalito, CA, USA, 2000.
33. Baghgar, M.; Labastide, J.A.; Bokel, F.; Hayward, R.C.; Barnes, M.D. Effect of Polymer Chain Folding on the Transition from H- to J-Aggregate Behavior in P3HT Nanofibers. *J. Phys. Chem. C* **2014**, *118*, 2229–2235. [CrossRef]
34. Spano, F.C. The Spectral Signatures of Frenkel Polarons in H- and J-Aggregates. *Accounts Chem. Res.* **2010**, *43*, 429–439. [CrossRef] [PubMed]
35. Parrenin, L.; Laurans, G.; Pavlopoulou, E.; Fleury, G.; Pecastaings, G.; Brochon, C.; Vignau, L.; Hadziioannou, G.; Cloutet, E. Photoactive Donor–Acceptor Composite Nanoparticles Dispersed in Water. *Langmuir* **2017**, *33*, 1507–1515. [CrossRef]
36. Wakim, S.; Beaupré, S.; Blouin, N.; Aich, B.-R.; Rodman, S.; Gaudiana, R.; Tao, Y.; Leclerc, M. Highly efficient organic solar cells based on a poly(2,7-carbazole) derivative. *J. Mater. Chem.* **2009**, *19*, 5351–5358. [CrossRef]
37. Parrenin, L.; Brochon, C.; Hadziioannou, G.; Cloutet, E.; Hadziiannou, G. Low Bandgap Semiconducting Copolymer Nanoparticles by Suzuki Cross-Coupling Polymerization in Alcoholic Dispersed Media. *Macromol. Rapid Commun.* **2015**, *36*, 1816–1821. [CrossRef] [PubMed]
38. Virgili, T.; Forni, A.; Cariati, E.; Pasini, D.; Botta, C. Direct Evidence of Torsional Motion in an Aggregation-Induced Emissive Chromophore. *J. Phys. Chem. C* **2013**, *117*, 27161–27166. [CrossRef]

Article

Determination of the Best Empiric Method to Quantify the Amplified Spontaneous Emission Threshold in Polymeric Active Waveguides

Stefania Milanese, Maria Luisa De Giorgi and Marco Anni *

Dipartimento di Matematica e Fisica "Ennio De Giorgi", Università del Salento, Via per Arnesano, 73100 Lecce, Italy; stefy_mil_94@hotmail.it (S.M.); marialuisa.degiorgi@unisalento.it (M.L.D.G.)
* Correspondence: marco.anni@unisalento.it

Received: 28 May 2020; Accepted: 24 June 2020; Published: 30 June 2020

Abstract: Amplified Spontaneous Emission (ASE) threshold represents a crucial parameter often used to establish if a material is a good candidate for applications to lasers. Even if the ASE properties of conjugated polymers have been widely investigated, the specific literature is characterized by several methods to determine the ASE threshold, making comparison among the obtained values impossible. We quantitatively compare 9 different methods employed in literature to determine the ASE threshold, in order to find out the best candidate to determine the most accurate estimate of it. The experiment has been performed on thin films of an homopolymer, a copolymer and a host:guest polymer blend, namely poly(9,9-dioctylfluorene) (PFO), poly(9,9-dioctylfluorene-cobenzothiadiazole) (F8BT) and F8BT:poly(3- hexylthiophene) (F8BT:rrP3HT), applying the Variable Pump Intensity (VPI) and the Variable Stripe Length (VSL) methods. We demonstrate that, among all the spectral features affected by the presence of ASE, the most sensitive is the spectral linewidth and that the best way to estimate the ASE threshold is to determine the excitation density at the beginning of the line narrowing. We also show that the methods most frequently used in literature always overestimate the threshold up to more than one order of magnitude.

Keywords: conjugated polymers; Amplified Spontaneous Emission; optically pumped laser; optical gain; active waveguides

1. Introduction

Earliest discoveries of stimulated emission from conjugated polymers, dating back to 1992 [1] and 1996 [2], gave rise to intense studies on the optical gain and lasing properties of these materials, aiming to develop optically and electrically pumped lasers. Conjugated molecules show unique properties such as wide chemical flexibility, high photoluminescence quantum efficiency (PLQE), high stimulated emission cross section and electrical conductivity [3–12]. Moreover, organic materials can be easily processed: they can be deposited from solution through simple techniques such as spin coating [13], drop casting [14], ink-jet printing [15,16], dip coating [17], solution shearing [18] and blade coating [19].

To date, optical gain and optically pumped lasing have been demonstrated in many families of conjugated molecules and with a wide range on different resonators [20,21] and a first indication of lasing effect under electrical pumping has been recently reported [22]. In particular the organic systems showing optical gain can be divided in the two big families of neat films or blends of active molecules [2,4–12,23–63], most of the times polymeric, and of blends between inert polymers and small molecules [30,64–73].

The neat films have the advantage to preserve the charge mobility, and are thus potentially interesting for the development of laser diodes, while the blends with inert polymers allow to avoid quenching effects due to intermolecular aggregation, and are thus particularly interesting for applications to optically pumped lasers.

Overall the actual performances of organic lasers are still not high enough to allow commercial applications, thus stimulating further research to develop novel active materials.

A typical initial experiment for the characterization of a new material for laser applications is the quantification of its Amplified Spontaneous Emission (ASE) properties. The active material is deposited in the form of thin films on quartz or glass substrates, thus obtaining a planar asymmetric waveguide. The photoluminescence (PL) spectra are acquired as a function of the excitation density and, if the material shows optical gain, for high enough excitation density the spontaneous emission is amplified during the propagation along the waveguide, allowing to observe the appearance of an ASE band in the PL spectra.

In order to compare different active materials and to determine the best candidate for a potential laser implementation, the ASE threshold, representing the minimum excitation density which allows light amplification, is typically used. Materials that show low ASE threshold are considered good laser active materials, and vice-versa.

Despite the importance of the ASE threshold value for the quantification of light amplification properties of materials and for their comparison, to date there is no consensus on the correct way to extract this parameter from the excitation density dependence of the PL spectra (Variable Pump Intensity, VPI, method). In order to have a complete picture of the current state of the art we focused our attention to experiments on the ASE properties of active polymers and we investigated 297 papers published to date, found by a research on Scopus with the string "Amplified Spontaneous Emission polymer" within "Article title, abstract, keywords" and written in English (see Figure 1) and manually removing the papers on inert polymer-dye blends and on non organic materials. The first interesting result (see Figure 1a) is that in 27.0% of the papers on ASE in polymers actually there is not any estimate of the threshold, while in 21.1% of the papers only qualitative values are provided. This means that overall in about one half of the papers the ASE threshold is not present at all, or in the best case only a rough evaluation is given. Focusing the attention on the 51.9% of the papers in which a procedure to determine the threshold is described, we identified the use of 9 different methods (see Figure 1b).

Most of these methods exploit the differences in the intensity increase with the excitation density and in the linewidth between the spontaneous emission and the ASE.

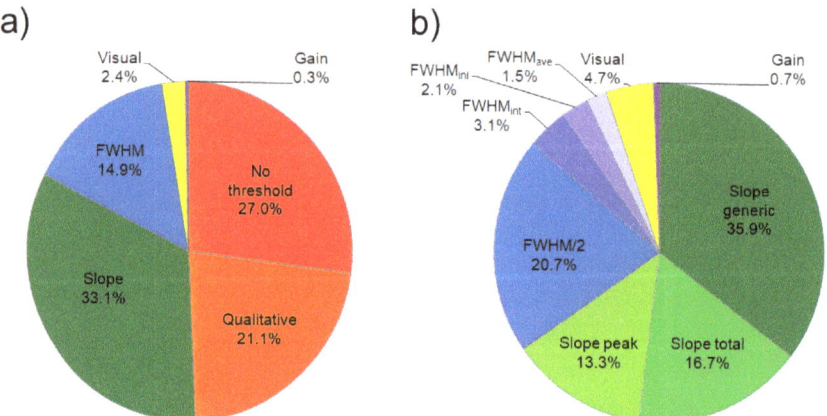

Figure 1. (a) Pie chart showing the percentage of papers on Amplified Spontaneous Emission (ASE) in polymers that quantify the ASE threshold in different ways. (b) Pie chart of the distribution of the main methods to quantify the ASE threshold.

In particular, ASE is characterized by an intensity increase with the excitation density stronger than the spontaneous emission and its spectral lineshape is typically much narrower. Thus, the transition between the excitation density regime dominated by the the spontaneous emission and the one in which ASE prevails can be identified by observing the variation of the emission intensity and of the spectral linewidth as a function of the excitation density.

About 65.9% percent of the papers thus determine the ASE threshold as the excitation density at which the plot of the emission intensity vs. the excitation density shows a slope increase [23–29], looking at the PL total intensity (area under the PL spectrum) in 16.7% of the papers, at the ASE peak intensity in the 13.3%, or without specifying which "intensity" has been investigated in the majority of the cases (35.9%).

About 27.4% percent of the papers instead find out the ASE threshold from the excitation density dependence of the spectral linewidth, by considering the peak Full Width at Half Maximum (FWHM), that typically shows a constant value at low excitation density, when only spontaneous emission is present, a narrowing when ASE sets in and a further constant lower value when ASE dominates. Several different criteria are used, like the excitation density at which the FWHM attains one half of the value it has at low density [33,35–38] (FWHM/2, used in 20.7% of the papers), starts to decrease (even if without a clear definition) [39–43] ($FWHM_{nar}$, 2.1%), reaches the average value between the values assumed at low and at high excitation density [44,45] ($FWHM_{ave}$, 1.5%) or as the excitation density at which the extrapolation of the data representing the FWHM decrease takes the value corresponding to the low excitation density [24,46] ($FWHM_{cros}$, 3.1%).

A different approach exploits the lineshape variation due to the appearance of the ASE band, identifying the ASE threshold as the minimum excitation density at which the ASE band becomes visible [49–56] (Visual, 4.7%).

A last method, apparently more rigorous but also less diffused, defines the ASE threshold as the energy density at which the net gain of the active medium becomes zero, similarly to what happens in a laser cavity [74] (gain, 0.7%).

Even if all these methods are based on some effect related to the appearance of the ASE, the different criteria and experimental methods obviously affect the inferred values, making almost impossible any meaningful comparison of the obtained thresholds and, more importantly, opening the problem of understanding which is the most correct way to determine the ASE threshold value.

In this paper we report on a detailed quantitative comparison between all the main methods to determine ASE threshold, in order to quantify the dependence on the used method and to determine the best one in terms of reliability of the ASE value and of ease of application.

Our analysis is performed on thin films of three different conjugated polymers whose emission covers the entire visible range, namely poly(9,9-dioctylfluorene) (PFO), poly(9,9-dioctylfluorene-cobenzothiadiazole) (F8BT) and a blend of F8BT and regio regular Poly(3-hexylthiophene) (rrP3HT) (F8BT:rrP3HT). These polymers show good solubility in standard organic solvent, good film forming properties and efficient ASE [12,36,38,41,45,55], and can be thus considered good prototype materials for homopolymers (PFO), copolymers (F8BT) and host guest blends (F8BT:rrP3HT).

We demonstrate that, for all the samples, the lowest value of the ASE threshold is obtained by determining the excitation density at which the FWHM of the PL emission peak starts to decrease. We also demonstrate that the two most widespread methods, which consider the output intensity slope variation or the FWHM halving, used in about 86.6% of the published papers that quantify the threshold, provide similar thresholds, but these values are between 2 and 14 times higher than the best threshold estimate. These results are evidence that, among all the spectral features depending on the ASE appearance, the most sensitive one is the FWHM decrease, that is used in only 2.1% of the papers. On the contrary our results suggest that in about 98% of the papers on ASE properties of conjugated polymers the threshold values are overestimated, up to more than one order of magnitude. These results are expected to be of general validity and can provide a useful starting base for the correct quantification of the ASE threshold in polymeric active waveguides.

2. Results

All the investigated methods to determine the ASE threshold are based on the excitation density dependence of the PL spectra. In order to probe all the samples in comparable pumping conditions we initially evaluated the minimum excitation density necessary to observe a PL lineshape variation, due to the ASE band appearance. This excitation density value defines the first ASE threshold estimation (called visual in the following) and allows to fix the excitation density range to perform the VPI measurements. For all the samples we acquired 25 PL spectra at different excitation densities between about 1/10 and about 10 times the visual threshold.

The PL spectra of the PFO sample show (see Figure 2a), at low excitation density, the typical spontaneous emission spectrum of the PFO glassy phase with the 0–0 transition peak at about 425 nm, followed by 0–1 and 0–2 vibronic replicas at about 443 and 480 nm [55]. As the excitation density increases the lineshape is unchanged, up to an excitation density value of 21 $\mu J cm^{-2}$ at which the spectrum shows a shoulder at about 450 nm (visual threshold). At higher excitation densities a clear ASE band peaked at 450 nm is observed, due to amplification of the 0–1 spontaneous emission, progressively dominating the PL spectra.

A progressive blue shift of the ASE band is observed with the excitation density, already observed in other samples of organic waveguides showing ASE, and typically ascribed to the competition between ASE and intermolecular energy migration within the disordered density of states [75–77].

The ASE band shows an intensity growth stronger than the spontaneous emission one and a lower linewidth, thus allowing to exploit the excitation density dependence of the emission intensity and of the spectral linewidth in order to determine the ASE threshold.

Concerning the line narrowing we observe that the FWHM is typically used in order to quantify the spectral linewidth. The obtained values (see Supplementary Materials for details) show (see Figure 2b) an almost constant value of about 16 nm up to about 14 $\mu J\,cm^{-2}$, a progressive decrease up to about 35 $\mu J\,cm^{-2}$ and a constant value of about 4 nm at higher excitation densities. In order to have a quantitative description of this behavior, we performed a best fit with a constant function up to 14 $\mu J\,cm^{-2}$ and with a linear decrease between 21 $\mu J\,cm^{-2}$ and 27 $\mu J\,cm^{-2}$.

From the best fit curves (see Figure 2b) we determined the threshold as the excitation density at which:

- the linewidth reaches one half of its constant value at low excitation density (FWHM/2);
- the linewidth reaches the average value between the high one at low excitation density and the low one at high excitation density ($FWHM_{ave}$);
- the linewidth starts to decrease ($FWHM_{nar}$);
- the extrapolations of the two best fit lines cross ($FWHM_{cros}$).

The intervals of the best fit functions have been determined by changing the constant term in the fit function within one standard deviation, as obtained by the fit procedure. These intervals were then used to estimate the uncertainty of the threshold values.

Concerning the $FWHM_{nar}$ threshold we considered the first point below the lower constant error line, representing the first point which deviates from the best line fit for more than one standard deviation. We thus estimated the threshold as the average between the excitation densities of this point and of the one immediately before, using their semidispersion as maximum error and converting it to statistical error. The obtained values, also reported in Table 1, are (26.7 ± 1.6) $\mu J\,cm^{-2}$, (25.0 ± 1.6) $\mu J\,cm^{-2}$, (20.4 ± 1.7) $\mu J\,cm^{-2}$ and (13.4 ± 0.4) $\mu J\,cm^{-2}$ for FWHM/2, $FWHM_{ave}$, $FWHM_{cros}$ and $FWHM_{nar}$, respectively.

The ASE threshold has been also determined exploiting the slope increase, due to the ASE presence, of the intensity raise with the excitation density (see Figure 2c). In this case we performed two different linear fits for the data before and after the slope increase, thus determining the threshold as the excitation density at which the two best fit lines cross. Three different values have been determined from the data of the total integrated intensity (I_{TOT}), of the peak intensity at the ASE peak wavelength

(I_{peak}) and of the total integrated intensity of the ASE band (see Supplementary Materials for the details of the procedure used to separate the integrated ASE intensity from the total integrated intensity) (I_{ASE}), namely (24.8 ± 1.5) µJ cm^{-2}, (27.22 ± 0.51) µJ cm^{-2} and (27.8 ± 1.1) µJ cm^{-2}, respectively from I_{TOT}, I_{ASE} and I_{peak} plots (Table 1).

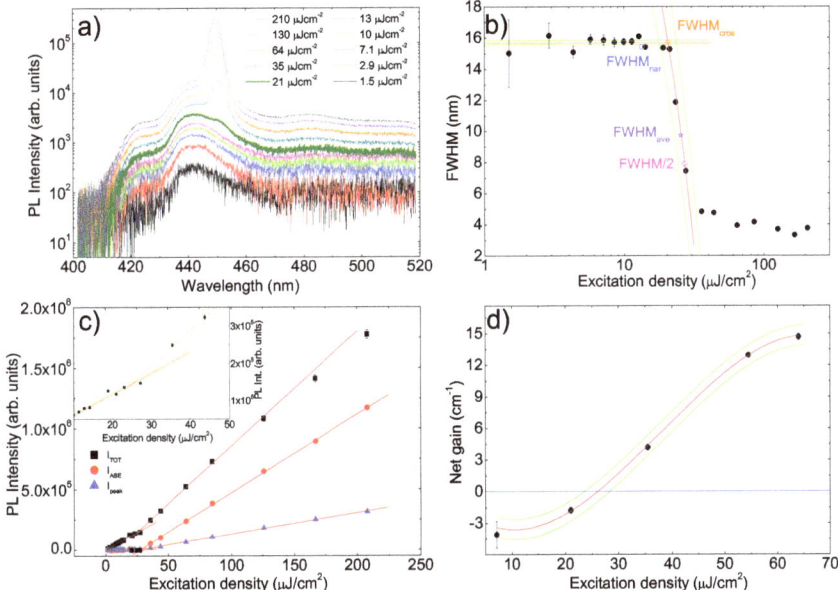

Figure 2. (a) Excitation density dependence of the photoluminescence (PL) spectra of the poly(9,9-dioctylfluorene) (PFO) sample. The thicker line evidences the first spectrum in which the lineshape is modified by the ASE presence. Only 10 spectra of the 25 acquired ones are shown for clarity. (b) Excitation density dependence of the PL spectra Full Width at Half Maximum (FWHM). The red lines are the best fit curves and the green lines the limits of the uncertainty range. The colored empty symbols represent the 4 threshold values extracted by $FWHM_{nar}$, $FWHM_{cros}$, $FWHM/2$ and $FWHM_{ave}$ methods. (c) Excitation density dependence of the integrated PL intensity (I_{TOT}) of the integrated ASE intensity (I_{ASE}) and of the intensity at the ASE band peak wavelength (I_{peak}). The red lines are the best fit curves. Inset: magnification of the I_{TOT} data evidencing the crossing of the best fit lines used to determine the threshold and the uncertainty bands. (d) Excitation density dependence of the net gain. The blue line evidences the zero.

Table 1. ASE threshold values obtained with all the methods and for all the investigated samples.

Method	ASE Threshold (µJ cm^{-2})		
	PFO	F8BT	F8BT:rrP3HT
Visual	~21	~85	~52
I_{TOT}	24.8 ± 1.5	N/A	148 ± 10
I_{ASE}	27.2 ± 0.5	142 ± 15	135.2 ± 5.6
I_{peak}	27.8 ± 1.1	145 ± 20	151.3 ± 7.7
FWHM/2	26.7 ± 1.6	167.9 ± 2.0	108.6 ± 7.8
$FWHM_{ave}$	25.0 ± 1.6	158.0 ± 2.0	104.6 ± 7.8
$FWHM_{cros}$	20.4 ± 1.7	64.9 ± 2.4	66.5 ± 7.9
$FWHM_{nar}$	13.4 ± 0.4	12.03 ± 0.41	28.1 ± 3.0
Gain	26.1 ± 1.4	39.5 ± 6.1	N/A

Finally we determined the waveguide net gain by measuring the PL spectra at fixed excitation density and with stripe length varied between 0 and 4 mm in steps of 0.1 mm (VSL method). The measurements have been performed at five different excitation densities: below, close to and above the visual ASE threshold.

Assuming uniform gain along the stripe, the PL intensity dependence on the stripe length is given by [78]:

$$I(\lambda, g', l) = \frac{I_0(\lambda)}{g'(\lambda)} \left[e^{g'(\lambda)l} - 1 \right] \tag{1}$$

where l is the stripe length, g' is the net optical gain coefficient and I_0 represents the spontaneous emission intensity per unity length. The experimental data (see Figure S2b) show an almost exponential increase with the stripe length, followed by a slower growth at high values of the stripe length. This effect is related to gain saturation, that is not included in the model. For this reason the data affected by gain saturation have been excluded in the fitting.

At each excitation density the net gain value g', given by the difference between the gain and the losses ($g' = g - \alpha$), has been obtained from the best fit with equation 1 of the intensity increase with the stripe length, at each wavelength (see Figure S2b). The best fit values of the net gain g' at the ASE peak wavelength of 450 nm at all the investigated excitation densities show (see Figure 2d) a progressive increase with the excitation density, starting from negative values for the two lowest excitation densities. This dependence has been reproduced with a third degree polynomial fit, obtaining a last estimate of the ASE threshold of (26.1 ± 1.4) µJ cm^{-2}, by determining the excitation density at which the net gain becomes 0, evidencing that the gain compensates the losses.

A similar experiment has been performed on the F8BT film. The PL spectra show, at low excitation density (see Figure 3a), the presence of a single broad band, with a shoulder at about 536 nm, a main peak at about 561 nm and a further shoulder at about 593 nm due to the 0–0, 0–1 and 0–2 transitions, respectively. As the excitation density increases the lineshape does not change up to 85 µJ cm^{-2} (visual threshold) at which a shoulder at about 573 nm starts to be visible. At higher excitation density a clear ASE band appears, with a peak wavelength of 573 nm [12,58].

In order to determine the ASE threshold from the spectral line narrowing we determined, from all the acquired spectra, the PL FWHM (see Figure 3b). The obtained results show a more gradual transition from the high value below the ASE threshold to the low value well above the threshold with respect to PFO, evidencing a slower increase of the ASE relative contribution to the total emission. From the best fit with a constant up to 11 µJcm^{-2} and a linear decrease between 100 µJ cm^{-2} and 210 µJ cm^{-2} we determined the threshold values: FWHM$_{nar}$ = (12.03 ± 0.41) µJ cm^{-2}, FWHM$_{cros}$ = (64.9 ± 2.4) µJ cm^{-2}, FWHM/2=(168.0 ± 2.0) µJ cm^{-2} and FWHM$_{ave}$ = (158.0 ± 2.0) µJ cm^{-2} (see also Table 1).

Concerning the excitation density dependence of the emission intensity we observe that the integrated ASE intensity and the ASE peak intensity show the typical kink related to the ASE appearance (see Figure 3c). This behavior is instead not observed for the total integrated emission intensity, that shows a gradual increase in all the investigated range, despite the presence of the ASE contribution (above the threshold). In order to understand this anomalous result we observe that, at the highest explored excitation density of 1200 µJ cm^{-2}, the integrated ASE intensity is about only 0.39 times the total intensity. This clearly evidences that, despite the progressive increase of the ASE contribution, even at the highest excitation density the total emission is mainly due to the spontaneous emission that has a clearly lower peak intensity, but also a much larger linewidth. Thus, the appearance of the ASE band in the spectra does not result in a clear variation of the integrated intensity growth, hidden by the dominating contribution of the spontaneous emission.

The ASE threshold has been thus determined only from the best fit with linear functions of the data of the integrated ASE intensity and of the ASE peak intensity, obtaining a value of (142 ± 15) µJ cm^{-2} and (145 ± 20) µJ cm^{-2}, respectively.

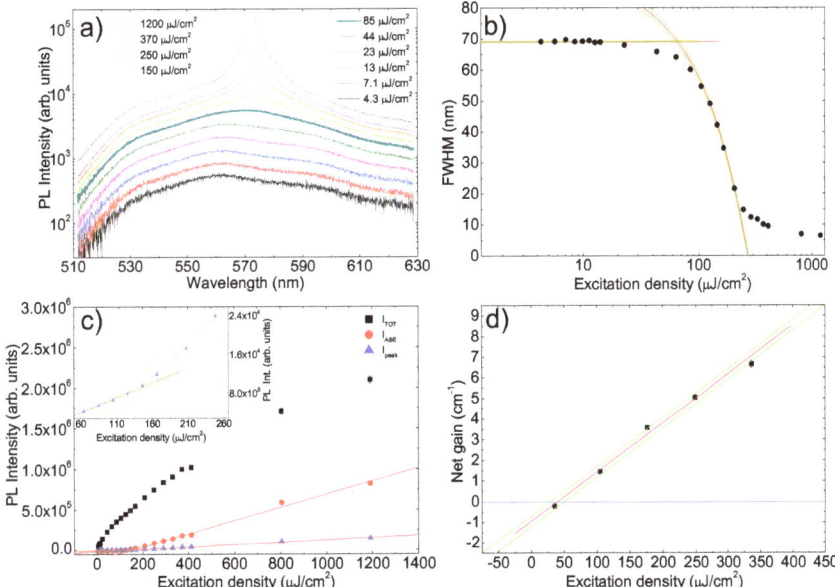

Figure 3. (**a**) Excitation density dependence of the PL spectra of the F8BT sample. The thicker line evidences the first spectrum in which the lineshape is modified by the ASE presence. Only 10 spectra of the 25 acquired ones are shown for clarity. (**b**) Excitation density dependence of the PL spectra FWHM. The red lines are the best fit curves and the green lines the limits of the uncertainty range. (**c**) Excitation density dependence of the integrated PL intensity (I_{TOT}) of the integrated ASE intensity (I_{ASE}) and of the intensity at the ASE band peak wavelength (I_{peak}). The red lines are the best fit curves. Inset: magnification of the I_{peak} data evidencing the crossing of the best fit lines used to determine the threshold and the uncertainty bands. (**d**) Excitation density dependence of the net gain. The blue line evidences the zero.

The last ASE estimate has been obtained from the excitation density dependence of the net gain at the ASE peak wavelength of 573 nm (see Figure 3d), showing an almost linear increase with the excitation density. The excitation density of zero net gain is $(39.5 \pm 6.1)\,\mu J\,cm^{-2}$.

The last investigated sample is the F8BT:rrP3HT blend film. The PL spectra show, below the ASE threshold, the typical linewidth of the rrP3HT emission (see Figure 4a), with a 0–0 line at about 640 nm and a 0–1 vibronic replica at about 666 nm. The absence of F8BT PL is due to the efficient F8BT→rrP3HT Förster Resonant Energy Transfer [36,38]. As the excitation density reaches 52 $\mu J\,cm^{-2}$ the relative intensity of the 0–1 line starts to increase (visual ASE threshold) and, at higher excitation density, a clear ASE band peaked at about 668 nm is observed. In this sample the ASE band shows a weak modulation, well visible above 130 $\mu J\,cm^{-2}$, likely due to random lasing assisted by scattering from morphological irregularities in the film [59–62].This attribution has been confirmed by SEM measurements showing (see Figure S3c,d) the lack of thickness uniformity in the film. In particular the thickness is about 1.25 μm close to the substrate edge, and progressively decreases to about 340 nm close to the film center. This effect is likely related to the deposition from a hot solution and to the solution temperature decrease during the spin coating, leading to a higher thickness toward the edges.

Similarly to the F8BT films, the spectral FWHM shows a gradual transition from the constant value of about 89 nm well below threshold to about 8.5 nm well above it (see Figure 4b). Concerning the emission intensity increase with the excitation density we observed a clear slope variation, related to the ASE appearance, in the total integrated intensity, the ASE integrated intensity and the ASE peak intensity (see Figure 4c). Following the already described procedures we thus estimated the ASE threshold values reported in Table 1.

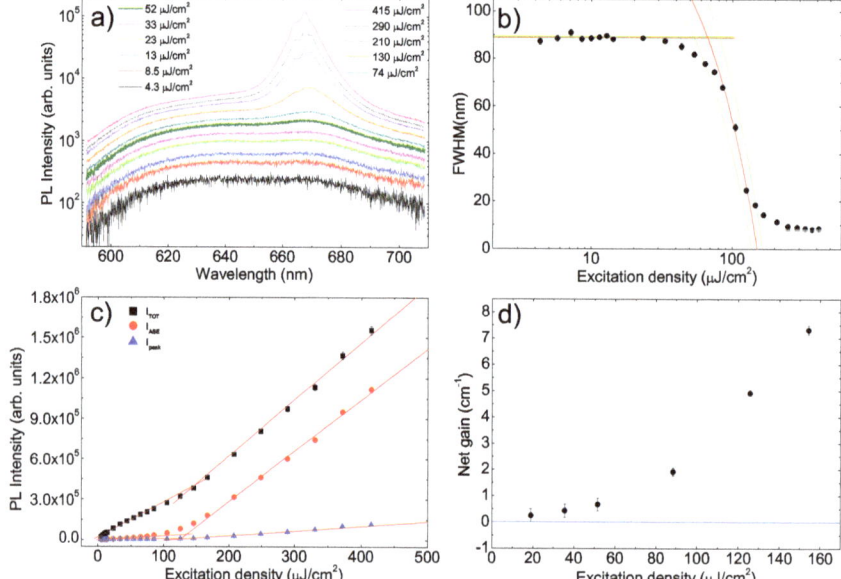

Figure 4. (**a**) Excitation density dependence of the PL spectra of the F8BT:rrP3HT sample. The thicker line evidences the first spectrum in which the lineshape is modified by the ASE presence. Only 11 spectra of the 25 acquired ones are shown for clarity. (**b**) Excitation density dependence of the PL spectra FWHM. The red lines are the best fit curves and the green lines are the limits of the uncertainty range. (**c**) Excitation density dependence of the integrated PL intensity (I_{TOT}) of the integrated ASE intensity (I_{ASE}) and of the intensity at the ASE band peak wavelength (I_{peak}). The red lines are the best fit curves. (**d**) Excitation density dependence of the net gain showing the lack of negative values. The blue line evidences the zero.

On the contrary, the excitation density dependence of the net gain at the ASE peak wavelength (see Figure 4d) shows a progressive increase with the excitation density, but it never shows negative gain values, even at the lowest input energy density that is 1.5 times below the lowest quantitative estimate of the ASE threshold and about 3 times below the visual ASE threshold. The lack of transition between negative and positive values of the net gain prevented an evaluation of the ASE threshold from the $g' = 0$ condition.

3. Discussion

All the obtained ASE values are determined by exploiting the effects of the ASE appearance on several spectral features, like linewidth, lineshape and output intensity, or by relating the ASE to the presence of positive net gain. In this section we will compare the values extracted from the all the investigated methods in order to determine the most reliable one and thus the best method to correctly quantify the ASE threshold.

As a preliminary step we observe that the ASE threshold is the minimum value of the excitation density that induces the ASE presence. By definition of minimum all the experimental values, determining the excitation density at which some effect related to ASE becomes observable, cannot be lower than the real threshold. For this reason, among all the experimental values, the lowest value will be the one closest to the real threshold, and will thus provide the most reliable estimate of the threshold.

The plot of the threshold values as a function of the used method (see Figure 5a) allows to observe a similar trend for all the three investigated polymers evidencing that the relative values extracted by different methods are mostly independent of the specific material.

In particular, for all the investigated samples, the lowest ASE threshold is always obtained by the $FWHM_{nar}$ method. This allows us to conclude that the quantity most sensitive to the ASE presence is the spectral linewidth and that the beginning of the line narrowing allows to determine the lowest threshold value. For this reason the method $FWHM_{nar}$ can be considered the best one for a reliable determination of the ASE threshold value.

It is important to observe that, rather surprisingly, the ASE threshold is determined by the $FWHM_{nar}$ method only in about 2.1% of the papers quantifying the ASE threshold, evidencing that in the remaining 97.9% the reported thresholds are systematically overestimated.

The sensitivity of the FWHM decrease to the ASE appearance can be clearly evidenced by observing that in all the samples the line narrowing is detectable at an excitation density much lower than the one that allows to observe a variation of the linewidth (given by the visual threshold). In particular the two samples with the highest relative ASE contribution to the emission, i.e., PFO and F8BT:rrP3HT, show a $FWHM_{nar}$ threshold about 2 times lower than the visual one, while in F8BT, that shows the lowest relative ASE intensity, the linewidth reduction starts at an excitation density 7 times smaller than the visual threshold.

Concerning the other methods the ones that provide ASE threshold closer to the $FWHM_{nar}$ are always the visual method and the $FWHM_{cros}$, both leading to similar ASE threshold values (with relative differences between almost negligible for PFO, and of about 30% for F8BT). This suggests that the excitation density determined by $FWHM_{cros}$ basically coincides with the one of ASE appearance in the spectra.

Moving to the most popular methods, we observe that the threshold values obtained with I_{tot}, I_{ASE} and I_{peak} are always comparable within about 10%. However these values are systematically much larger than the $FWHM_{nar}$ ones, with differences between about 2 times for PFO up to 12 times for F8BT.

The threshold values obtained by the FWHM/2 method are overall comparable with the ones extracted from the intensity slope increase, with relative differences between almost 0 for PFO and about 28% for F8BT:rrP3HT.

One should be aware that the methods based on the slope variation and the FWHM/2 one lead to a systematic overestimation of the threshold, that is dependent on the specific material and that can exceed one order of magnitude.

The threshold overestimation is intrinsic of the method definitions as both the intensity slope increase and the FWHM halving are obtained when the ASE contribution starts to dominate the emission, and not when ASE emission appears.

Finally we observe that the threshold values obtained from the $g' = 0$ condition are completely unpredictable and not reliable. This is due to the dependence of the gain values on the thickness uniformity; so irregularities result in a limited capability to determine the net gain with the accuracy (within fraction of cm^{-1}) necessary to correctly find the excitation density of 0 net gain. Our samples clearly show this effect, as PFO is the sample with the highest thickness uniformity and shows a threshold value obtained from the gain value comparable to the ones obtained from the intensity and the FWHM halving. F8BT, presenting some thickness fluctuations, is instead characterized by a clearly reduced value obtained from the gain. Finally the F8BT:rrP3HT film, that is the worst sample in terms of uniformity, never exhibits negative gain values.

Another important aspect to consider in order to compare the different methods is their ease of use, that can be evaluated from an estimate of the total working time that they need (see the Supplementary Materials for details and Figure 5b).

In this case we observe that the visual method is largely the fastest one, needing just a few minutes, and it is the most suitable for the investigation of the ASE threshold uniformity across the sample in short times. All the other methods based on VPI experiments are instead roughly comparable in term of working times, strongly suggesting that FWHM$_{nar}$ provides overall the best choice to combine correct values and reasonable measurement times. The only foresight in order to be able to determine the beginning of the line narrowing with a small error bar is to acquire a reasonable number of spectra at excitation densities below the visual threshold one. In our experiment the presence of 10 spectra below the visual threshold and starting from an excitation density about 10 times below it allowed to determine the FWHM$_{nar}$ threshold with a relative uncertainty below 10% in all the samples.

Moreover, by looking at the three methods based on the intensity growth analysis we suggest I$_{peak}$ as the best choice, as it allows to get a threshold value comparable with I$_{TOT}$ and I$_{ASE}$, but in clearly shorter times.

Finally, also working time considerations strongly suggest to avoid the use of gain measurements to quantify the threshold as, beyond the risk of completely unreliable values, the necessary time to get the threshold values are at least 4 times higher than the FWHM$_{nar}$ ones. It is also relevant to evidence that this time has been estimated for the determination of the gain at a single wavelength. In our experiment the maximum gain wavelength has been determined from the gain spectra including values at 75 different values around the ASE peak wavelength, considerably increasing the time to obtain each reported value.

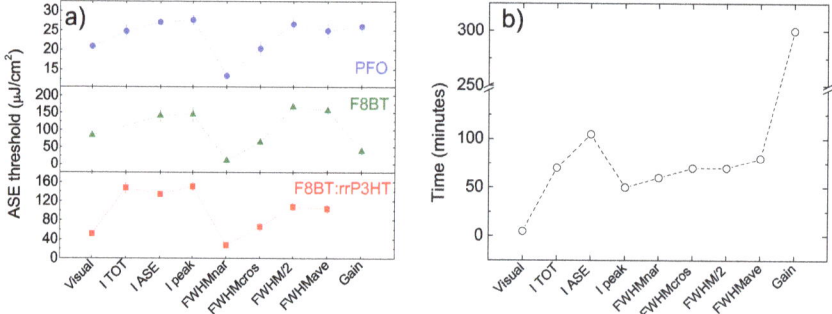

Figure 5. (**a**) ASE threshold values for PFO (top), F8BT (middle) and F8BT:rrP3HT (bottom). (**b**) Estimated working times for all the methods. All the lines are guides for the eyes.

4. Materials and Methods

4.1. Sample Preparation

PFO and F8BT were provided by ADS dyes (Canada) while rrP3HT was provided by Sigma Aldrich (now Merck KGaA, Germany). All the polymers have been used as received.

All the samples have been realized by spin coating from 15 mg/mL toluene solutions, with a rotation speed of 500 rpm for 5 s and 2500 rpm for 2 min, on glass substrates.

The PFO solution has been heated at 50° for few minutes, in order to avoid the formation of the β-phase [55,63].

The F8BT:rrP3HT blend has been prepared by mixing two 15 mg/mL toluene solutions of the individual materials with a relative concentration of 80:20 by weight, chosen in order to minimize ASE threshold [38]. In order to avoid rrP3HT aggregation the final solution has been heated at about 60° and the spin coating has been done from the hot solution [36,38].

4.2. VPI and VSL Measurements

All the samples have been excited by a LTB MNL 100 Nitrogen laser, delivering 3 ns pulses with a repetition rate of 10 Hz and a wavelength of 337 nm. The laser beam has been focused on the sample by a cylindrical lens, obtaining a rectangular pump stripe with 80 μm width. The stripe length has been varied through a micrometric slit up to 4 mm. The pump stripe has been placed at the edge of the film and the edge emitted radiation has been collected by a lens system coupled with an optical fiber connected to a computer-controlled Acton 750 spectrometer, and detected by an Andor Peltier cooled CCD. The spectral resolution was 0.5 nm. All the measurements have been performed at room temperature and, in order to avoid photodegradation effects, in vacuum (10^{-2} mbar).

The VPI measurements have been performed by keeping the stripe length fixed at 4 mm and by changing the excitation density with a continuously variable neutral filter. As a first step in the measurements we determined the excitation density at which the ASE band starts to be visible in the spectra by acquiring the PL spectra in real time while progressively increasing the excitation density. The VPI measurements have been then performed at 25 different excitation density values from 1/10 to 10 times this value.

The VLS measurements have been instead performed by keeping fixed the excitation density and by changing the pump stripe length from 0 mm to 4 mm in steps of 0.1 mm. The measurements have been performed at no less than 5 different values of the excitation density between 1/3 and 3 times the visual ASE threshold.

4.3. Thickness Measurements

The film thickness has been determined from Scanning Electron Microscopy (SEM) images in cross section. The images have been collected by a JEOL JSM-6480LV SEM, operated at 20 kV. In order to prevent charging effects the samples have been metalized by depositing a 10 nm-thick gold film on the surface by sputtering in Ar atmosphere at a pressure of 10^{-1} mbar, with a Quorum Technologies-Emitech K550x sputter coater.

5. Conclusions

In conclusion, we reported a quantitative comparison among various currently employed methods to determine the ASE threshold of organic active waveguides, in order to find the one that allows to determine the most accurate estimate of the ASE threshold.

We demonstrated that the spectral feature most sensitive to the ASE appearance is the spectral linewidth, and that the most reliable way to quantify the ASE threshold is to determine the excitation density at which the line narrowing starts. We also evidenced that the most common methods used in literature, determining the threshold from the slope variation of the intensity growth or from the FWHM halving, permit to determine the excitation regime at which the ASE starts to dominate the emission, but systematically overestimate the ASE threshold up to 14 times. Our results will be useful to correctly quantify the ASE threshold in polymeric waveguides, and to easily compare the ASE properties of different novel materials.

Supplementary Materials: The following are available online, Figure S1: Example of the variation of the peak wavelength with the excitation density for the PFO film, Figure S2: a: Example of the determination of the ASE integrated intensity for the PFO film. b: Intensity increase with the stripe length for the PFO film and best fit line. Figure S3: SEM image in cross section of the investigatd samples, Table S1: Estimated working time in minutes for the determination of the ASE threshold values for each method.

Author Contributions: Conceptualization and supervision, M.A.; writing—original draft preparation, S.M. S.M., M.L.D.G. and M.A. contributed to methodology, validation, formal analysis, investigation and writing—review and editing. All authors have read and agreed to the published version of the manuscript.

Funding: This research received no external funding.

Conflicts of Interest: The authors declare no conflict of interest.

References

1. Moses, D. High Quantum Efficiency Luminescence from a Conducting Polymer in Solution: A Novel Polymer Laser Dye. *Appl. Phys. Lett.* **1992**, *60*, 3215–3216. [CrossRef]

2. Hide, F.; Diaz-García, M.A.; Schwartz, B.J.; Andersson, M.R.; Pei, Q.; Heeger, A.J. Semiconducting Polymers: A New Class of Solid-State Laser Materials. *Science* **1996**, *273*, 1833–1836. [CrossRef]
3. Friend, R.H.; Gymer, R.W.; Holmes, A.B.; Burroughes, J.H.; Marks, R.N.; Taliani, C.; Bradley, D.D.C.; Dos Santos, D.A.; Brédas, J.L.; Lögdlund, M.; et al. Electroluminescence in Conjugated Polymers. *Nature* **1999**, *397*, 121–128. [CrossRef]
4. Frolov, S.V.; Gellermann, W.; Ozaki, M.; Yoshino, K.; Vardeny, Z.V. Cooperative Emission in π-Conjugated Polymer Thin Films. *Phys. Rev. Lett.* **1997**, *78*, 729–732. [CrossRef]
5. Tessler, N. Lasers Based on Semiconducting Organic Materials. *Adv. Mater.* **1999**, *11*, 363–370. [CrossRef]
6. Scherf, U.; Riechel, S.; Lemmer, U.; Mahrt, R. Conjugated Polymers: Lasing and Stimulated Emission. *Curr. Opin. Solid State Mater. Sci.* **2001**, *5*, 143–154. [CrossRef]
7. McGehee, M.D.; Heeger, A.J. Semiconducting (Conjugated) Polymers as Materials for Solid-State Lasers. *Adv. Mater.* **2000**, *12*, 1655–1668. [CrossRef]
8. Holzer, W.; Penzkofer, A.; Gong, S.H.; Blau, W.J.; Davey, A.P. Effective Stimulated Emission and Excited-State Absorption Cross-Section Spectra of Para-phenylene-ethynylene Polymers. *Opt. Quantum Electron.* **1998**, *30*, 1–14. [CrossRef]
9. Frolov, S.V.; Fujii, A.; Chinn, D.; Vardeny, Z.V.; Yoshino, K.; Gregory, R.V. Cylindrical Microlasers and Light Emitting Devices from Conducting Polymers. *Appl. Phys. Lett.* **1998**, *72*, 2811–2813. [CrossRef]
10. Tessler, N.; Denton, G.J.; Friend, R.H. Lasing from Conjugated Polymer Microcavities. *Nature* **1996**, *382*, 695–697. [CrossRef]
11. McGehee, M.D.; Diaz-García, M.A.; Hide, F.; Gupta, R.; Miller, E.K.; Moses, D.; Heeger, A.J. Semiconducting Polymer Distributed Feedback Lasers. *Appl. Phys. Lett.* **1998**, *72*, 1536–1538. [CrossRef]
12. Xia, R.; Heliotis, G.; Hou, Y.; Bradley, D.D.C. Fluorene-Based Conjugated Polymer Optical Gain Media. *Org. Electron.* **2003**, *4*, 165–177. [CrossRef]
13. Chua, L.L.; Ho, P.K.H.; Sirringhaus, H.; Friend, R.H. Observation of Field-Effect Transistor Behavior at Self-Organized Interfaces. *Adv. Mater.* **2004**, *16*, 1609–1615. [CrossRef]
14. Kim, D.H.; Han, J.T.; Park, Y.D.; Jang, Y.; Cho, J.H.; Hwang, M.; Cho, K. Single-Crystal Polythiophene Microwires Grown by Self-Assembly. *Adv. Mater.* **2006**, *18*, 719–723. [CrossRef]
15. Minemawari, H.; Yamada, T.; Matsui, H.; Tsutsumi, J.; Haas, S.; Chiba, R.; Kumai, R.; Hasegawa, T. Inkjet Printing of Single-Crystal Films. *Nature* **2011**, *475*, 364–367. [CrossRef]
16. Arias, A.C.; Ready, S.E.; Lujan, R.; Wong, W.S.; Paul, K.E.; Salleo, A.; Chabinyc, M.L.; Apte, R.; Street, R.A.; Wu, Y.; et al. All Jet-Printed Polymer Thin-Film Transistor Active-Matrix Backplanes. *Appl. Phys. Lett.* **2004**, *85*, 3304–3306. [CrossRef]
17. Rogowski, R.Z.; Dzwilewski, A.; Kemerink, M.; Darhuber, A.A. Solution Processing of Semiconducting Organic Molecules for Tailored Charge Transport Properties. *J. Phys. Chem. C* **2011**, *115*, 11758–11762. [CrossRef]
18. Becerril, H.A.; Roberts, M.E.; Liu, Z.; Locklin, J.; Bao, Z. High-Performance Organic Thin-Film Transistors through Solution-Sheared Deposition of Small-Molecule Organic Semiconductors. *Adv. Mater.* **2008**, *20*, 2588–2594. [CrossRef]
19. Wu, D.; Kaplan, M.; Ro, H.W.; Engmann, S.; Fischer, D.A.; De Longchamp, D.M.; Richter, L.J.; Gann, E.; Thomsen, L.; McNeill, C.R.; et al. Blade Coating Aligned, High-Performance, Semiconducting-Polymer Transistors. *Chem. Mater.* **2018**, *30*, 1924–1936. [CrossRef]
20. Anni, M.; Lattante, S. (Eds.) *Organic Lasers: Fundamentals, Developments and Applications*; Pan Stanford Publishing: Singapore, 2018.
21. Kuehne, A.J.C.; Gather, M.C. Organic Lasers: Recent Developments on Materials, Device Geometries, and Fabrication Techniques. *Chem. Rev.* **2016**, *116*, 12823–12864. [CrossRef]
22. Sandanayaka, A.S.D.; Matsushima, T.; Bencheikh, F.; Terakawa, S.; Potscavage, W.J.; Qin, C.; Fujihara, T.; Goushi, K.; Ribierre, J.C.; Adachi, C. Indication of current-injection lasing from an organic semiconductor. *Appl. Phys. Express* **2019**, *12*, 061010. [CrossRef]
23. McGehee, M.D.; Gupta, R.; Veenstra, S.; Miller, E.K.; Díaz-García, M.A.; Heeger, A.J. Amplified spontaneous emission from photopumped films of a conjugated polymer. *Phys. Rev. B* **1998**, *58*, 7035–7039. [CrossRef]
24. Laquai, F.; Keivanidis, P.E.; Baluschev, S.; Jacob, J.; Müllen, K.; Wegner, G. Low-Threshold Amplified Spontaneous Emission in thin Films of Poly(tetraarylindenofluorene). *Appl. Phys. Lett.* **2005**, *87*, 261917. [CrossRef]

25. Park, J.Y.; Srdanov, V.I.; Heeger, A.J.; Lee, C.H.; Park, Y.W. Amplified Spontaneous Emission from an MEH-PPV Film in Cylindrical Geometry. *Synth. Met.* **1999**, *106*, 35–38. [CrossRef]
26. Lee, T.W.; Park, O.O.; Choi, D.H.; Cho, H.N.; Kim, Y.C. Low-threshold blue amplified spontaneous emission in a statistical copolymer and its blend. *Appl. Phys. Lett.* **2002**, *81*, 424–426. [CrossRef]
27. Heliotis, G.; Xia, R.; Bradley, D.D.C.; Turnbull, G.A.; Samuel, I.D.W.; Andrew, P.; Barnes, W.L. Blue, Surface-Emitting, Distributed Feedback Polyfluorene Lasers. *Appl. Phys. Lett.* **2003**, *83*, 2118–2120. [CrossRef]
28. Chen, Y.; Herrnsdorf, J.; Guilhabert, B.; Kanibolotsky, A.L.; Mackintosh, A.R.; Wang, Y.; Pethrick, R.A.; Gu, E.; Turnbull, G.A.; Skabara, P.J.; et al. Laser action in a surface-structured free-standing membrane based on a π-conjugated polymer-composite. *Org. Electron.* **2011**, *12*, 62–69. [CrossRef]
29. Foucher, C.; Guilhabert, B.; Kanibolotsky, A.L.; Skabara, P.J.; Laurand, N.; Dawson, M.D. RGB and White-Emitting Organic Lasers on Flexible Glass. *Opt. Express* **2016**, *24*, 2273–2280. [CrossRef]
30. Pan, J.Q.; Yi, J.P.; Xie, G.; Lai, W.Y.; Huang, W. Enhancing Optical Gain Stability for a Deep-Blue Emitter Enabled by a Low-Loss Transparent Matrix. *J. Phys. Chem. C* **2018**, *122*, 21569–21578. [CrossRef]
31. Kim, D.H.; Sandanayaka, A.S.D.; Zhao, L.; Pitrat, D.; Mulatier, J.C.; Matsushima, T.; Andraud, C.; Ribierre, J.C.; Adachi, C. Extremely low amplified spontaneous emission threshold and blue electroluminescence from a spin-coated octafluorene neat film. *Appl. Phys. Lett.* **2017**, *110*, 023303. [CrossRef]
32. Mai, V.T.; Shukla, A.; Mamada, M.; Maedera, S.; Shaw, P.E.; Sobus, J.; Allison, I.; Adachi, C.; Namdas, E.B.; Lo, S.C. Low Amplified Spontaneous Emission Threshold and Efficient Electroluminescence from a Carbazole Derivatized Excited-State Intramolecular Proton Transfer Dye. *ACS Photonics* **2018**, *5*, 4447–4455. [CrossRef]
33. Holzer, W.; Penzkofer, A.; Schmitt, T.; Hartmann, A.; Bader, C.; Tillmann, H.; Raabe, D.; Stockmann, R.; Hörhold, H.H. Amplified spontaneous emission in neat films of arylene-vinylene polymers. *Opt. Quantum Electron.* **2001**, *33*, 121–150. [CrossRef]
34. Xia, R.; Lai, W.Y.; Levermore, P.A.; Huang, W.; Bradley, D.D.C. Low-Threshold Distributed-Feedback Lasers Based on Pyrene-Cored Starburst Molecules with 1,3,6,8-Attached Oligo(9,9-Dialkylfluorene) Arms. *Adv. Funct. Mater.* **2009**, *19*, 2844–2850. [CrossRef]
35. Lampert, Z.E.; Lappi, S.E.; Papanikolas, J.M.; Lewis Reynolds, C. Intrinsic optical gain in thin films of a conjugated polymer under picosecond excitation. *Appl. Phys. Lett.* **2013**, *103*, 033303. [CrossRef]
36. Anni, M.; Lattante, S. Amplified Spontaneous Emission Optimization in Regioregular Poly(3-hexylthiophene) (rrP3HT): Poly(9,9-dioctylfluorene-cobenzothiadiazole) (F8BT) Thin Films through Control of the Morphology. *J. Phys. Chem. C* **2015**, *119*, 21620–21625. [CrossRef]
37. Lampert, Z.E.; Papanikolas, J.M.; Lappi, S.E.; Reynolds, C.L. Intrinsic gain and gain degradation modulated by excitation pulse width in a semiconducting conjugated polymer. *Opt. Laser Technol.* **2017**, *94*, 77–85. [CrossRef]
38. Xia, R.; Stavrinou, P.N.; Bradley, D.D.C.; Kim, Y. Efficient Optical Gain Media Comprising Binary Blends of Poly(3-hexylthiophene) and Poly(9,9-dioctylfluorene-co-benzothiadiazole). *J. Appl. Phys.* **2012**, *111*, 123107. [CrossRef]
39. Schweitzer, B.; Wegmann, G.; Giessen, H.; Hertel, D.; Bässler, H.; Mahrt, R.F.; Scherf, U.; Müllen, K. The optical gain mechanism in solid conjugated polymers. *Appl. Phys. Lett.* **1998**, *72*, 2933–2935. [CrossRef]
40. Wegmann, G.; Schweitzer, B.; Hertel, D.; Giessen, H.; Oestreich, M.; Scherf, U.; Müllen, K.; Mahrt, R. The dynamics of gain-narrowing in a ladder-type π-conjugated polymer. *Chem. Phys. Lett.* **1999**, *312*, 376–384. [CrossRef]
41. Heliotis, G.; Bradley, D.D.C. Light Amplification and Gain in Polyfluorene Waveguides. *Appl. Phys. Lett.* **2002**, *81*, 415–417. [CrossRef]
42. Gupta, R.; Stevenson, M.; Heeger, A.J. Low Threshold Distributed Feedback Lasers Fabricated from Blends of Conjugated Polymers: Reduced Losses through Förster Transfer. *J. Appl. Phys.* **2002**, *92*, 4874–4877. [CrossRef]
43. Azuma, H.; Kobayashi, T.; Shim, Y.; Mamedov, N.; Naito, H. Amplified spontaneous emission in α-phase and β-phase polyfluorene waveguides. *Org. Electron.* **2007**, *8*, 184–188. [CrossRef]
44. Kim, Y.C.; Lee, T.W.; Park, O.O.; Kim, C.Y.; Cho, H.N. Low-Threshold Amplified Spontaneous Emission in a Fluorene-Based Liquid Crystalline Polymer Blend. *Adv. Mater.* **2001**, *13*, 646–649. [CrossRef]
45. Lattante, S.; Cretí, A.; Lomascolo, M.; Anni, M. On the Correlation between Morphology and Amplified Spontaneous Emission Properties of a Polymer: Polymer Blend. *Org. Electron.* **2016**, *29*, 44–49. [CrossRef]

46. Li, J.Y.; Laquai, F.; Wegner, G. Amplified spontaneous emission in optically pumped neat films of a polyfluorene derivative. *Chem. Phys. Lett.* **2009**, *478*, 37–41. [CrossRef]
47. Abdel-Awwad, M.; Luan, H.; Messow, F.; Kusserow, T.; Wiske, A.; Siebert, A.; Fuhrmann-Lieker, T.; Salbeck, J.; Hillmer, H. Optical amplification and photodegradation in films of spiro-quaterphenyl and its derivatives. *J. Lumin.* **2015**, *159*, 47–54. [CrossRef]
48. Brouwer, H.J.; Krasnikov, V.V.; Pham, T.A.; Gill, R.E.; Hadziioannou, G. Stimulated emission from vacuum-deposited thin films of a substituted oligo(p-phenylene vinylene). *Appl. Phys. Lett.* **1998**, *73*, 708–710. [CrossRef]
49. Anni, M.; Alemanno, M. Temperature dependence of the amplified spontaneous emission of the poly(9,9-dioctylfluorene) β phase. *Phys. Rev. B* **2008**, *78*, 233102. [CrossRef]
50. Anni, M.; Perulli, A.; Monti, G. Thickness Dependence of the Amplified Spontaneous Emission Threshold and Operational Stability in poly(9,9-dioctylfluorene) Active Waveguides. *J. Appl. Phys.* **2012**, *111*, 093109. [CrossRef]
51. Anni, M. Operational lifetime improvement of poly(9,9-dioctylfluorene) active waveguides by thermal lamination. *Appl. Phys. Lett.* **2012**, *101*, 013303. [CrossRef]
52. Chang, S.J.; Liu, X.; Lu, T.T.; Liu, Y.Y.; Pan, J.Q.; Jiang, Y.; Chu, S.Q.; Lai, W.Y.; Huang, W. Ladder-type poly(indenofluorene-co-benzothiadiazole)s as efficient gain media for organic lasers: Design, synthesis, optical gain properties, and stabilized lasing properties. *J. Mater. Chem. C* **2017**, *5*, 6629–6639. [CrossRef]
53. Lattante, S.; De Giorgi, M.L.; Pasini, M.; Anni, M. Low threshold Amplified Spontaneous Emission properties in deep blue of poly[(9,9-dioctylfluorene-2,7-dyil)-alt-p-phenylene] thin films. *Opt. Mater.* **2017**, *72*, 765–768. [CrossRef]
54. Zhang, Q.; Liu, J.; Wei, Q.; Guo, X.; Xu, Y.; Xia, R.; Xie, L.; Qian, Y.; Sun, C.; Lüer, L.; et al. Host Exciton Confinement for Enhanced Förster-Transfer-Blend Gain Media Yielding Highly Efficient Yellow-Green Lasers. *Adv. Funct. Mater.* **2018**, *28*, 1705824. [CrossRef]
55. Anni, M. Dual band amplified spontaneous emission in the blue in Poly(9,9-dioctylfluorene) thin films with phase separated glassy and β-phases. *Opt. Mater.* **2019**, *96*, 109313. [CrossRef]
56. Anni, M. Poly[2-methoxy-5-(2-ethylhexyloxy)-1,4-phenylenevinylene] (MeH-PPV) Amplified Spontaneous Emission Optimization in Poly(9,9-dioctylfluorene(PFO):MeH-PPV Active Blends. *J. Lumin.* **2019**, *215*, 116680. [CrossRef]
57. Virgili, T.; Anni, M.; De Giorgi, M.L.; Borrego Varillas, R.; Squeo, B.M.; Pasini, M. Deep Blue Light Amplification from a Novel Triphenylamine Functionalized Fluorene Thin Film. *Molecules* **2019**, *25*, 79. [CrossRef]
58. Anni, M. Photodegradation Effects on the Emission Properties of an Amplifying Poly(9,9- dioctylfluorene) Active Waveguide Operating in Air. *J. Phys. Chem. B* **2012**, *116*, 4655–4660. [CrossRef]
59. Sznitko, L.; Mysliwiec, J.; Miniewicz, A. The Role of Polymers in Random Lasing. *J. Polym. Sci. Part B Polym. Phys.* **2015**, *53*, 951–974. [CrossRef]
60. Anni, M.; Rhee, D.; Lee, W.K. Random Lasing Engineering in Poly-(9-9dioctylfluorene) Active Waveguides Deposited on Wrinkles Corrugated Surfaces. *ACS Appl. Mater. Int.* **2019**, *11*, 9385–9393. [CrossRef]
61. Frolov, S.V.; Vardeny, Z.V.; Yoshino, K.; Zakhidov, A.; Baughman, R.H. Stimulated emission in high-gain organic media. *Phys. Rev. B* **1999**, *59*, R5284–R5287. [CrossRef]
62. Anni, M.; Lattante, S.; Cingolani, R.; Gigli, G.; Barbarella, G.; Favaretto, L. Far-field emission and feedback origin of random lasing in oligothiophene dioxide neat films. *Appl. Phys. Lett.* **2003**, *83*, 2754–2756. [CrossRef]
63. Anni, M. The role of the β-phase content on the stimulated emission of poly(9,9-dioctylfluorene) thin films. *Appl. Phys. Lett.* **2008**, *93*, 023308. [CrossRef]
64. Yamashita, K.; Kuro, T.; Oe, K.; Yanagi, H. Low Threshold Amplified Spontaneous Emission from Near-Infrared Dye-Doped Polymeric Waveguide. *Appl. Phys. Lett.* **2006**, *88*, 241110. [CrossRef]
65. Navarro-Fuster, V.; Calzado, E.M.; Ramirez, M.G.; Boj, P.G.; Henssler, J.T.; Matzger, A.J.; Hernandez, V.; Lopez Navarrete, J.T.; Diaz-Garcia, M.A. Effect of ring fusion on the amplified spontaneous emission properties of oligothiophenes. *J. Mater. Chem.* **2009**, *19*, 6556–6567. [CrossRef]
66. Kazlauskas, K.; Kreiza, G.; Bobrovas, O.; Adomeniene, O.; Adomenas, P.; Jankauskas, V.; Jursenas, S. Fluorene- and benzofluorene-cored oligomers as low threshold and high gain amplifying media. *Appl. Phys. Lett.* **2015**, *107*, 043301. [CrossRef]

67. Lahoz, F.; Oton, C.J.; Capuj, N.; Ferrer-González, M.; Cheylan, S.; Navarro-Urrios, D. Reduction of the amplified spontaneous emission threshold in semiconducting polymer waveguides on porous silica. *Opt. Express* **2009**, *17*, 16766–16775. [CrossRef]
68. Navarro-Fuster, V.; Calzado, E.M.; Boj, P.G.; Quintana, J.A.; Villalvilla, J.M.; Diaz-Garcia, M.A.; Trabadelo, V.; Juarros, A.; Retolaza, A.; Merino, S. Highly photostable organic distributed feedback laser emitting at 573 nm. *Appl. Phys. Lett.* **2010**, *97*, 171104. [CrossRef]
69. Calzado, E.M.; Villalvilla, J.M.; Boj, P.G.; Quintana, J.A.; Diaz-Garcia, M.A. Tuneability of amplified spontaneous emission through control of the thickness in organic-based waveguides. *J. Appl. Phys* **2005**, *97*, 093103. [CrossRef]
70. Liu, J.; Qian, Y.; Wei, Q.; Zhang, Q.; Xie, L.; Lee, C.; Kim, H.; Kim, Y.; Xia, R. Deep Blue Laser Gain Medium Based on Triphenylamine Substituted Arylfluorene With Improved Photo-Stability. *IEEE J. Sel. Top. Quantum Electron.* **2016**, *22*, 15–20. [CrossRef]
71. Munoz-Marmol, R.; Zink-Lorre, N.; Villalvilla, J.M.; Boj, P.G.; Quintana, J.A.; Vaszquez, C.; Anderson, A.; Gordon, M.J.; Sastre-Santos, A.; Fernandez-Lazaro, F.; et al. Influence of Blending Ratio and Polymer Matrix on the Lasing Properties of Perylenediimide Dyes. *J. Phys. Chem. C* **2018**, *122*, 24896–24906. [CrossRef]
72. Calzado, E.M.; Villalvilla, J.M.; Boj, P.G.; Quintana, J.A.; Gomez, R.; Segura, J.L.; Diaz-Garcia, M.A. Effect of Structural Modifications in the Spectral and Laser Properties of Perylenediimide Derivatives. *J. Phys. Chem. C* **2007**, *111*, 13595. [CrossRef]
73. Cerdán, L.; Costela, A.; García-Moreno, I.; García, O.; Sastre, R. Waveguides and Quasi-Waveguides based on Pyrromethene 597-Doped Poly(methyl methacrylate). *Appl. Phys. B* **2009**, *97*, 73–83. [CrossRef]
74. Kumar, G.A.; Riman, R.E.; Banerjee, S.; Kornienko, A.; Brennan, J.G.; Chen, S.; Smith, D.; Ballato, J. Infrared fluorescence and optical gain characteristics of chalcogenide-bound erbium cluster-fluoropolymer nanocomposites. *Appl. Phys. Lett.* **2006**, *88*, 091902. [CrossRef]
75. Pisignano, D.; Anni, M.; Gigli, G.; Cingolani, R.; Zavelani-Rossi, M.; Lanzani, G.; Barbarella, G.; Favaretto, L. Amplified spontaneous emission and efficient tunable laser emission from a substituted thiophene-based oligomer. *Appl. Phys. Lett.* **2002**, *81*, 3534–3536. [CrossRef]
76. Ryu, G.; Xia, R.; Bradley, D.D.C. Optical gain characteristics of β-phase poly(9,9-dioctylfluorene). *J. Phys. Condens. Matter* **2007**, *19*, 056205. [CrossRef]
77. Conwell, E.M. *Organic Electronic Materials*; Springer: Berlin/Heidelberg, Germany, 2001; pp. 170–174.
78. Dal Negro, L.; Bettotti, P.; Cazzanelli, M.; Pacifici, D.; Pavesi, L. Applicability Conditions and Experimental Analysis of the Variable Stripe Length Method for Gain Measurements. *Opt. Commun.* **2004**, *229*, 337–348. [CrossRef]

© 2020 by the authors. Licensee MDPI, Basel, Switzerland. This article is an open access article distributed under the terms and conditions of the Creative Commons Attribution (CC BY) license (http://creativecommons.org/licenses/by/4.0/).

MDPI
St. Alban-Anlage 66
4052 Basel
Switzerland
Tel. +41 61 683 77 34
Fax +41 61 302 89 18
www.mdpi.com

Molecules Editorial Office
E-mail: molecules@mdpi.com
www.mdpi.com/journal/molecules

www.ingramcontent.com/pod-product-compliance
Lightning Source LLC
LaVergne TN
LVHW070708100526
838202LV00013B/1051